Green Energy and Technology

Edited By

Hatim Machrafi

Université Paris 6
France
&
Université de Liège
Belgium

CONTENTS

CHAPTERS

FOREWORD

The large evolution of the energy consumption during the last century is correlated with the development of large cities, a strong increase of the human transportation and the large exchanges of mineral and food resources. At the same time, the large part of the energy that is needed is produced by combustion systems, which give us a very flexible energy freedom, while the storage is just connected with the extraction of fossil carbon sources (oil, coal or gas).

In 2008, the European Economic Community opens a new challenge by the European Energy set plans (20 % of ENR, 20 % of increasing energy efficiency, 20 % of carbon dioxide capture) in order to open the way for sustainable development concerning energy resources.

Despite the energy production losses of about 60 % of the energy content of the primary sources, and in any way the second law of the thermodynamics point out the technical difficulties of energy efficiency, this new management is starting with more complex industrial processes which open the field for "green energy processes".

The feeling of this new management of energy sources is built on energy and mass balance of the process in order to measure firstly the energy efficiency, but at the same time also the byproducts of the process such as ashes, carbon dioxides, HAP, nitrous oxides, sulfur oxides, heavy metals, mercury…, and at the end ***the water energy*** consumption of the process for each MWh produced.

By doing so, the real cost of a new process is clearer, because it takes into account at the same time both the energy production and – in agreement with the European regulation – the recycling wastes and by-products. Besides hazard effects, the impact on health can also be identified, so that we can speak about green energy processes.

Taking into account the evolution of the energy sources for sustainable development, we have three kinds of energy: fossil carbon resources, which are limited but give us a very flexible energy storage, due to the high level of energy content by liter of raw material, renewable energy (solar, wind, geothermal sources, biomass) and nuclear energy sources (U_{235} fission or H_2 fusion in the future).

All of these are developed but we have to point out the large difference in term of financial cost (CAPEX), management and utilities, flexibility, waste treatment, recycling, energy storage, ***power size, and electrical*** network and social agreement.

The future of our civilization needs a large creativity supported by a strong engineering education to manage the ***mixed energetic network***. The sustainable energy concept has to take into account all the best working new processes but at the same time the water consumption, the climate, the human activities, the industrial development and the ***agricultural*** techniques. That is why this book ***opens*** three parts to give an overview of energy efficiency and environmental impact, renewable energy and energy for transportation.

The first part is mainly focused on environmental processes, including hydrogen and heavy oil sources and allows us obtaining the leading knowledge for a better efficiency of the processes.

The second part is about renewable energy sources such as solar, wind, and biofuels. It is focused on the large development of wind turbines, photovoltaic techniques, thermal solar techniques, and presents the large plants which are built to day around the world. However, the question of the energy storage and the flexibility of these nanotechniques need more development to explain how to develop a ***mixed energetic network*** for a continuous energy demand. At last the biofuel is one of the more complex energies: in one way, we have large sources of biomass from wood, by product of agricultural activities, algae and animal waste, but today the main industries working in that field use corn or sugar which are in competition with food resources.

The last part is about energy for transportation, we have to remember that it represents 40 % of the energy consumption and controls the international trade. The evolution of the efficiency of energy from 20 to 40 % is one of the technical developments, with the new synfuel from FT, hydrogen, fuel cells. It is an evolution for the transportation process which is starting.

So, that the field of "green energy technology" be the starting point of creativity for those who enjoy to work in the field of new technologies.

Prof. J. Amouroux

ENSCP/UPMC LGPPTS - E-MRS
France

PREFACE

This new book presents recent developments, state-of-the-art and progresses in the field of energy where efforts are done in order to improve the usability of energy systems, reducing their environmental impact. The book aims at providing researchers, academics, engineers and advanced students information and points of discussion, a platform for future improvements in green energy. Both theoretical and applied aspects are treated in this book. Many illustrations and mathematical equations as well as practical on-the-field applications are incorporated. This book aims at contributing to the increasing interest in reducing the environmental impact of energy as well as its further development.

Three parts are considered. The first part treats different energy applications and the efforts that are done in order to improve their impact on the environment. It deals with bio-energy, well-to-wheel analyses and heavy oils. Concerning bio-energy, the costs are evaluated under strict regulations imposed by a climate stabilization target, where a certain case is taken as example. The well-to-wheel energy analyses take into consideration a comprehensive study on greenhouse gases and certain criteria that concern the emissions in the case of hydrogen based fuel-cells.

The second part deals with different kinds of renewable energy sources. The chapters discuss bioenergy, wind energy and solar energy. The use of biofuels is critically discussed. A real application and evaluation of wind energy is presented, by means of designing and optimalizing a wind power system. A detailed discussion about the generation of electricity by means of different kinds of solar energy is performed.

The third part puts emphasis on alternative energy processes for transport utilities. In this part, is discussed, amongst others, the Homogeneous Charge Compression Ignition combustion mode. Advantages and inconveniences are proposed. Wide and comprehensive studies on this combustion mode are presented. Both a large synthesis of available work and experimental results are used in an effort to discuss these advantages and inconveniences in order to propose possible solutions. Some examples are treated elaborately. Hydrogen-based combustion is discussed and its efficiency evaluated in detail. Elaborate studies and examples give a rather complete vision of the probable use of hydrogen in internal combustion engines and the different aspects that are to be examined are treated in detail. The commercialization of fuel-cells are investigated, giving trends and possible developments that can be envisaged. Finally, a philosophy of controlling the auto-ignition in an HCCI engine is presented. This philosophy is somewhat general and the outlines can also be used for the abovementioned transportation modes.

A rather general overview is given in this book, starting from different visions on reducing the impact of energy on the environment (green energy) and continuing on how this can be achieved (green technology). The different results and the available literature that is treated in various chapters in this book show that many researchers are investigating the efficient use of energy whilst looking for ways to decrease its impact on the environment. This shows the importance of such research and the work that remains to be done.

Hatim Machrafi

Université Paris 6
France
&
Université de Liège
Belgium

List of Contributors

Abourriche Abdelkrim

Laboratoire Matériaux, Procédés, Environnement et Qualité, École Nationale des Sciences, Appliquées, B.P. 63, 46000 Safi, Morocco
E-mail: krimabou@hotmail.com

Abourriche Abdelmjid

Université Hassan II, Faculté des Sciences Ben M'sik, B.P. 7955 Casablanca, Morocco
E-mail: a.abourriche@univh2m.ac.ma

Bandyopadhyay Santanu

Department of Energy Science and Engineering, Indian Institute of Technology, Bombay, Powai, Mumbai, 400076, India
E-mail: santanu@me.iitb.ac.in

Birot Marc

Université de Bordeaux, Institut des Sciences Moléculaires, CNRS-UMR 5255, 351 cours de la Libération, F-33405 Talence, France
E-mail: m.birot@ism.u-bordeaux1.fr

Glasnovic Zvonimir

Faculty of Chemical Engineering and Technology, Dept. of Measurement and Process Control, Univ. of Zagreb, Savska cesta16, HR-10000 Zagreb, Croatia
E-mail: zvonglas@fkit.hr

Hannache Hassan

Laboratoire des Matériaux Thermostructuraux, Faculté des Sciences Ben M'sik, B.P. 7955 Casablanca, Morocco
E-mail: hannache.hassan@caramail.com

Huang Zhijia

School of Civil Engineering, Anhui University of Technology, Maanshan, Anhui, 243002 China
E-mail: huangzhijia99@hotmail.com

Kazim Ayoub

Department of Mechanical Engineering, United Arab Emirates University, P.O. Box 17555, Al-Ain, United Arab Emirates
E-mail: akazim@uaeu.ac.ae

Komninos Neofytos P.

School of Mechanical Engineering, National Technical University of Athens (NTUA), 9 Heroon Polytechniou St., Zografou Campus, 15780 Athens, Greece
E-mail: nkom@central.ntua.gr

Liu Haifeng

State Key Laboratory of Engines, Tianjin University, Tianjin 300072, China
E-mail: haifeng0515@gmail.com

Machrafi Hatim

ENSCP, Université Paris 6, 11, rue Pierre et Marie Curie, 75005 Paris, France and Université de Liège, Thermodynamics of Irreversible Phenomena, Allée du 6-Août, 17, 4000, Liège, Belgium
E-mail: hatim-machrafi@enscp.fr or H.Machrafi@ulg.ac.be

Margeta Jure

Faculty of Civil Engineering and Architecture, Univ. of Split, Matice Hrvatske 15, 21000 Split, Croatia
E-mail: Jure.Margeta@gradst.hr

Narodoslawsky Michael

Institute of Process Engineering, Technical University Graz, Inffeldgasse 21B, A-8010 Graz, Austria
E-mail: braunegg@glvt.tu-graz.ac.at

Naslain Roger

Université Bordeaux 1/CNRS/CEA/SAFRAN, Laboratoire des Composites Thermostructuraux (LCTS), 3 allée de la Boétie, F-33600 Pessac, France
E-mail: admin@lcts.u-bordeaux.fr

Oumam Mina

Laboratoire des Matériaux Thermostructuraux, Faculté des Sciences Ben M'sik, B.P. 7955 Casablanca, Morocco
E-mail: m.oumam@univh2m.ac.ma

Pailler René

Université Bordeaux 1/CNRS/CEA/SAFRAN, Laboratoire des Composites Thermostructuraux (LCTS), 3 allée de la Boétie, F-33600 Pessac, France
E-mail: pailler@lcts.u-bordeaux.fr

Roy Anindita

Department of Energy Science and Engineering, Indian Institute of Technology, Bombay, Powai, Mumbai, 400076, India
E-mail: anindita@iitb.ac.in

Sierens Roger

Ghent University, Department of Flow, Heat and Combustion Mechanics, Sint-Pietersnieuwstraat 41, B-9000 Gent, Belgium
E-mail: Roger.Sierens@UGent.be

Stoeglehner Gernot

University of Natural Resources and Applied Life Sciences, Department of Spatial, Landscape and Infrastructure Sciences, Institute of Spatial Planning and Rural Development, Peter-Jordan-Strasse 82, A-1190 Wien, Austria
E-mail: gernot.stoeglehner@boku.ac.at

Takeshita Takayuki

Transdisciplinary Initiative for Global Sustainability, The University of Tokyo, Japan
E-mail: takeshita@ir3s.u-tokyo.ac.jp

Verhelst Sebastian

Ghent University, Department of Flow, Heat and Combustion Mechanics, Sint-Pietersnieuwstraat 41, B-9000 Gent, Belgium
E-mail: sebastian.verhelst@ugent.be

Wallner Thomas

Energy Systems Division, Argonne National Laboratory, Building 362, 9700 South Cass Avenue, Argonne, IL 60439-4815, USA
E-mail: twallner@anl.gov

Yao Mingfa

State Key Laboratory of Engines, Tianjin University, Tianjin 300072, China
E-mail: y_mingfa@tju.edu.cn

Zheng Zhaolei

State Key Laboratory of Engines, Tianjin University, Tianjin 300072, China & Key Laboratory of Low-grade Energy Utilization Technologies and Systems, Ministry of Education, Chongqing University, Chongqing 400030, China

E-mail: zhengzhaolei2002@yahoo.com.cn

INTRODUCTION

Energy is indispensable in present society. All depend on a constant and reliable source of energy, whether it be for transport, industrial or home applications. The use of such energy sources can present some inconveniences, such as source depletion, pollution or health problems. The different kinds of energy use one may think of are electricity, mechanical conversion or heating. Fossil fuels are the main energy sources that are used for these kinds of energies. However, the transport sector and the industry, one of the primary energy consumers, contribute a lot to the pollution of the atmosphere. Many ways can be opted in order to reduce the impact of pollution on the environment. One may think of renewable energy sources or new, alternative energy conversion processes that pollute less or a combination of these two. These three ways of rational energy use, reducing altogether its impact on the environment, are the main thought of this book.

Even if there is a range of energy use, renewable energy sources are generally to be used for three purposes: power generation, energy conversion for transport and the generation of heat and electricity (keeping in mind, the production of cold as well). The sources, in this case, are meant to be inexhaustible, such as the sun, the wind, the geothermic heat, the flowing of water, or also replaceable such as fuels extracted from plants. The combustion of fossil fuels generates greenhouse gases as well as other harmful pollutants. Renewable energy, however, has the great potential to produce less greenhouse gases and other pollutants. Caution needs also to be made when considering renewable energy sources from biological origin, especially if it is alimentary. The competition with the food industry needs to be avoided. Some renewable energy sources do not reduce the carbondioxide level, but are said to be carbon neutral, as for instance biomass. The carbondioxide production when using biomass is balanced by its absorption when producing the organic material. Such an energy source needs to be studied thoroughly in order to avoid a higher carbondioxide production than its absorption by carefully examining the whole life-cycle (or well-to-wheel process in case of transport applications) of such an energy source.

As said before, different renewable energy sources can be thought of, principally of solar (photovoltaic cells for electricity production or solar thermal systems for heating), wind (wind turbines for electricity generation or conventional windmill for water pumping), water (hydro electric, wave and tidal systems for electricity generation), biomass (combustion applications of the produced gas, electricity generation or heat production), biofuel (mainly for combustion for transport purposes) and geothermal (using the temperature of the earth for heat and/or electricity generation) nature. Some major energy uses are depicted in Fig. **1**.

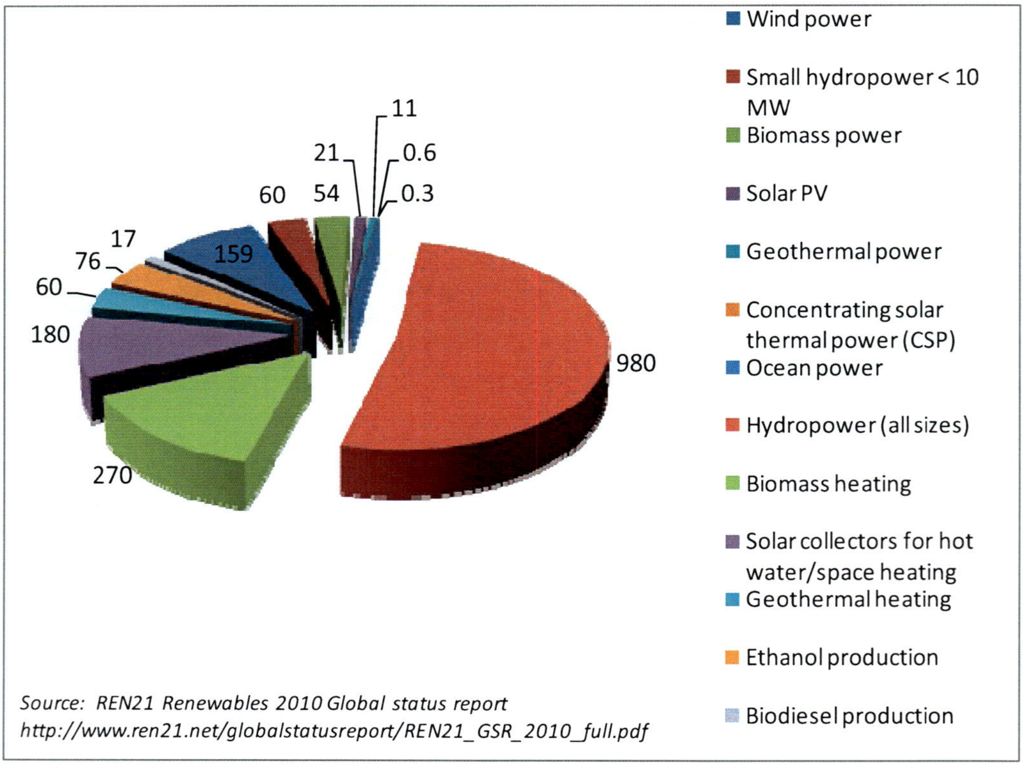

Figure 1: The energy sources in the world: renewable versus total.

The aforementioned six renewable energies have the main characteristic that they are natural phenomena. However, their use does not depend only on the natural resources, but also on the technologies that are needed to make these energies usable. Using solar energy has great potential. Solar radiation, which provides the energy, can be converted by photovoltaics or heat engines in order to generate electricity, heating or cooling. Besides the technologies that can convert solar radiation into useful energy, one may also think of using the warmth of solar radiation in an intelligent way by arranging building spaces or choosing building material in such a way that the sunlight is well captured and evenly distributed. Wind energy needs wind turbines in order to be useful as a renewable energy. Since the power output from a wind turbine depends strongly on the wind speed, wind energy will gain much success in areas where there are strong winds at an average constant rate. One may think of high altitudes such as plateaus, the situation of hills or offshore facilities. Another renewable energy source, water, has an interesting capacity of storing energy and using it when needed. The much greater density of water (with respect of air) permits producing a considerable amount of energy with even a small flow. Plants and the biological domain can offer great advantages if properly used in order not to affect the nature. On one side, there is biomass, relying on the capturing of energy of the sun by plants by means of photosynthesis. When these plants are used for combustion purposes, the stored energy is then released. Biomass can thus be considered as the natural storage of solar energy. On the other side, there is biofuel from different kind of natural sources such as trees, grass, sugar, starch crops or vegetable oils, to mention a few. In most cases, the biofuel is used as an additive to either gasoline or diesel, attaining in some cases reduced levels of particulates, CO and hydrocarbons from internal combustion engines. The last example of renewable energy, mentioned before, is geothermal energy. This type of energy is obtained by the heat of the earth itself. This heat can be used for making steam out of water, which can be used directly in a steam turbine, creating electricity. Another form of geothermal energy is hot water coming from within the earth that can be used through heat exchangers alimenting another turbine circuit. Of course, this depends largely on the closeness of hot water and regions where the heat is closer to the earth's surface.

These are some important examples of the various green energies that are used and/or are in development. Another side is the existing technology for the green energies and the (economic/commercial) feasibility of the processes. Many renewable energy technologies show an increasing growth in the past few years. These forms of energy can be expensive. However, as time passes, renewable energy can become cheaper. This can be understood by the fact that once the facilities for using renewable energy are built, the renewable energy source is free (*i.e.* sun, wind). Furthermore, the improvement of the existing technologies renders such processes more efficient and less costly. The latter is also achieved if these technologies will be used at a larger scale (the material costs decrease if the production is increased). This has the consequence that many other new renewable energy technologies are being developed and others still under research.

As can be seen, renewable energies and the corresponding needed technologies, in other words, green energy and technology, are of great interest. Already some developments are performed and others need yet to be made. This book aims at discussing and presenting some important results in this field concerning recent developments and an up-to-date synthesis of parts of the studied green energy technologies. This book is divided in three parts. The first part treats how green energy can reduce the impact on the environment. The second part deals with renewable energy sources. The third part shows various applications of green energy on the transport sector. The chapters in these three parts discuss a wide variety of green energy technologies, their applications as well as feasibility studies. As such, it presents an overall look of the present state-of-the-art in this field and of new results, both of which present a critical and promising look on the present and future use of green energy and technology.

1

CHAPTER 1

Cost-Optimal Use of Bioenergy Under a Stringent Climate Stabilization Target

T. Takeshita*

Transdisciplinary Initiative for Global Sustainability, The University of Tokyo, Japan

Abstract: Using a global energy model describing the bioenergy sector in detail, this chapter examines the cost-optimal use of modern bioenergy over the period 2010-2100 under a 400 ppmv CO_2 stabilization constraint and its potential contribution to satisfying this stringent constraint. The following three main results are obtained. First, it is cost-optimal to use modern bioenergy largely to generate heat and replace direct coal use until around 2040. As second-generation bioenergy conversion technologies and CO_2 capture and storage (CCS) technologies become mature in the second half of the century, it becomes cost-optimal to produce biofuels and electricity using these technologies. All biomass gasification-based conversion technologies are combined with CCS (called BECCS) from 2060. Second, introducing modern bioenergy, particularly the strategy of negative CO_2 emissions provided by BECCS, makes a substantial contribution to stabilizing the atmospheric CO_2 concentration at 400 ppmv in 2100 and is a robust future technology option under such a stringent climate stabilization constraint. However, from around 2060, bioenergy supply potentials place a severe limit on the amount of modern bioenergy produced. Third, under the 400 ppmv CO_2 stabilization constraint, BECCS holds a large share of the global amount of CCS throughout the time horizon and offers great flexibility in the timing of CO_2 reductions, whose value is estimated to be as high as $13.3 trillion in constant 2000 US dollars. A significant portion of the CO_2 capture is implemented in now-developing regions, implying the importance of the effective transfer of CCS technologies to now-developing regions for achieving stringent climate stabilization targets.

Keywords: Global warming, energy, supply, energy supply security, cost optimal use, bioenergy, energy policy, stringent climate constraint, CO_2 neutral, global energy system, optimization, model, alternative fuel, sensitivity analysis, stabilization, target.

INTRODUCTION

In general, the two major challenges for current world energy policy are to

***Address correspondence to T. Takeshita:** Transdisciplinary Initiative for Global Sustainability, The University of Tokyo, Japan; Tel: +81-3-5841-8576; Fax: +81-3-5841-1545; E-mail: takeshita@ir3s.u-tokyo.ac.jp

Hatim Machrafi (Ed)

improve energy supply security and to address environmental concerns associated with energy conversion and use [1]. From an energy supply security point of view, introducing alternative fuels to petroleum products could play an important role because of the world's heavy dependence on oil, the uneven distribution of oil resources, and a possible peak in conventional oil production before 2030 [2]. From an environmental point of view, the challenge is much more difficult to achieve. Global warming, which is mainly caused by energy use and resulting generation of CO_2, has become one of the most critical issues for all humankind. The Intergovernmental Panel on Climate Change (IPCC) Fourth Assessment Report (AR4) [3] states that using the "best estimate" assumption of climate sensitivity, limiting global mean temperature increases to 2–2.8°C above the pre-industrial level, at equilibrium, would require CO_2 emissions to peak before 2020.

Among various energy policy tools to tackle these problems, substituting modern bioenergy for fossil fuels is a very promising candidate for three reasons. First, there is a high degree of substitutability between modern bioenergy and each type of fossil fuels, because it can be converted into a variety of modern energy carriers such as heat, electricity, liquid biofuels (*i.e.* biodiesel, bioethanol, methanol, dimethyl ether (DME), and Fischer-Tropsch (FT) synfuels), and gaseous biofuels (*i.e.* biogas and hydrogen). Second, biofuels can provide a domestic rather than imported source of transportation fuels for many countries [1]. Even if imported, biofuels would likely come from regions other than those producing petroleum. Hence, expanding the use of modern bioenergy instead of fossil fuels could help meet the policy goal of enhancing energy supply security. Third, modern bioenergy is "CO_2 neutral" since the carbon it emits into the atmosphere when burned is offset by the carbon plants absorb from the atmosphere while growing [1]. Moreover, it should be emphasized that the combination of bioenergy and CO_2 capture and storage (CCS) can yield negative CO_2 emissions (*i.e.* a net removal of CO_2 from the atmosphere) because the CO_2 put into storage comes from biomass and the biomass absorbs CO_2 as it grows [4]. These suggest that modern bioenergy has a possibility of greatly reducing CO_2 emissions throughout the fuel cycle.

The aim of this chapter is to examine the cost-optimal use of modern bioenergy over the period 2010-2100 under a stringent climate stabilization constraint which might ensure the avoidance of dangerous climate change and to assess its potential contribution to such a climate stabilization regime. This is done by using a regionally disaggregated global energy model with 70 regions (REDGEM70), which is characterized by a detailed technological representation [5]. In this model simulation study, a special focus is placed on the potential role and value of the combination of bioenergy conversion technologies and CCS (called BECCS hereafter) in satisfying such a stringent climate constraint. Also, sensitivity analysis is performed to test the robustness of the findings.

The rest of the chapter proceeds as follows. First, the structure of the REDGEM70 model is outlined and then key input data and assumptions for the model are given. The model simulation results and discussion are presented subsequently, finishing with the conclusions.

MODEL STRUCTURE

REDGEM70 is a bottom-up type, global energy systems optimization model formulated as an intertemporal linear programming problem. With a 5% discount rate, the model is designed to determine the optimal energy strategy from 2010 to 2100 at 10-year time steps for each of 70 world regions so that total discounted energy system costs are minimized under constraints on the satisfaction of exogenously given useful energy or energy service demands, the availability of primary energy resources, the market penetration rate of new technologies, and so forth. The 70 regions of the model are categorized into "energy production and consumption regions" and "energy production regions." The whole world is first divided into the 48 energy production and consumption regions to which future useful energy or energy service demands are allocated. The 22 energy production regions, which are defined as geographical points, are then distinguished from the energy production and consumption regions to represent the geographical characteristics of the areas endowed with large amounts of primary energy resources. Such a detailed regional disaggregation enables the explicit consideration of the regional characteristics in terms of energy resource supply, energy demands, and geography. The model has a full flexibility in when and where CO_2 emissions reductions are achieved to stabilize the atmospheric CO_2 concentration at a given level.

REDGEM70 is also characterized by a detailed description of the whole energy system, from primary energy supply through energy conversion to final energy consumption, as illustrated in Fig. **1**. In the downstream part of the model, future useful energy demand trajectories are given exogenously for each of the industrial and residential/commercial sectors and decomposed by demand category, whereas future energy service demand trajectories (expressed in passenger-km or tonne-km) given for each of 13 transportation modes. For each end-use demand category, the possibility of price-induced energy conservation, substitutability among final energy carriers (*e.g.,* high-quality energy carriers can be used for a wide range of applications), and changes in efficiency and costs associated with final energy substitution are considered in the model. Various transport technologies, which refer to possible combinations of propulsion systems and transportation fuels, are characterized by parameters such as energy intensity, capital cost, and operating and maintenance cost, and their cost-optimal mix is endogenously determined for each transportation mode.

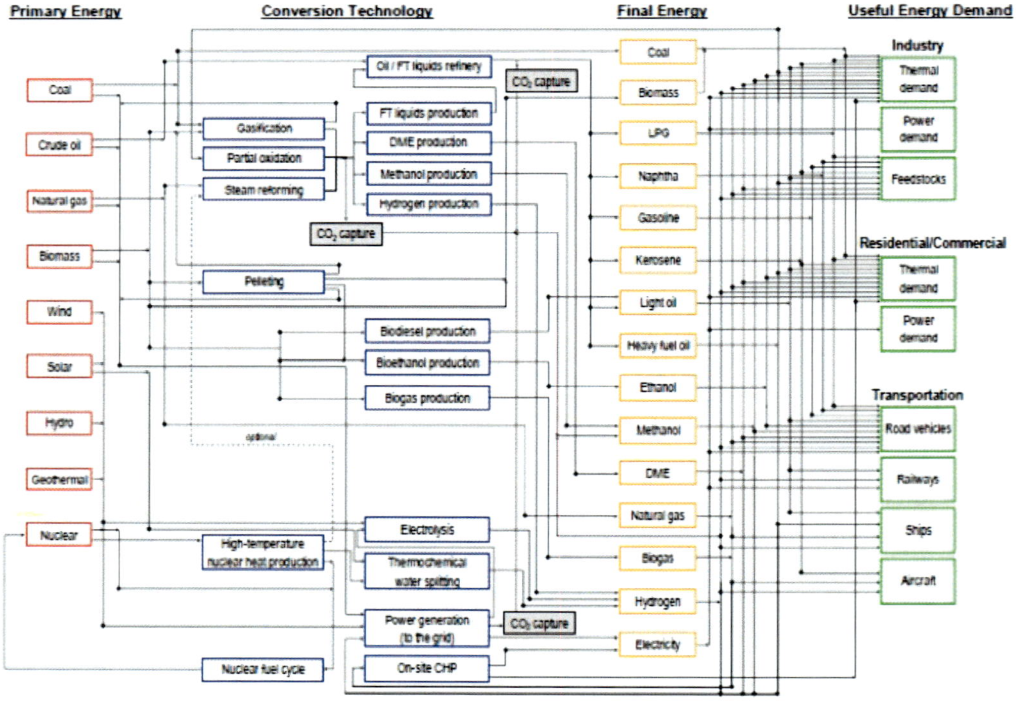

Figure 1: Schematic representation of the structure of REDGEM70.

On the supply side, REDGEM70 considers the entire supply chain of final energy carriers, which includes primary energy production, interregional energy transportation, coastal storage (in the case of sea transportation), conversion into secondary energy, intraregional secondary energy distribution, and final energy supply at retail sites (*e.g.,* refueling). To represent the economics of each of these final energy supply chain stages in a realistic manner, the model considers the capital and operating costs separately at each stage of the fuel supply chain (excluding resource extraction) by treating the corresponding infrastructure explicitly. Note that final energy carriers are not always supplied in this order: a wide variety of final energy supply patterns can be selected in the model. The model treats the interregional transportation of 10 types of energy carriers and CO_2 between representative cities/sites in the 70 model regions and is able to identify its cost-optimal evolution path (for a more detailed description of how to model their interregional transportation, see [5]). Furthermore, the model considers the difference in the cost of local secondary energy distribution not only by energy carrier, but also by time point, region, and end-use sector. The siting characteristics of secondary energy production plants are also taken into account in the model. That is, although centralized secondary energy production plants using fossil fuels can be constructed near demand centers, the location of those using renewable or nuclear energy (except for wood pellets) is constrained by resource availability or safety concerns. Thus, the cost associated with intraregional secondary energy transportation to demand centers is added for these supply options. To make such modeling possible, the spatial structure of an energy production and consumption region is represented in detail in the model by consideration of the distribution of energy system components in this type of model regions.

REDGEM70 considers a number of promising energy conversion technologies. In particular, the model comprehensively includes technologies for producing alternative energy carriers such as synthetic fuels (*i.e.* hydrogen, methanol, DME, and FT synfuels) and conventional biofuels (*i.e.* bioethanol, biodiesel, and biogas). The bioenergy sector is described in great detail in the model. The structure of the bioenergy submodel is schematically illustrated in Fig. **2**. For biomass resources, the model considers not only plantation biomass such as energy crops, which are defined as fast-growing trees (*e.g.,* hybrid poplars and willows) in the model, modern fuelwood, sugar crops, grain crops, and oilseed crops, but also waste biomass. Given the amount of excess cropland that can be used for energy purposes, the model determines its optimal allocation among different plantation-based crop biomass productions based on crop yields per hectare of land, crop supply costs, and characteristics of conversion technologies available. The characteristics of bioenergy conversion technologies are specified for each feedstock in the light of the unique properties of biomass feedstocks.

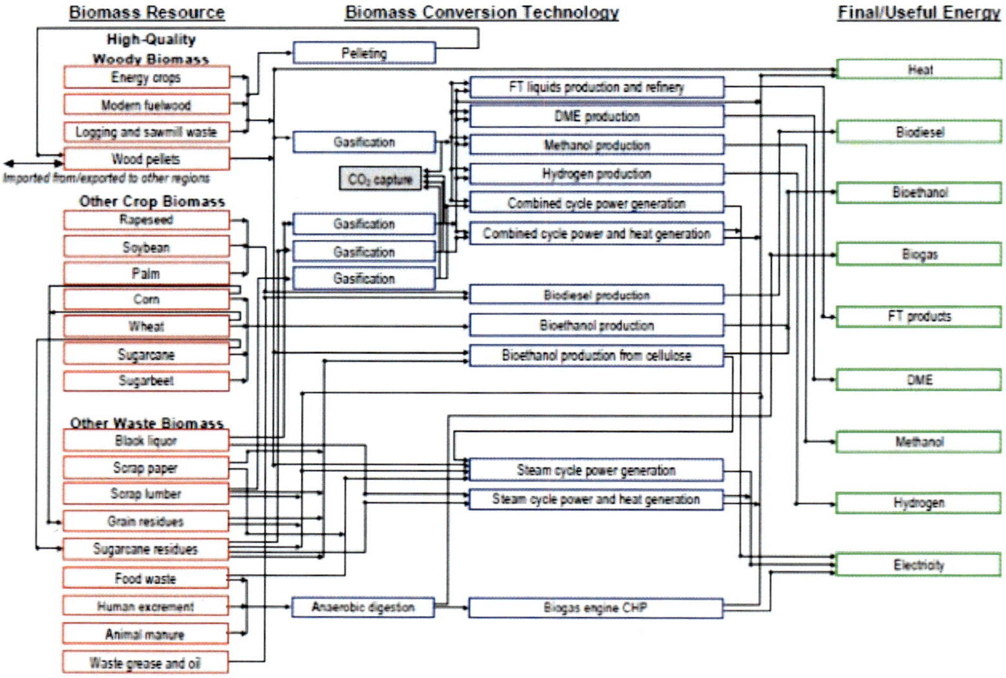

Figure 2: Schematic representation of the structure of the bioenergy submodel of REDGEM70.

In REDGEM70, the CO_2 generated from power plants (excluding those used for on-site combined heat and power production (CHP) and biomass-fired steam cycle power production), synthetic fuels production plants (excluding those used for converting stranded gas and decentralized small-scale hydrogen production), and oil/FT refinery plants can be captured for subsequent sequestration in geologic formations or methanol synthesis. As shown in Fig. **2**, the combination of biomass gasification-based conversion technologies and CCS is included as a

BECCS option in the model, which removes CO_2 from the atmosphere. For CO_2 capture technologies, the model considers that physical absorption is the preferred method at high CO_2 concentrations (*e.g.,* in the synthesis gas (syngas) produced by the gasification process), while chemical absorption is the preferred method at low CO_2 concentrations (*e.g.,* in flue gas of thermal power plants). The geological CO_2 sequestration options included are enhanced oil recovery, enhanced coalbed methane recovery, depleted natural gas well disposal, and deep aquifer disposal. Parameters for CO_2 capture and its transportation to storage sites are specified for each combination of conversion technology and feedstock.

DATA AND ASSUMPTIONS

Scenario and Driving Forces

Future trajectories for scenario driving forces such as population, gross domestic product measured in purchasing power parities (GDP_{ppp}), and end-use demands are based on the "Middle Course" case B developed by the International Institute for Applied Systems Analysis (IIASA) and the World Energy Council (WEC) [6]. Because end-use demand projections were originally for 11 world regions used in [6], they were disaggregated into the 48 energy production and consumption regions of REDGEM70. This was done by using the country- or state-level statistics/estimates (and projections if available) on population, GDP_{ppp}, geography, energy use by type, and transportation activity by mode, and by taking into account the underlying storyline of the case B that regional diversity might be somewhat preserved throughout the 21st century.

For non-energy-related CO_2 emissions, the future global CO_2 emissions trajectory from cement production and gas flaring was taken from the IPCC Special Report on Emissions Scenarios B2 scenario [7] in the light of its comparability with the IIASA/WEC case B [7, 8], while that from deforestation was taken from [9].

Primary Energy Resource Availability

Tables **1** and **2** show the input data for plantation biomass and waste biomass, respectively. Monetary values are all expressed in constant 2000 US dollars. The estimates of the availability of various biomass resources and excess cropland are those that can be used for energy purposes without conflicting with other biomass uses such as the production of food, paper, lumber, and traditional fuelwood. They were estimated for each of 11 world regions and each time point assuming that biomass is produced in a sustainable way so that biomass-derived energy carriers can be regarded as CO_2 neutral (for details see [9]). Fig. **3** shows the resulting global supply potential of bioenergy by resource type and by region. Throughout this chapter, an 11-region classification is identical to that of the joint IIASA/WEC study [6].

On the other hand, assumptions on the availability and extraction cost of fossil fuel resources and uranium resources were derived from [11] and [12], respectively. Fossil fuel resource availability is limited to the categories labeled as "resource base" in [11]. The potentials and supply cost of solar, wind, hydro, and geothermal energy sources were taken from [13]. These resource availability estimates were then allocated to the 70 model regions by using country-, state-, or site-level statistics/estimates.

Table 1: Data for plantation biomass[a].

Biomass Type	**Supply Cost[b] (US$$_{2000}$/toe)**	**Feedstock Yield[c] (tonne/ha/year)**	**Energy Content of Feedstock (toe/tonne-feedstock)**
Energy crops	76.4-277.1[d]	10.0-20.0	0.358
Modern fuelwood	220.6-862.2[d]	3.5-18.8	0.358
Rapeseed (in the EU)	206.2	3.7-4.5	0.669
Soybean (in the US)	166.4	3.2-3.9	0.436
Palm (in Malaysia and Indonesia)	137.1	4.0-4.9	0.939
Corn (in the US and China[e])	150.9	7.2-8.8	0.326
Wheat (in the EU and Canada)	284.0	6.6-8.1	0.334
Sugarcane (in Brazil)	132.2	68.7-83.8	0.121
Sugarcane (in South Asian countries, China[f], southeastern Asian countries, Australia, the US[g], South Africa, Egypt, and the other Latin American countries)	264.5	56.0-68.3	0.121
Sugarbeet (in the EU)	98.7	51.1-62.3	0.452

[a] *Source:* [10]. toe = tonnes of oil equivalent.

[b] The supply cost for biomass resources includes the costs for their harvest, intraregional transportation, storage, and processing.

[c] The range of these parameter values denotes that crop production per hectare for these crops is assumed to improve at 1% per year over the period 2000-2020, except that the range of the parameter values for energy crops and modern fuelwood denotes the difference by region and by time point.

[d] The range of these parameter values denotes the difference by region and by resource category.

[e] Excludes southern coastal China.

[f] Limited to southern regions (*i.e.* Guangxi, Guangdong, and Yunnan provinces).

[g] Limited to southern regions (*i.e.* Florida, Louisiana, Texas, and Hawaii).

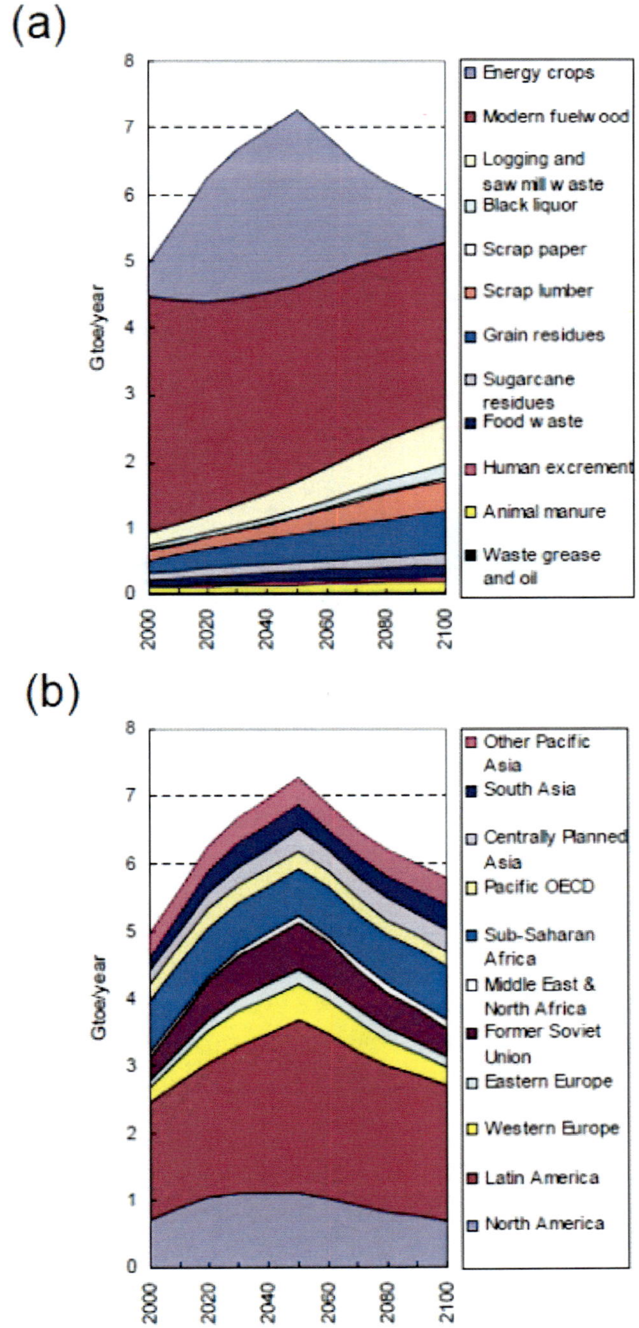

Figure 3: Global supply potential of bioenergy by resource type (a) and by region (b). This estimate was made assuming that all the available excess cropland is allocated to producing energy crops.

Table 2: Data for waste biomass[a].

Biomass Type	Supply Cost[b] ($US\$_{2000}$/toe)
Logging and sawmill waste	77.0-154.0[c]
Black liquor	0
Scrap paper	0
Scrap lumber	35.9
Grain residues	154.0
Sugarcane residues	0 (bagasse) 102.6 (trash)
Food waste	0
Human excrement	0
Animal manure	0
Waste grease and oil	0

[a] *Source:* [10].

[b] The supply cost for biomass resources includes the costs for their harvest, intraregional transportation, storage, and processing.

[c] The range of this parameter value denotes the difference by resource category.

Energy Conversion and Transportation Technologies

Tables **3-5** show the input data for power generation technologies (excluding those using solar, wind, hydro, and geothermal energy sources), synthetic fuels production technologies, and conventional biofuels production technologies, respectively. The actual model inputs for plant capital costs vary by region by applying a region-specific plant location factor, whose values are shown in Table **6**. The cost reduction and performance improvement in the future were estimated for each technology option based on the concept of technology clusters [16]: technologies are grouped into clusters depending on their technological proximity, and then a similar trend is assumed for the future cost and performance of technologies belonging to the same cluster. It is important to note that gasification-based conversion technologies are assumed to have a relatively large potential for technological improvements in the future [10].

The lower heating value (LHV) efficiency of direct heat generation from modern bioenergy is assumed to be 85% [10]. In the model, biomass feedstocks available for synthetic fuels production are limited to high-quality woody biomass (including wood pellets) and black liquor because of the necessity to produce a clean syngas with low levels of contaminants [14]. Among several gasifier designs, the directly-heated, pressurized, air/oxygen-blown, fluidized-bed gasification process developed by the Institute of Gas Technology (IGT) was chosen as the woody biomass gasifier, while the pressurized, oxygen-blown, entrained flow gasification process (*i.e.* the Chemrec process) was chosen as the black liquor gasifier. When CO_2 capture is an objective, pressurized, oxygen-blown gasification processes such as the IGT process are preferred because of the higher CO_2 partial pressures [17].

Table 3: Data for power generation technologies[a,b].

Technologies	**Capital Cost[c] (US$$_{2000}$/kW)**	**Conversion Efficiency[c] (%, LHV basis)**	**Maximum Capacity Factor (%)**	**Technical Lifetime (years)**
Coal-fired power generation				
Coal-fired steam cycle	1322	40.7-50.0	85	40
IGCC	1763-1433	42.2-55.0	85	40
Coal-fired IGCC-SOFCs[d]	1984	56.0-60.0	85	30
Oil-fired power generation	937	44.0-52.0	85	30
Natural gas-fired power generation				
Natural gas-fired steam cycle/NGCC	827-551	47.0-65.0	85	30
NGCC-SOFCs[d]	882	66.0-70.0	85	30
Nuclear power generation				
Light water reactors	2424-1929	34.0	85	40
Fast breeder reactors [d]	3637-2893	40.0	85	40
Biomass-fired power generation				
Biomass-fired steam cycle				
using wood chips	1653	23.0-34.0	85	30
using wood pellets	1579	27.3-34.0	85	30
using grain residues	1873	23.0-34.0	85	30
using sugarcane residues (a uniform mixture of bagasse and trash)	1653	20.4-30.2 (electricity) 18.5 (heat)	85	30
using black liquor	3891	12.2-13.0 (electricity) 61.0 (heat)	85	30
using municipal wastes	1433	21.0-31.0	85	30
Biomass IGCC[e]				

Table 3: cont….

using wood chips	2109-1726	36.0-45.0	85	30
using wood pellets	2023-1655	38.1-47.1	85	30
using sugarcane residues (a uniform mixture of bagasse and trash)	2452-2006	31.2-37.9 (electricity) 18.5 (heat)	85	30
using black liquor	1771-1450	24.6-28.7 (electricity) 45.7 (heat)	85	30
Biogas CHP using a gas engine	2821	35.0-40.0 (electricity) 84.0 (total efficiency)	85	20
Hydrogen-fired power generation using a gas turbine	1984-1818	51.4-65.0	85	30
Methanol-fired power generation using a gas turbine	2094-1929	44.9-60.0	85	30
DME-fired power generation using a gas turbine	1482-992	49.0-62.0	85	30
CHP by stationary fuel cells				
Hydrogen-fueled PEMFCs used for residential/commercial applications	12893-1289	40.9-50.0 (electricity) 40.0 (heat)	50	15
Natural gas-fueled PEMFCs used for residential/commercial applications	14546-1455	32.7-40.0 (electricity) 40.0 (heat)	50	15
Hydrogen-fueled SOFCs used for residential/commercial applications[e]	12232-1223	31.7-42.8 (electricity) 50.9-47.2 (heat)	50	10
Natural gas-fueled SOFCs used for residential/commercial applications[e]	12508-1251	40.7-55.0 (electricity) 41.9-35.0 (heat)	50	10
Hydrogen-fueled SOFCs used for industrial applications[e]	9918-992	36.6-49.4 (electricity) 46.0-40.6 (heat)	85	10
Natural gas-fueled SOFCs/MCFCs used for industrial applications[e]	10138-1014	47.0-63.5 (electricity) 35.6-26.5 (heat)	85	10

[a] *Source:* The data for power plants using biomass feedstocks are from [10].

[b] IGCC = integrated gasification combined cycle; NGCC = natural gas combined cycle; SOFC = solid oxide fuel cell; PEMFC = proton exchange membrane fuel cell; MCFC = molten carbonate fuel cell.

[c] These ranges denote the assumed evolution of the parameter values. The left-side value of the range denotes the value in 2000, while its right-side value denotes the value in 2100. All cost numbers are based on North American manufacture and construction.

Table 3: cont….

[d] Assumed to be available from 2030.
[e] Assumed to be available from 2010.

Table 4: Data for synthetic fuels production technologies.[a]

Technologies	Capital Cost[b] (US$$_{2000}$/toe-fuel/year)	Conversion Efficiency[b] (%, LHV basis)	Electric Power Required (MWh/toe -feedstock)
Centralized large-scale hydrogen production			
Coal gasification	1037-749	64.0-72.5	0.074
Partial oxidation of fuel oil	695-614	71.0-72.8	0.035
Steam methane reforming of natural gas	398-352	79.2-81.2	0.293
High-quality woody biomass gasification[c]	1029-744	60.6-68.7	0.582
Black liquor gasification[c, d]	1454-1050	53.6-60.8 (fuel) 38.3 (heat)	0.014
Electrolysis of water using grid AC power	970-856	68.2-69.9	-
Electrolysis of water using solar, wind, hydro, or nuclear DC power on a stand-alone basis	745-658	71.0-72.8	-
Steam methane reforming of natural gas using nuclear heat[e]	484-426[f]	70.0-71.8[g]	0.652[g]
Iodine-sulfur thermochemical water splitting cycle using nuclear heat[h]	807-645[f]	45.0-51.0	0
Iodine-sulfur thermochemical water splitting cycle using solar heat[h]	2958-2362	56.0-63.4	0
Decentralized small-scale hydrogen production			
Steam methane reforming of natural gas	1791-1580	72.8-74.6	0.330
Electrolysis of water using grid AC power	2179-1923	68.2-69.9	-
Methanol production (full conversion concept)			
from coal	1490-1077	57.7-65.4	0.279
from natural gas	728-642	70.0-71.8	0.161
from high-quality woody biomass[c]	1576-1138	54.5-61.7	0.255
from black liquor[c, d]	2226-1608	55.7-63.1 (fuel) 37.9 (heat)	-0.244
from hydrogen and carbon dioxide	302	89.7	1.481
DME production (full conversion concept)[i]			
from coal[c]	1632-1179	55.2-62.5	0.345
from natural gas[c]	798-704	71.0-72.8	0.069

Table 4: cont….

from high-quality woody biomass[c]	1726-1246	52.2-59.1	0.314
from black liquor[c, d]	2312-1670	56.5-64.0 (fuel) 37.9 (heat)	-0.223
Raw FT liquids production (full conversion concept)[j]			
from coal	1539-1112	47.3-53.6	0.087
from natural gas	784-691	61.6-63.2	0
from high-quality woody biomass[c]	1873-1353	46.0-52.1	0.177
from black liquor[c, d]	2700-1951	44.3-50.2 (fuel) 37.9 (heat)	-0.223

[a] *Sources:* [10, 14].
[b] These ranges denote the assumed evolution of the parameter values. The left-side value of the range denotes the value in 2000, while its right-side value denotes the value in 2100. All cost numbers are based on North American manufacture and construction.
[c] Assumed to be available from 2010.
[d] A mill steam deficit that occurs is assumed to be compensated using imported woody biomass (for details see [10]).
[e] Assumed to be available from 2020.
[f] The capital cost of a high temperature gas-cooled reactor (including the intermediate loop) is excluded. It is estimated at $US\$_{2000}$ $469/kW_{th}$.
[g] The energy input (feedstock) into the process is defined as natural gas and nuclear heat combined.
[h] Assumed to be available from 2030.
[i] Single-step DME synthesis is considered here.
[j] Assumptions on the refinery process streams for raw FT liquids are described in [14].

Table 5: Data for conventional biofuels production technologies[a].

Technologies	**Capital Cost[b] ($US\$_{2000}$/toe-fuel/year)**	**Conversion Efficiency[b] (%, LHV basis)**	**External Process Energy (toe/toe-feedstock)**
Bioethanol production			
from high-quality woody biomass[c]	1713-1230[d]	47.0-57.3	0
from wood pellets[c]	1469-1055[d]	54.8-66.9	0
from corn (in the US and China)	704[d]	56.0	Fossil fuels of 0.280
from wheat (in the EU and Canada)	1348[e]	55.0	Fossil fuels of 0.446
from sugarcane (in Latin American countries, South Asian countries, China, southeastern Asian countries, Australia, the US, South Africa, and Egypt)[f]	280[g]	38.0	0
from sugarbeet (in the EU)	921	12.0	Fossil fuels of 0.067
from cellulosic waste biomass[c]	1966-1412[d]	38.0-46.4	0

Table 5: cont….

Biodiesel production			
from rapeseed (in the EU)	312	29.0	Fossil fuels of 0.096 and co-produced glycerine
from soybean (in the US)	374	29.0	Fossil fuels of 0.096 and co-produced glycerine
from palm (in Malaysia and Indonesia)	302	29.0	Fossil fuels of 0.096 and co-produced glycerine
from waste grease and oil	468[d]	29.0	Fossil fuels of 0.096 and co-produced glycerine
Biogas production			
in now-developed countries	305	50.0	0
in now-developing countries	506-305	50.0	0

[a] *Source:* [10].
[b] These ranges denote the assumed evolution of the parameter values. The left-side value of the range denotes the value in 2000, while its right-side value denotes the value in 2100.
[c] Assumed to be available from 2010. Electricity can be co-generated at a rate of $0.092kW_e/kW_{ethanol}$.
[d] Based on North American manufacture and construction.
[e] Based on European manufacture and construction.
[f] A bagasse surplus of $0.088kW/kW_{ethanol}$ can be used for electricity co-generation.
[g] Based on Brazilian manufacture and construction.

Table 6: Location factors for plant capital costs in the years 2000, 2050, and 2100.[a]

Region	For Power Plants			For the Other Types of Plants		
	2000	**2050**	**2100**	**2000**	**2050**	**2100**
North America	1	1	1	1	1	1
Western Europe	0.993	0.995	0.996	0.987	0.991	0.993
Pacific OECD	1.001	1.000	0.999	1.002	1.000	0.999
Eastern Europe	0.931	0.957	0.978	0.878	0.924	0.961
Former Soviet Union	0.930	0.956	0.976	0.877	0.922	0.958
Latin America	0.942	0.954	0.969	0.898	0.920	0.946
Middle East & North Africa	0.936	0.941	0.956	0.887	0.896	0.922
Sub-Saharan Africa	0.898	0.907	0.943	0.821	0.837	0.900
Centrally Planned Asia	0.902	0.936	0.959	0.827	0.888	0.928
South Asia	0.891	0.914	0.954	0.808	0.848	0.919
Other Pacific Asia	0.934	0.957	0.968	0.884	0.924	0.944

[a] *Source:* [15].

The transportation of oil or FT products by tanker is generally the cheapest means of interregional energy transportation. Other liquid fuels such as bioethanol, methanol, and DME are slightly more expensive to transport due to their relatively low volumetric energy density. In contrast, the interregional transportation of electricity and hydrogen is very costly. These points make a sharp contrast with the difference between synthetic fuels production costs. For biomass transportation, high-quality woody biomass (*i.e.* energy crops, modern fuelwood, and logging and sawmill waste) is assumed to be pelletized before transportation to increase its volumetric energy density. Even so, the transportation costs of solid fuels are higher than for liquid fuels. A detailed description of the estimated data for interregional energy transportation costs (including the costs and energy requirements needed for compression, liquefaction, and regasification) was given by [5, 14]. On the other hand, the trend of the interregional transportation costs by energy carrier is almost true for the intraregional distribution costs and retail supply costs (on an energy basis) by energy carrier. For these cost estimates, see [5, 18].

CCS Technologies

The capital costs of CO_2 capture systems were estimated for each type of new energy conversion plants based on two assumptions. First, the capital cost of a CO_2 capture system per unit of CO_2 capture capacity is assumed to scale with CO_2 capture capacity with a scaling factor of 0.608 (the scaling factor for a CO_2 capture system, using data from [19], was estimated to be 0.608). Second, for typical modern energy conversion plants, the assumed plant input capacity is 1632 MW_{feed} for coal, 704 MW_{feed} for natural gas, and 344 MW_{feed} for modern bioenergy [10]. REDGEM70 also includes the option to retrofit CO_2 capture onto energy conversion technologies for which CCS is assumed to be applicable to newly built plants. The cost of retrofitting CO_2 capture onto existing power plants was estimated from [17, 20]. In the case of synthetic fuels production plants, the cost of retrofitting CO_2 capture onto existing plants is assumed to be the same as that of installing it in new plants. The reason is that the cost of CO_2 capture in the production of synthetic fuels is mainly due to CO_2 drying and compression because separation of CO_2 is already carried out as part of their production processes even if CO_2 is vented [17, 19, 21, 22]. This means that synthetic fuels production plants can be regarded as being inherently ready for the addition of CO_2 capture equipment.

As a result, the incremental capital costs of CO_2 capture in terms of $US\$_{2000}$ per tonne of carbon captured per year were estimated to be 240.3 for power generation from new biomass gasification combined cycle (BGCC) plants, 395.4 for power generation from existing BGCC plants, 62.2 for hydrogen production from woody biomass, 75.9 for methanol production from woody biomass, 89.4 for DME production from woody biomass, and 73.1 for raw FT liquids production from woody biomass, respectively, if commercially-mature (N^{th} plant) CO_2 capture plant designs are assumed. The resulting estimates (for details see [10]) indicate that the incremental capital costs of CO_2 capture are higher for woody biomass-based plants than for coal-based plants by a factor of about 2.1, but are lower for woody biomass-based plants than for natural gas-based plants. This trend can be explained by two factors. First, a low-cost physical absorption process can be used to capture CO_2 from biomass gasification-based conversion plants, whereas CO_2 is captured using a relatively costly chemical absorption process in natural gas-based plants. Second, because of the high carbon content of biomass and the low efficiency of bioenergy conversion technologies, biomass-based plants produce more capturable CO_2 than natural gas-based plants.

Taking into account the result for the evolution of the share of power plants with CO_2 capture in total power generation capacities presented in [23], a constraint is imposed on the growth rate of CO_2 capture capacities in the model. That is, an annual growth rate of 1.1% was assumed for the share of energy conversion plants equipped with CCS in their total capacities (according to results from [23], the share of coal-fired power plants with CO_2 capture in total coal-fired power generation capacities was 23.8% in 2070 and 65.1% in 2080).

For intraregional CO_2 transportation, the cost of transporting CO_2 captured from fossil-based plants to a storage site was estimated to be $US\$_{2000}$ 20.2/t-C, assuming that captured CO_2 is transported through 250 km of pipeline [16] and that CO_2 pipeline "backbones" would be developed by which a number of centralized energy conversion plants are connected with storage sites [24]. In contrast, due to the dispersed nature of biomass feedstocks, transportation of CO_2 captured from biomass-based plants to a storage site is assumed to suffer from diseconomies of small scale, the cost of which was estimated to lie in the range of 49.5-80.1 $US\$_{2000}$/t-C using the relationship between the CO_2 pipeline transportation cost and CO_2 flow rate given by [22, 24]. For the costs of transporting CO_2 interregionally, see [5].

The global CO_2 storage capacity in geologic formations has been estimated to be 998 Gt-C at costs for CO_2 storage and monitoring in the range of 11.8-713.8 US$_{2000}$/t-C stored [15, 17]. The regional distribution of the CO_2 storage capacity of deep aquifers is based on the country-level estimates and geographic information system data reported by [25], whereas that of the CO_2 storage capacity of oil/gas fields and coalbeds is based on the country-, state-, and site-level estimates of the availability of related fossil energy resources (*e.g.*, conventional natural gas for the potential of depleted natural gas well disposal). The estimated regional CO_2 storage capacity is characterized by an abundant worldwide availability of low-cost aquifer storage options.

Climate Stabilization Policy

Because this chapter focuses on the potential contribution of modern bioenergy to avoiding dangerous climate change, an appropriate constraint is imposed on the atmospheric CO_2 concentration. A CO_2 stabilization target of 400 ppmv was chosen for the simulation: the atmospheric concentration of CO_2 is assumed to be stabilized at 400 ppmv in 2100, which has been assumed to assure stabilization of greenhouse gas (GHG) concentrations at 490 ppmv CO_2-equivalent by 2100 [3]. The reason for this choice is because the IPCC AR4 [3] states that avoidance of many key vulnerabilities requires temperature change in 2100 to be below 2.6°C above pre-industrial levels and estimates that achieving the CO_2 stabilization target of 400 ppmv would be a sufficient condition for limiting the global mean temperature change below 2.6°C above pre-industrial levels, using a best estimate climate sensitivity of 3°C. Similar to [26], overshoots are allowed before 2100 in REDGEM70.

RESULTS AND DISCUSSION

Cost-Optimal Use of Modern Bioenergy and its Contribution to Climate Stabilization

Fig. **4** shows the evolution of the global primary energy production for the 400 ppmv case. If the CO_2 stabilization constraint is not imposed, fossil fuels continue to dominate the global primary energy supply and the contribution of modern bioenergy to it is almost negligible (at most 3.3% in 2080). In contrast, the global primary energy structure becomes diversified in the 400 ppmv case, where natural gas, renewables (mainly modern bioenergy), and nuclear energy gain shares at the expense of coal and oil. It should be noted that modern bioenergy's share of total global primary energy supply increases substantially under the CO_2 stabilization target of 400 ppmv, which accounts for approximately 25% in 2060 and becomes one of the pillars of the primary energy supply structure.

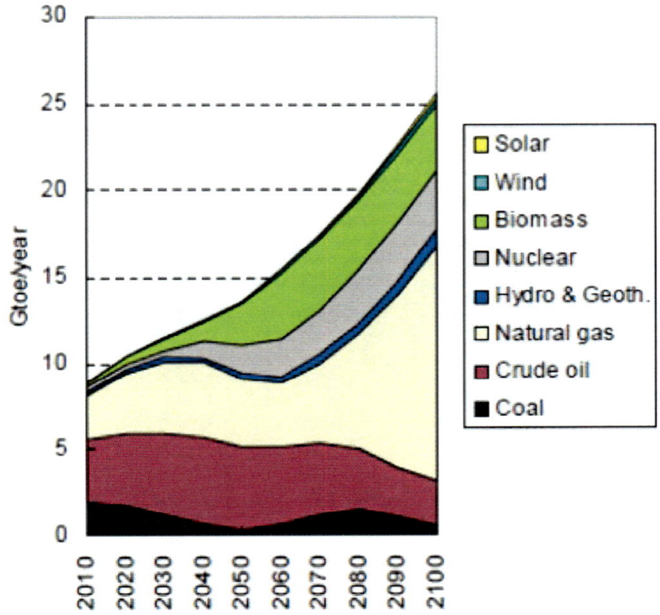

Figure 4: Global primary energy production in the 400 ppmv case. Primary energy production was calculated by using the direct equivalent method for all nonthermal uses of renewables and nuclear energy.

The supply-demand balance for biomass resources and excess cropland is shown in Table **7** for the 400 ppmv case. It can be seen that almost all of the biomass resources and excess cropland are used for energy purposes from 2060 onward. This result means that the global supply of modern bioenergy cannot increase and remains virtually unchanged from 2060 in the 400 ppmv case (see Fig. **4**) because bioenergy supply potentials place a limit on the amount of modern bioenergy that can be introduced and because global bioenergy supply potentials show a declining trend from 2050. The results in Fig. **4** and Table **7** indicate that the large-scale introduction of modern bioenergy is a critical, cost-effective strategy for meeting the stringent climate stabilization constraint.

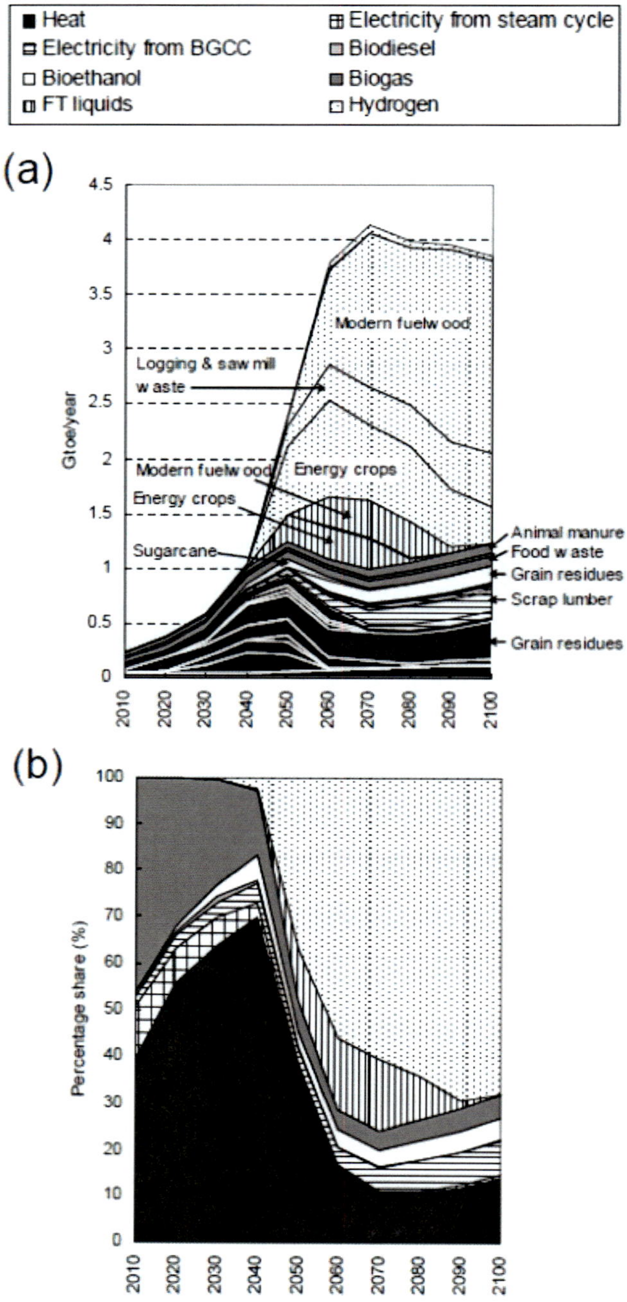

Figure 5: Total global modern biomass-derived final energy consumption by final energy carrier and by supply source in the 400 ppmv case (a) and percentage share of each modern biomass-derived final energy carrier in total global modern biomass-derived final energy consumption in the 400 ppmv case (b).

Table 7: Supply-demand ratio for biomass resources and excess cropland in the 400 ppmv case.

	2010	2020	2030	2040	2050	2060	2070	2080	2090	2100
Biomass resources	15.6	11.9	8.9	5.8	2.3	1.2	1.0	1.0	1.0	1.0
Excess cropland	-[a]	95.4	21.4	7.7	1.5	1.0	1.0	1.0	1.0	1.0

[a] The denominator, *i.e.* the demand, is equal to zero.

Figs. **5(a)** and **(b)** show the total global modern biomass-derived final energy consumption by final energy carrier and by supply source and the share of each modern biomass-derived final energy carrier in total global modern biomass-derived final energy consumption, for the 400 ppmv case, respectively. The share of modern biomass-derived final energy consumption in total global final energy consumption (excluding traditional biomass use) climbs up from 3.1% in 2010 to 29.1% in 2060 and declines thereafter, which is caused by the downward trend of global bioenergy supply potentials from 2050, but remains over 20% in the second half of the century. This supports the view that modern bioenergy would likely play an important role in the stabilization of the atmospheric CO_2 concentration at 400 ppmv in 2100. As shown in these figures, it is cost-optimal to use a dominant part of biomass resources to produce heat and energy carriers other than electricity. Nuclear energy and natural gas, both of which increase their shares in the 400 ppmv case, are two main sources for electricity generation in this CO_2 constrained case: nuclear represents more than nearly half of total global electricity generation in the second half of the century in this case. In short, modern bioenergy and nuclear would contribute to achieving the stringent climate stabilization target in such a manner that they are used for different applications.

Furthermore, there is a clear difference in the cost-optimal strategy for introducing modern biomass-derived final energy carriers into the energy market under the 400 ppmv CO_2 stabilization constraint between the first and second half of the century. Until around 2040, it is cost-optimal to use modern bioenergy largely to generate heat and replace direct coal use. The share of modern biomass-derived heat in total global modern biomass-derived final energy consumption increases up to 69.9% in 2040. The first reason is that, because direct heat generation from modern bioenergy is much more energy efficient than the other modern bioenergy conversion technologies, more CO_2 reductions can be achieved per toe of biomass feedstock or per hectare of cropland if modern bioenergy is used to generate heat and replace coal than if it is used to produce liquid or gaseous biofuels replacing oil or natural gas. This aspect is important in that bioenergy supply costs are relatively high and that the supply-demand balance for modern bioenergy becomes tighter over time. The second reason is that, because later CO_2 emissions reductions are usually cheaper due to technological progress and discounting, modern bioenergy is not preferably used for gasification-based conversion (which can achieve deep cuts in global emissions by combining it with CCS) until later time periods. This result is comparable to that by [27].

On the other hand, as the cost and performance of the so-called "second-generation bioenergy conversion technologies" (*e.g.,* cellulose-to-ethanol production technologies and biomass gasification-based conversion technologies) and CCS technologies are considerably improved in the second half of the century, biofuels and electricity produced using these technologies gain share at the expense of direct heat generation from modern bioenergy. Although developing infrastructure for hydrogen distribution and refueling requires huge investment, hydrogen production from modern bioenergy is preferred because hydrogen production with CCS can sequester larger amounts of CO_2 (the final product, hydrogen, contains no carbon), because the incremental capital costs of CO_2 capture are much lower for hydrogen production than for electricity generation, and because biomass-to-hydrogen production is more energy efficient than electricity generation from modern bioenergy. Hydrogen is used mainly for the following applications: (1) CHP by stationary SOFCs in the industrial sector, (2) FC light-duty vehicles, and (3) liquid hydrogen aircraft.

As regards other modern biomass-derived final energy carriers, FT synfuels, electricity from BGCC, cellulose-derived bioethanol, and biogas account for a visible share in the second half of the century in the 400 ppmv case. Biodiesel is produced from all the available amount of waste grease and oil from 2030, but its small resource potential makes the share of biodiesel negligible. According to the result for the cost-optimal use of biogas, its direct use as a replacement for natural gas is favorable. Grain-derived bioethanol and oilseed-derived biodiesel are not selected mainly because of their high feedstock costs, low biofuels yields per hectare of cropland, and use of fossil fuel as a process energy source.

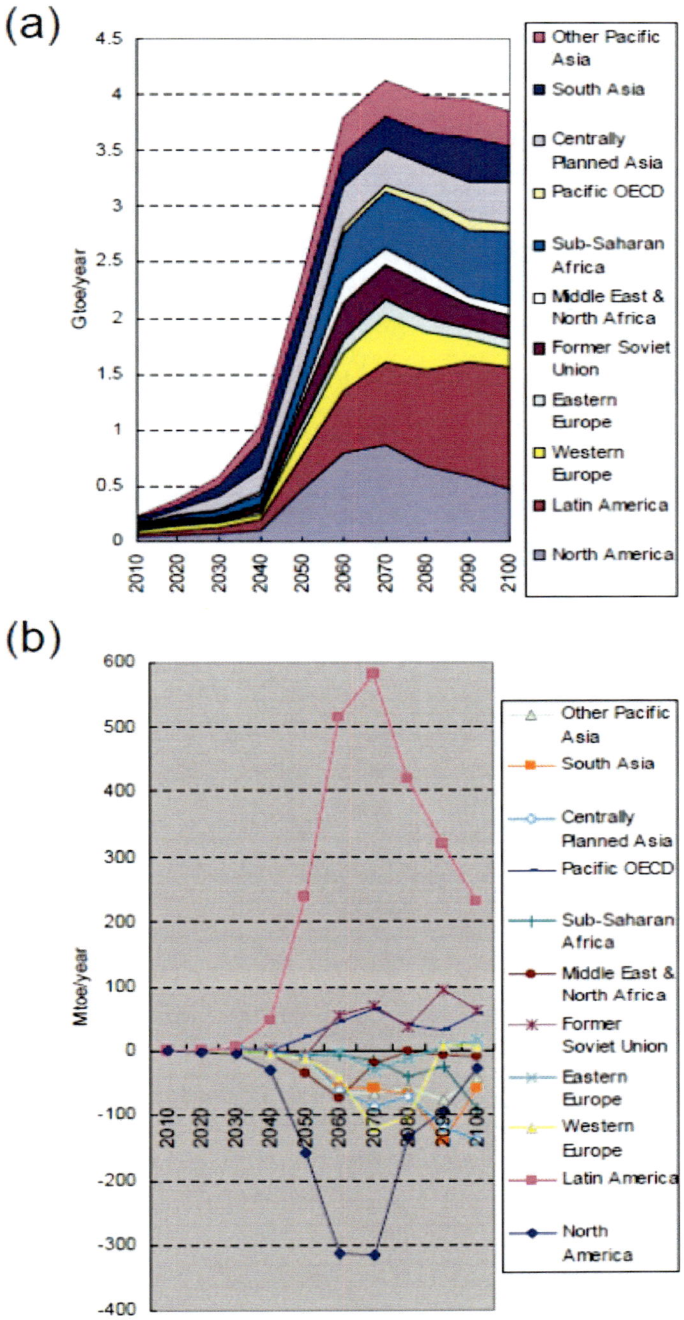

Figure 6: Total global modern biomass-derived final energy consumption by region in the 400 ppmv case (a) and net trade of modern biomass-derived secondary energy carriers for 11 world regions in the 400 ppmv case (b).

To provide some insights into regional differences in the importance and role of modern biomass-derived energy carriers, Fig. **6(a)** and **(b)** show the regional breakdown of total global modern biomass-derived final energy consumption and the net trade of modern biomass-derived secondary energy carriers for 11 world regions, for the 400 ppmv case, respectively. Fig. **6(a)** shows that the degree of regional difference in the amount of modern biomass-derived final energy consumption is obviously lower than the degree of regional difference in bioenergy supply potentials. This leads to the result that modern biomass-derived energy carriers become one of the important sources of final energy consumption in each of these world regions in the 400 ppmv case. As shown in Fig. **6(b)**, this can be accomplished by an extensive interregional trade of modern biomass-derived secondary energy carriers.

Table **8** shows the global penetration rate of CO_2 capture in all types of plants to which CCS is applicable and in biomass gasification-based conversion plants for the 400 ppmv case. As can be seen in this table, the CO_2 capture is introduced increasingly over time and the penetration rate reaches 100% on a global basis in 2070. All the captured CO_2 is sequestered in geologic formations. It should be emphasized that all of the biomass gasification-based conversion technologies (*i.e.* electricity generation from BGCC, biomass-to-hydrogen production, and biomass-to-FT synfuels production) are combined with CCS from 2060 onward in the 400 ppmv case. This implies that the benefit of negative CO_2 emissions may outweigh the diseconomies of small scale associated with BECCS under the stringent climate stabilization constraint.

Table 8: Global penetration rate of CO_2 capture in plants to which CCS is applicable in the 400 ppmv case.

	2010	2020	2030	2040	2050	2060	2070	2080	2090	2100
Biomass gasification-based conversion plants (%)	0	0	0	26.0	90.0	100	100	100	100	100
All types of plants to which CCS is applicable (%)	0	3.5	12.9	28.5	79.5	98.8	100	100	100	100

Fig. **7** shows the global amount of CO_2 captured for geological storage by source of CO_2 capture for the 400 ppmv case. In this figure, the global amount of CO_2 captured from biomass gasification-based conversion plants is disaggregated into 11 world regions. Two findings are worth noting. First, BECCS holds a large share of the global amount of CO_2 captured for storage throughout the time horizon, whose share is 77.9% in 2050 and 37.9% in 2100. This decline reflects the downward trend of global bioenergy supply potentials from 2050 and the resulting decline in the share of modern biomass-derived final energy consumption. Second, a significant portion of the CO_2 capture from biomass is implemented in now-developing regions. The explanation for this is that these regions, especially Latin America and tropical central Africa, are rich in biomass resources suitable for gasification-based conversion. Interestingly, this holds true not only for BECCS but also for CCS from all types of plants to which it is applicable. The share of developing regions in the global amount of CO_2 captured for geological storage is in the range of 60.2-66.6% in the second half of the century. This result suggests that the effective transfer of CCS technologies to now-developing regions is crucially important for achieving stringent climate stabilization targets.

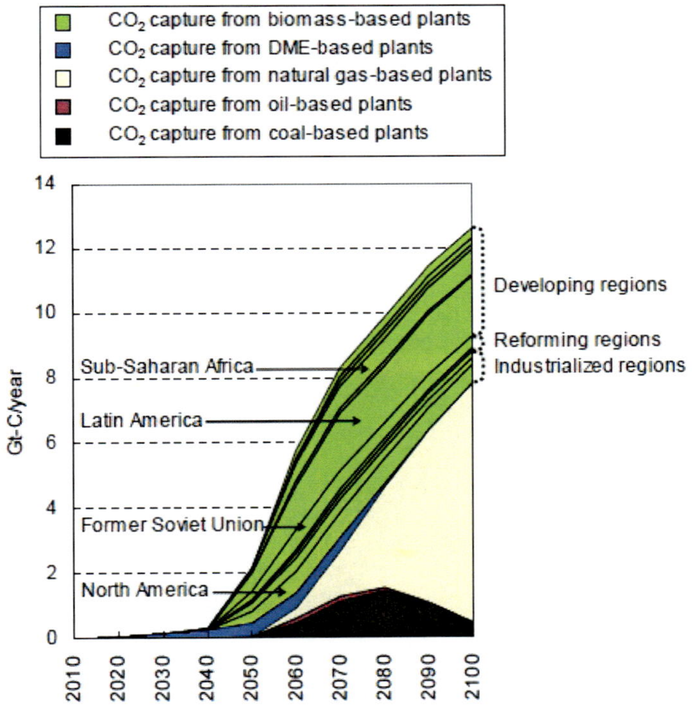

Figure 7: Global amount of CO_2 captured for geological storage by source of CO_2 capture in the 400 ppmv case. The global amount of CO_2 captured from biomass gasification-based conversion plants is disaggregated into 11 world regions.

Fig. **8** shows the cost-optimal global CO_2 emissions trajectories obtained with and without the 400 ppmv CO_2 stabilization constraint and the contributions of the following abatement options to meeting this atmospheric CO_2 stabilization target: price-induced energy demand reductions including energy efficiency improvements on the supply side, fuel switching among fossil fuels, fuel switching to non-fossil fuels, and CCS. It is readily apparent that the introduction of modern bioenergy, particularly the strategy of negative CO_2 emissions provided by BECCS, makes a substantial contribution to stabilizing the atmospheric CO_2 concentration at 400 ppmv in 2100. The contributions of fuel switching to modern bioenergy and CCS from it combined remain almost the same over the period to 2040, then rise sharply to 57.5% in 2060, and thereafter decrease to 29.4% in 2100. It can, therefore, be inferred that such a net negative emissions strategy might be essential for achieving very stringent climate stabilization targets.

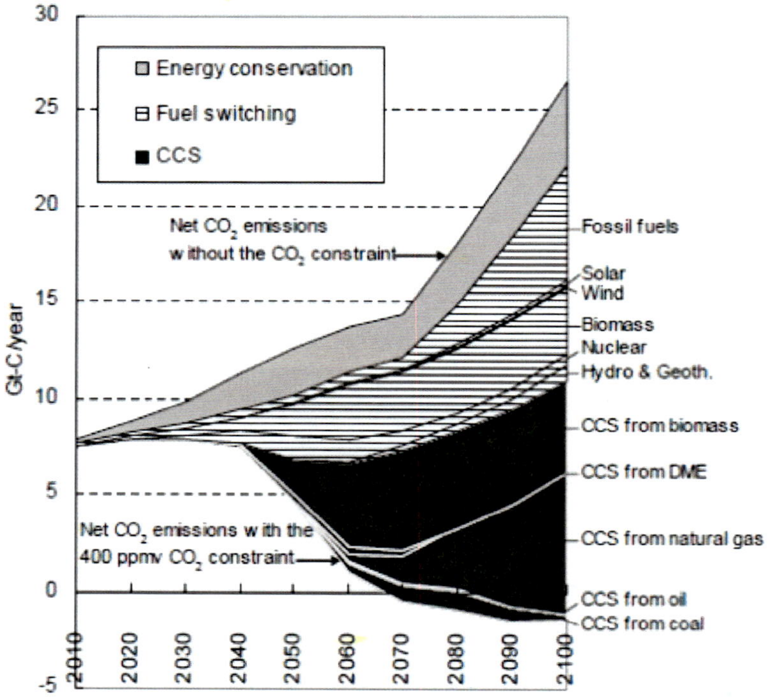

Figure 8: Global CO_2 emissions trajectories obtained with and without the 400 ppmv CO_2 stabilization constraint and contributions of the three main options to meeting this atmospheric CO_2 stabilization target.

Value and Potential Risk of BECCS

Now, the value of the strategy of negative CO_2 emissions provided by BECCS is examined by simulating the CO_2 400 ppmv stabilization case without the availability of BECCS and by simulating the CO_2 400 ppmv stabilization case where the future global CO_2 emissions trajectory is constrained to be the same as that derived from the CO_2 550 ppmv stabilization case until a certain point in time. Also, through these analyses, potential concern and risk associated with the deployment of BECCS are tried to be identified.

The simulation results show that if BECCS is available, it is feasible to stabilize the atmospheric CO_2 concentration at 400 ppmv in 2100 with the global CO_2 emissions trajectory constrained to be the same as that derived from the CO_2 550 ppmv stabilization case until 2040 (see Fig. **9**). This figure also shows that BECCS can be viewed as a potential rapid-response prevention strategy for abrupt climate change. In the case where the global CO_2 emissions trajectory is constrained to be the same as that derived from the CO_2 550 ppmv stabilization case until 2050 and beyond, it is infeasible to do so even if BECCS is available. On the other hand, the simulation results show that if BECCS is unavailable, it is feasible to stabilize the atmospheric CO_2 concentration at 400 ppmv in 2100 with the global CO_2 emissions trajectory constrained to be the same as that derived from the CO_2 550 ppmv stabilization case only until 2020.

These results have important implications for decision making on the deployment of BECCS. The results indicate that the large-scale deployment of BECCS could make it possible to respond to CO_2 stabilization targets that are strengthened from 550 ppmv (*i.e.* the moderate target) to 400 ppmv (*i.e.* the stringent target) by 2040 in a flexible manner. This implies that if international agreements on climate change were delayed, the large-scale deployment of BECCS could enable the stabilization of the atmospheric CO_2 concentration at 400 ppmv in 2100 as long as such agreements would be reached and implemented by around 2040. In this sense, it is reasonable to say that BECCS offers great flexibility in the timing of CO_2 emissions reductions under the CO_2 stabilization target of 400 ppmv.

To provide information on the amount worth investing in research, development, and demonstration on BECCS technologies, the net present value of BECCS is estimated under the 400 ppmv CO_2 stabilization constraint following the approach proposed by [26]: it is calculated as the difference in total discounted energy system costs with and without BECCS under the CO_2 constraint. As a result, BECCS is estimated to have a net present value of approximately US$$_{2000}$ 13.3 trillion under the 400 ppmv CO_2 stabilization constraint, which is in the same order as that in [26]. This value is almost the same as the US GDP for 2008 (in constant 2000 US dollars).

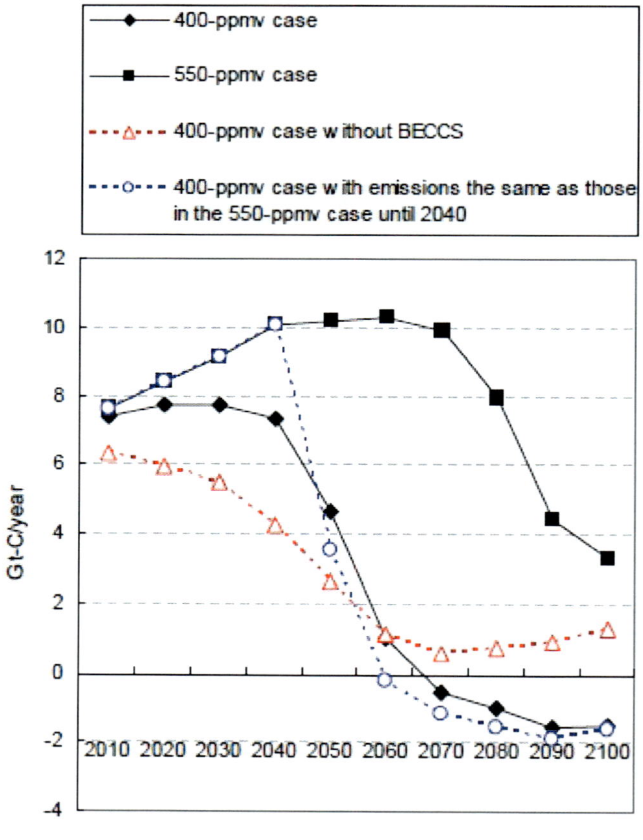

Figure 9: Global CO_2 emissions trajectories in the 400 ppmv case, the 550 ppmv case, the 400 ppmv case without BECCS, and the 400 ppmv case constrained to keep global CO_2 emissions the same as those in the 550 ppmv case until 2040.

However, the strategy of delayed emissions reductions enabled by deploying BECCS on a large scale results in a temporal overshoot of the stabilization target before the eventual date of stabilization. It is true that such a delayed response strategy might provide a significant economic value and benefit from better technologies and higher CO_2 fertilization rates from natural systems at later points in time, but the associated rate of warming from overshoots would increase the risk of exceeding critical climate thresholds [3]. In the simulation result of the CO_2 400 ppmv stabilization case without BECCS, a temporal overshoot of the atmospheric CO_2 concentration is kept below 6.1 ppmv level over the simulation period (which can be achieved at the incremental cost described above). However, in the simulation results of the CO_2 400 ppmv stabilization cases with the availability of BECCS, a temporal overshoot of the atmospheric CO_2 concentration becomes no longer negligible: its temporal overshoot is approximately 40 ppmv in the 400 ppmv case and approximately 55 ppmv in the 400 ppmv case constrained to keep global CO_2 emissions the same as those in the 550-ppmv case until 2040. To summarize, it is necessary that the value of delaying CO_2 emissions reductions be weighted against potential risks caused by such an overshoot.

Sensitivity Analysis

To assess the robustness of the findings, particularly regarding the importance of modern biomass-derived energy carriers, sensitivity analysis was performed for the following four parameters: (1) bioenergy-related costs (*i.e.* supply costs of bioenergy and capital costs of bioenergy conversion technologies); (2) the availability of BECCS; (3) the leakage rate for CO_2 stored in geologic formations; and (4) the atmospheric CO_2 stabilization target. In the first step of the sensitivity analysis, as a result of literature survey, a possible upper bound is assumed to be a 50% increase in reference supply costs of bioenergy, whereas it is assumed to be a 30% increase in reference capital costs of bioenergy conversion technologies. REDGEM70 was then run under the assumption that both cost parameters take their upper bound values. In the second step, the model was run under the assumption that BECCS is unavailable. In the third step, the model was run using annual leakage rates of 0.1%, 0.5%, 1.0%, and 1.5%. The leakage of geologically stored CO_2 was modeled based on previous studies on this issue (*e.g.*, [28]): leakage is assumed to be proportional to the cumulative amount of CO_2 stored at any point in time. For simplicity and due to data limitations, the same leakage rate was applied to all the CO_2 sequestration reservoirs over the simulation period, although it would certainly differ not only by type of reservoir but also by site and time. In the fourth step, the model was run under an atmospheric CO_2 constraint equal to 450 ppmv, 500 ppmv, and 550 ppmv.

Fig. **10** shows the total global modern biomass-derived final energy consumption as a function of the changes in the four parameters. The result of the first sensitivity analysis shows that the competitiveness of modern biomass-derived energy carriers is virtually unaffected by these cost parameter changes: even if these two cost parameters take their upper bound values, total global modern biomass-derived final energy consumption throughout the time horizon decreases only by 4.3%. On the other hand, Fig. **10(a)** indicates that if BECCS is unavailable, modern biomass-derived energy carriers become more competitive and their total final consumption throughout the time horizon increases by a factor of approximately 1.2 at the global level. The reason for this is that the global energy system is forced to rely more on fuel switching to non-fossil fuels to meet the stringent CO_2 stabilization constraint as a result of the lack of BECCS with large potential for CO_2 emissions reductions. Moreover, the unavailability of BECCS means that a rapid response to the stringent climate stabilization target is impossible, which leads to the early introduction of CO_2-neutral fuels such as modern biomass-derived energy carriers.

For the impact of imperfect retention of stored CO_2, total global modern biomass-derived final energy consumption increases especially until 2060 as the leakage rate rises. This is because if the leakage of CO_2 from sequestration reservoirs were inevitable, earlier CO_2 abatement actions would be required to compensate for the impact of the leakage under the CO_2 stabilization target of 400 ppmv. The result of the fourth sensitivity analysis (see Fig. **10(c)**) shows that among the four parameters explored here, the atmospheric CO_2 stabilization target has the largest impact on the competitiveness of modern biomass-derived energy carriers, which implies that concern over climate change is a driving force for their introduction.

Another interesting result of the sensitivity analysis worth emphasizing is the robustness of BECCS's competitiveness to the leakage rate for CO_2 stored in geologic formations. Although not shown here, the result of the third sensitivity analysis shows that the leakage rate for CO_2 stored in geologic formations has

(a)

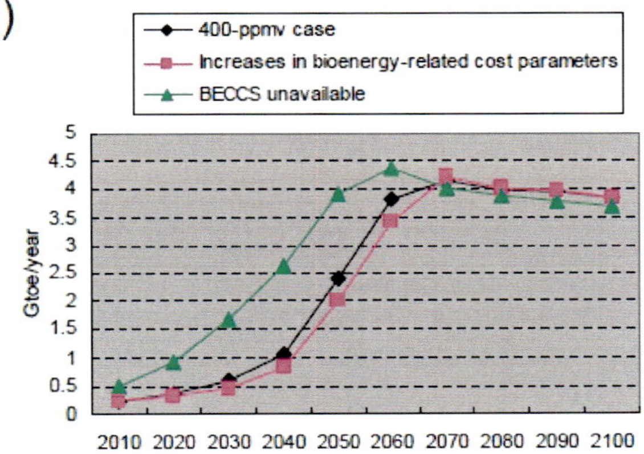

(b)

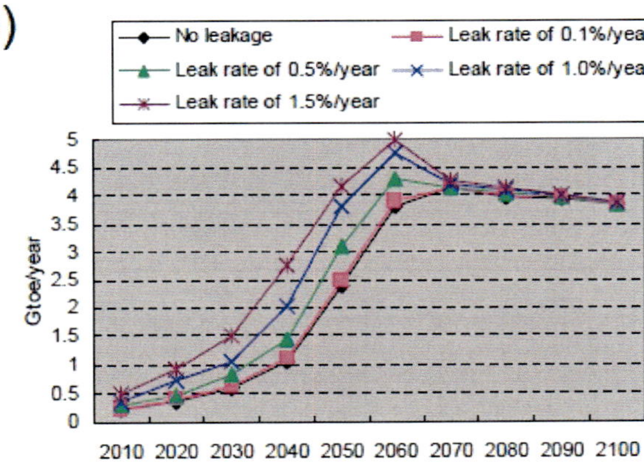

(c)

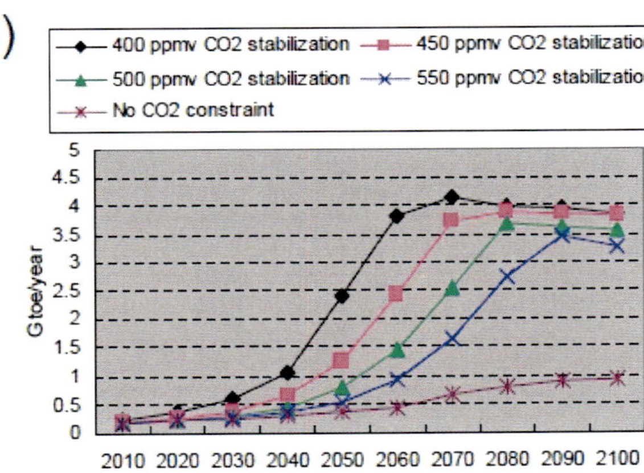

Figure 10: Total global modern biomass-derived final energy consumption as a function of bioenergy-related cost parameters and the availability of BECCS (a), the leakage rate for geologically stored CO_2 (b), and the atmospheric CO_2 stabilization target (c).

little effect on the global amount of CO_2 captured from biomass gasification-based conversion plants, *i.e.* BECCS's cost-competitiveness, whereas that of CCS as a whole decreases with the leakage rate. This implies that BECCS remains important under the CO_2 stabilization target of 400 ppmv regardless of whether or not the retention of CO_2 stored in geologic formations is perfect.

Taken all together, these results suggest that introducing modern biomass-derived energy carriers is a robust future technology option under the stringent climate stabilization constraint. It has also been confirmed that the findings obtained in the results remain almost unchanged on condition that the atmospheric CO_2 concentration is to be stabilized at 400 ppmv in 2100 and that BECCS is available.

CONCLUSIONS

In this chapter, the regionally disaggregated global energy model with 70 regions, which is characterized by a detailed description of the bioenergy sector, was used to examine the cost-optimal use of modern bioenergy over the period 2010-2100 under the 400 ppmv CO_2 stabilization constraint and to assess its potential contribution to satisfying such a stringent climate stabilization constraint. The model used here explicitly takes into account land competition with other biomass uses such as food supply and lumber production and addresses the sustainability issues on the harvest of various biomass resources. The major findings and implications obtained can be summarized as follows.

First, there is a clear difference in the cost-optimal strategy for introducing modern biomass-derived final energy carriers under the 400 ppmv CO_2 stabilization constraint between the first and second half of the century. Until around 2040, it is cost-optimal to use modern bioenergy largely to generate heat and replace direct coal use because direct heat generation from modern bioenergy is much more energy efficient and expected to achieve larger CO_2 reductions than converting it into liquid or gaseous forms. As second-generation bioenergy conversion technologies and CCS technologies become mature in the second half of the century, biofuels and electricity produced using these technologies gain share at the expense of modern biomass-derived heat. All biomass gasification-based conversion technologies are combined with CCS from 2060, which enables the net CO_2 removal from the atmosphere.

Second, the introduction of modern bioenergy, particularly the strategy of negative CO_2 emissions provided by BECCS, makes a substantial contribution to stabilizing the atmospheric CO_2 concentration at 400 ppmv in 2100. The contributions of fuel switching to modern bioenergy and CCS from it combined to meeting the 400 ppmv CO_2 stabilization constraint remain over 29.4% in the second half of the century, with a peak of 57.5% in 2060. As a result, modern bioenergy and modern biomass-derived energy carriers become one of the important primary and final energy sources, respectively, in the second half of the century under the CO_2 stabilization target of 400 ppmv. This result holds true in the case of changes in the value of parameters explored here so long as the atmospheric CO_2 concentration is to be stabilized at 400 ppmv in 2100, which suggests that introducing modern biomass-derived energy carriers is a robust future technology option under stringent climate stabilization constraints. However, neither modern bioenergy's share of total primary energy supply nor modern biomass-derived energy carriers' share of total final energy consumption can increase from 2060 because bioenergy supply potentials place a severe limit on the amount of modern bioenergy produced. Identical results are observed for each of 11 world regions.

Third, BECCS holds a large share of the global amount of geologically stored CO_2 throughout the time horizon under the CO_2 stabilization target of 400 ppmv. In this case, BECCS's competitiveness is virtually unaffected by the imperfect retention of CO_2 stored in geologic formations, while it has a negative impact on the competitiveness of CCS as a whole. It is also shown that BECCS offers great flexibility in the timing of CO_2 emissions reductions and that this flexibility is estimated to have a significant economic value. However, the strategy of delayed emissions reductions enabled by BECCS results in a temporal overshoot of the atmospheric CO_2 concentration, which would increase the risk of exceeding critical climate thresholds. Therefore, it is necessary that the value of delaying CO_2 emissions reductions be weighted against potential risks caused by such a temporal overshoot. From a regional perspective, a significant portion of the CO_2 capture is implemented in now-developing regions, implying that the effective transfer of CCS technologies to now-developing regions is crucially important for achieving stringent climate stabilization targets.

Finally, limitations of this model analysis need to be recognized when interpreting the results. In the real world, the motivations to aggressively introduce modern bioenergy are complex and multidimensional, which include the desire to address not only concerns over climate change but also issues related to energy supply security, local environmental pollution, rural development and poverty alleviation, the improvement of trade balances, and the expansion of the agricultural sector [29]. On the other hand, such issues as the direct and indirect impacts on land-use changes and related GHG emissions, biodiversity, food supply security, and increased water consumption can be considered as the limiting factors for bioenergy development [29]. The model used here does not take into account all these issues, and thus the "cost-optimal" strategy for introducing modern bioenergy derived from the model is not optimal in terms of all these issues. Nevertheless, the findings of this chapter provide important information useful for decision making on the commercial development of bioenergy.

REFERENCES

[1] International Energy Agency (IEA), *Biofuels for transport: an international perspective*. Paris: IEA, 2004.

[2] S. Sorrell, J. Speirs, R. Bentley, A. Brandt, and R. Miller, *Global oil depletion*. London: UK Energy Research Centre, 2009.

[3] B. Metz, O. Davidson, P. Bosch, R. Dave, and L. Meyer, Eds., *Climate change 2007: mitigation*. New York: Cambridge University Press, 2007.

[4] K. Mollersten, J. Yan, and J.R. Moreira, "Potential market niches for biomass energy with CO_2 capture and storage - opportunities for energy supply with negative CO_2 emissions", *Biomass Bioenergy*, vol. 25, pp. 273-285, 2003.

[5] T. Takeshita, K. Yamaji, and Y. Fujii, "Prospects for interregional energy transportation in a CO_2-constrained world", *Environ. Econ. Policy Stud.*, vol. 7, pp. 285-314, 2006.

[6] N. Nakicenovic, A. Grubler, and A. McDonald, Eds., *Global energy: perspectives*. Cambridge: Cambridge University Press, 1998.

[7] Intergovernmental Panel on Climate Change (IPCC), *Special report on emissions scenarios*. Cambridge: Cambridge University Press, 2000.

[8] K. Riahi and R.A. Roehrl, "Greenhouse gas emissions in a dynamics-as-usual scenario of economic and energy development", *Tech. Forecasting and Soc. Change*, vol. 63, pp. 175-205, 2000.

[9] H. Yamamoto, J. Fujino, and K. Yamaji, "Evaluation of bioenergy potential with a multi-regional global-land-use-and-energy model", *Biomass Bioenergy*, vol. 21, pp. 185-203, 2001.

[10] T. Takeshita, "A strategy for introducing modern bioenergy into developing Asia to avoid dangerous climate change", *Appl. Energy*, vol. 86 (Suppl. 1), pp. 222-232, 2009.

[11] H-H. Rogner, "An assessment of world hydrocarbon resources", *Annu. Rev. Energy Environ.*, vol. 22, pp. 217-262, 1997.

[12] Organization for Economic Co-operation and Development (OECD) Nuclear Energy Agency (NEA) and International Atomic Energy Agency (IAEA), *Uranium 2001: resources, production and demand.* Paris: OECD, 2002.

[13] K. Akimoto, T. Tomoda, Y. Fujii, and K. Yamaji, "Assessment of global warming mitigation options with integrated assessment model DNE21", *Energy Econ.*, vol. 26, pp. 635-653, 2004.

[14] T. Takeshita and K. Yamaji, "Important roles of Fischer-Tropsch synfuels in the global energy future", *Energy Policy*, vol. 36, pp. 2791-2802, 2008.

[15] T. Takeshita and K. Yamaji, "Potential contribution of coal to the future global energy system", *Environ. Econ. Policy Stud.*, vol. 8, pp. 55-88, 2006.

[16] K. Riahi, L. Barreto, S. Rao, and E.S. Rubin, "Towards fossil-based electricity systems with integrated CO_2 capture: implications of an illustrative long-term technology policy", in 7th International Conference on Greenhouse Gas Control Technologies (GHGT-7), 2004.

[17] Intergovernmental Panel on Climate Change (IPCC), *Special report on carbon dioxide capture and storage*. Cambridge: Cambridge University Press, 2005.

[18] T. Takeshita, "Exploring the cost-effective strategy for introducing hydrogen by considering hydrogen supply infrastructure in detail", in International Energy Workshop 2009, 2009.

[19] E.D. Larson and R. Tingjin, "Synthetic fuel production by indirect coal liquefaction", *Energy Sustain. Develop.*, vol. 7, pp. 79-102, 2003.

[20] J.P. Ciferno, *CO_2 capture-ready coal power plants*. Pittsburgh, PA: U.S. Department of Energy, National Energy Technology Laboratory, 2008.

[21] J.J. Marano and J.P. Ciferno, *Life-cycle greenhouse-gas emissions inventory for Fischer-Tropsch fuels*. Gaithersburg, MD: Energy and Environmental Solution, LLC, 2001.

[22] R.H. Williams and E.D. Larson, "A comparison of direct and indirect liquefaction technologies for making fluid fuels from coal", *Energy Sustain. Develop.*, vol. 7, pp. 103-129, 2003.

[23] K. Riahi, E.S. Rubin, and L. Schrattenholzer, "Prospects for carbon capture and sequestration technologies assuming their technological learning", *Energy*, vol. 29, pp. 1309-1318, 2004.

[24] International Energy Agency (IEA), *Prospects for CO_2 capture and storage*. Paris: IEA, 2004.

[25] Research Institute of Innovative Technology for the Earth (RITE), *Study project for implementation planning of the New Earth program*. Tokyo: New Energy and Industrial Technology Development Organization, 2002 (in Japanese).

[26] C. Azar, K. Lindgren, E.D. Larson, and K. Mollersten, "Carbon capture and storage from fossil fuels and biomass - costs and potential role in stabilizing the atmosphere", *Climatic Change*, vol. 74, pp. 47-79, 2006.

[27] M. Grahn, C. Azar, K. Lindgren, G. Berndes, and D. Gielen, "Biomass for heat or as transportation fuel? a comparison between two model-based studies", *Biomass Bioenergy*, vol. 31, pp. 747-758, 2007.

[28] J.J. Dooley and M.A. Wise, "Retention of CO_2 in geologic sequestration formations: desirable levels, economic considerations, and the implications for sequestration R&D", in: J. Gale and Y. Kaya, Eds., *Greenhouse Gas Control Technologies*. Oxford: Elsevier Science, 2003.

[29] J. Yan and T. Lin, "Biofuels in Asia", *Appl. Energy*, vol. 86 (Suppl. 1), pp. 1-10, 2009.

CHAPTER 2

Well-to-Wheel Energy, Greenhouse Gases and Criteria Pollution Emissions Evaluation of Hydrogen Based Fuel-Cell Vehicle Pathways in Shanghai

Z. Huang*

School of Civil Engineering Anhui University of Technology Maanshan, Anhui, China

Abstract: Due to high energy efficiency and zero emissions, some believe fuel cell vehicles (FCVs) could revolutionize the automobile industry by replacing internal combustion engine technology, and could be boosted and boomed in China first. However, hydrogen infrastructure is one of the major barriers. Because different H2 pathways have very different energy and emissions effects, the well-to-wheels analyses are necessary for adequately evaluating fuel/vehicle systems. The pathways used to supply H2 for FCVs must be carefully examined by their WTW energy use, GHGs emissions, total criteria pollutions emissions, and urban criteria pollutions emissions.

Ten hydrogen pathways in Shanghai have been simulated. The results include well-to-wheels energy use, GHGs emissions, total criteria pollutions and urban criteria pollutions.

A fuel-cycle model developed at Argonne National Laboratory – called the Greenhouse gases, Regulated Emissions, and Energy use in Transportation (GREET) model – was used to evaluate well-to-wheels energy and emissions impacts of hydrogen pathways in this study. Because GREET model has no coal and naphtha-based hydrogen pathways, four hydrogen pathways (No. 5-8) computer program were added to GREET in the research. To analyze uncertain impacts, commercial software, Crystal BallTM, is used to conduct Monte Carlo simulations. Instead of the point estimates, the results of this study were probability distributions.

Through the research, the following conclusions can be achieved:

(1) All the pathways have significant reduction in WTW petroleum use, except two H2 pathways from naphtha, which achieve about 20% reduction in WTW petroleum.

(2) All the pathways have significant reduction in WTW urban criteria pollutions emissions, except two H2 pathways from coal, which offer significant increase in WTW urban SO_x emissions.

***Address correspondence to Z. Huang:** School of Civil Engineering Anhui University of Technology Maanshan, Anhui, China; Tel: +86 555 2487046; Fax: +86 555 2400862; E-mail: huangzhijia99@hotmail.com

Hatim Machrafi (Ed)

(3) The NG-based H2 pathways have best WTW energy efficiencies, and the electrolysis H2 pathways have worst WTW energy efficiencies. The WTW energy efficiencies of H2 pathways from naphtha and coal are between NG-based pathways and electrolysis pathways. The pathways from naphtha have higher energy efficiencies than the pathways from coal. Only four pathways (G NG C, G NG R, G N C, and L NG C) offer WTW energy benefits, and the other six pathways consume more WTW energy than baseline-conventional gasoline vehicles.

(4) Changes in WTW GHGs emissions have nearly identical results with changes in WTW energy use.

(5) For WTW total criteria pollutions emissions, all pathways can achieve significant reduction in WTW total VOCs and CO. the other criteria pollutions emissions-NOx, PM10, and SOx, have certainly reduction in NG and crude oil-based H2 pathways, but have significant increase in electrolysis and coal-based pathways.

Keywords: Wheel-to-wheel, energy, greenhouse gas, pollution, pollution criteria, emission, emission evaluation, hydrogen, fuel cell, fuel cell analysis, life cycle analysis, calculation, modeling, energy consumption, pathways, vehicle.

INTRODUCTION

Due to China's ongoing high rate of economic growth, the number of vehicles is expected to also grow significantly over time. The increase is expected to also increase demands on infrastruture, including roads, enengy, and environment. This study intends to project possible impacts on energy and the environment under different scenarios.

Gasoline and diesel are traditonal fuels of internal combustion engine vehicles. These fuels come from petroleum, but China's petroleum resources is not rich and the reserves per person are only one tenth of the world average. Domestic petroleum production was 0.16 billion metric tons and net imported petroleum was 80 million metric tons in 2000, therefore, the reliance on imported oil was 33% [1]. If there are 10 private vehicles per hundred people in 2030, then there would be 130 million more private vehicles by then. Suppose the average oil consumption of the private vehicle is one ton per year, then there is demand for 130 million more metric tons petroleum in 2030 than in 2000. Since private vehicles consume 80% of domestic petroleum production and there is little chance for demostic petroleum pruduction to increase, the reliance on imported oil will rise to over 55%. This is unacceptable for China since it would likely threaten national energy safety and political stability. China must therefore diversify its vehicle energy sources in order to reduce reliance on imported oil.

Motor vehicles are a major source of urban air pollution. In 1999, in downtown Shanghai, motor vehicles accounted for 86% of total CO emissions, 96% of total volatile organic compound (VOC) emissions, and 56% of total nitrogen oxide (NO_x) emissions [2]. In recent years Shanghai has done more work to improve air quality: *e.g.,* stationary pollution sources (chimneys) in downtown were removed; the moving pollution sources (vehicles) were subjected to new emissions code. Even with these controls on petroleum based vehicles, the number of vehicles grows continuously, and the air quality in Shanghai is under national standards. Controlling emissions cannot thoroughly solve the issue of pollution by vehicles.

Due to high efficiencies and zero emissions, fuel cell vehicles are undergoing vigorous research and development (R&D) efforts at major automobile companies worldwide. Hydrogen can be produced from all of the primary energy sources, such as natural gas, coal, petroleum, and electrolysis. Therefore, hydrogen based fuel cell vehicles can abate vehicle reliance on petroleum and urban air pollution. In fact, some believe that FCVs could revolutionize the automobile industry by replacing internal combustion engine technology. Besides investing in FCV R&D efforts, governments and private industries are actively investing to understand market barriers for vehicles and create fuel infrastructure so that the introduction of FCVs will be successful when the technology is ready for the mass market.

When H_2 is used to fuel FCVs, water and electricity are the main products. However, production of H_2 can generate a significant amount of emissions and incur significant energy losses. To provide a complete evaluation of energy and emissions impacts of FCVs powered with different fuels, energy use and emissions from well to pumps (for fuel) and from pumps to wheels (for vehicles) must be taken into account.

Hydrogen is used in a number of industrial applications, with ammonia production accounting for 62.4% of the world's hydrogen, and refining and methanol production consuming 24.3% and 8.7%, respectively [3]. Hydrogen can be produced from natural gas, coal, naphtha, and electrolysis. Production can occur in either a central plant or a refueling station. The hydrogen on board FCVs can be either a gaseous or a liquid state. Therefore, the energy and environmental effects of their use can differ. To thoroughly examine the energy and environment effects of different hydrogen pathways, the researcher must evaluate their full fuel cycle energy and environmental effects. In this study, we use the fuel cycle model GREET (Greenhouse gases, Regulated Emissions, and Energy use in Transportation) developed at Argonne National Laboratory to estimate per-mile, fuel-cycle energy use and emissions.

METHODS OF FUEL CYCLE ANALYSIS

The GREET Model

A WTW analysis of a vehicle/fuel system covers all stages of the fuel cycle – from energy feedstock recovery (wells) to energy delivered at vehicle wheels (wheels). A WTW analysis is also called a fuel-cycle analysis in the transportation fuels area and a life-cycle analysis (LCA) for consumer products. Since 1995, with funding from the US Department of Energy (DOE), Wang (Argonne National Laboratory) [4-10] has been developing the Greenhouse gases, Regulated Emissions, and Energy use in Transportation (GREET) model as an analytical tool for use by researchers and practitioners to estimate WTW energy use and emissions associated with transportation fuels and advanced technology vehicles.

Fig. **1** presents stages and activities covered in GREET WTW simulations of vehicle/fuel systems. A WTW analysis includes the feedstock (or primary energy), fuel, and vehicle operation stages. The feedstock and fuel stages together are called "well to pump"(WTP) or "upstream" stages, and vehicle operation stages is called the "pump to wheels"(PTW) or "downstream" stages. In GREET, WTW energy and emission results are presented for each of the three stages shown in Fig. **1**.

The GREET model can calculate Btu-per-mile (Btu/mi) energy use and grams-per-mile (g/mi) emissions by taking into account energy use and emissions of fuel combustion and non-combustion sources. The model calculates total energy use, fossil energy use, and petroleum use. It includes emissions of three major GHGs

(CO_2, CH_4 and N_2O) and five criteria pollutions (VOCs, CO, NO_x, PM_{10}, and SO_x). Table **1** lists output items from the GREET model.

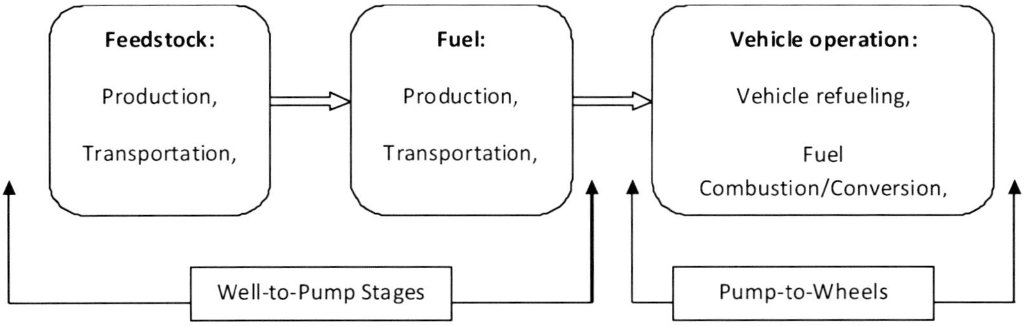

Figure 1: WTW stages of vehicle/fuel systems covered in the GREET model.

Table 1: Output items from the GREET model.

Category	**Output Item**	**Remarks**
Energy (Btu/mi)	Total energy use	
	Fossil energy use	
	Petroleum use	
Greenhouse Gases	CO_2	GHGs are converted into
(g/mi)	CH_4	CO_2-equivalent emissions
	N_2O	with their GWPs
	VOC (optional)	
	CO (optional)	
	NO_X (optional)	
Criteria Pollution	VOC	These emissions are
(g/mi)	CO	seperated into total and
	NO_X	urban emissions
	PM_{10}	
	SO_X	

Calculation Logic

Fig. **2** is a simplified diagram showing calculation logic for energy use and emissions for each upstream stage. The model uses assumptions about shares of fuel combustion technologies, energy efficiencies, total and urban emission shares, and shares of process fuels. For a given stage, energy use by fuel type is calculated on the basis of energy efficiencies and process fuels shares. We then calculate emissions by using energy use by fuel type, emission factors by fuel type, and combustion technology shares. Finally, urban emissions are estimated from total emissions and a spilt of facility locations between urban and non-urban locations. For CO_2 and SO_2 emissions, GREET takes carbon-balance and sulfur-balance approach.

Fig. **3** shows the way in which GREET simulates transportation of feedstock and fuels. First, we determined the types and shares of transportation modes (*i.e.* ocean tankers, pipelines, barges, rail and trucks), the types and shares of process fuels (residual oil, diesel fuels, natural gas, electricity, *etc.*), energy intensity, transportation distance, emission factor, and urban emission shares to be used for transporting a given feedstock and fuel. Second, we calculated the energy use, total emissions, and urban emissions by mode. Finally, we added together the energy use and emissions of all transportation mode to be used for transporting the given feedstock and fuel.

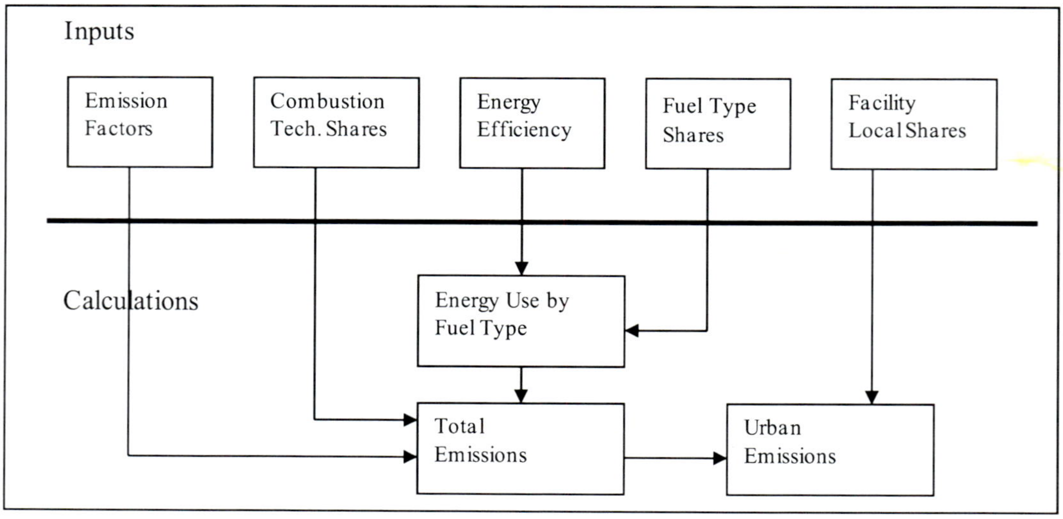

Figure 2: Calculation logic for upstream stages energy use and emissions.

Calculation of Energy Use

Total energy use (at the primary energy level), fossil energy use, and petroleum use can be calculated by GREET model. Total energy includes fossil energy and renewable energy such as solar energy, hydropower energy. The total energy use can explain the vehicle fuel efficiency from well to wheel. Fossil energy includes coal, natural gas, and petroleum. Different fossil energy has different energy efficiency and emissions. Petroleum reserves are less than other fossil energy, and traditional vehicles' fuels come from it. Therefore, this study has interest in total energy use, fossil energy use, and petroleum use. The model can estimate the amount of fossil fuel and petroleum displaced because of using alternative transportation fuels and advanced vehicle technologies instead of conventional vehicles fueled with gasoline.

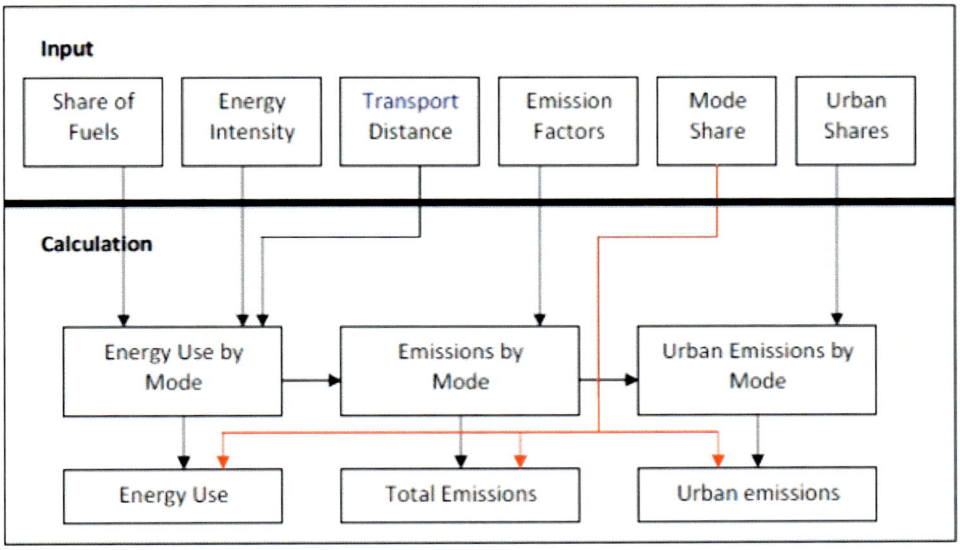

Figure 3: Calculation logic for well-to-pump energy use and emissions of transportation fuels.

Upstream Stage Energy Use

(1) Direct consumption of process fuel *i*

The consumption of process fuel *i* for a particular stage are calculated by using the following formulas:

$$DEP_i = \frac{1}{\eta} \times S_{fuel,i}, \text{ or} \quad \textbf{(1)}$$

$$DEP_i = \left(\frac{1}{\eta} - 1\right) \times S_{fuel,i} \quad \textbf{(2)}$$

where

DEP_i: consumption of process fuel *i* (Btu/10^6Btu of fuel throughput);

$S_{fuel,\ i}$: share of process fuel *i* out of all process fuels consumed during the stage ($\sum_i S_{fuel,i} = 1$) and

η: energy efficiency for a given stage (defined as [energy output]/[energy input] for the stage).

(2) Primary energy level consumption of process fuel *i*

Because producing fuel consumed energy, and recovering feedstock of fuel also consumes energy, the following equation is used to calculate consumption of process fuel *i* in primary energy level:

$$PEP_i = \sum_i DEP_i \left(1 + \left(k_{1,i} \times PEP_{feedstock,i} + PEP_{fuel,i}\right)/1000000\right) \quad \mathbf{(3)}$$

where

PEP_i: consumption of process fuel *i* in primary energy level (Btu/10^6Btu of fuel throughput);

$PEP_{feedstock,\ i}$: primary energy level consumption of feedstock recovery for fuel *i* (Btu/10^6Btu of feedstock output for fuel *i*);

$PEP_{fuel,\ i}$: primary energy level consumption of fuel production for fuel *i* (Btu/10^6Btu of fuel output for fuel *i*) and

$k_{1,\ i}$: loss factor of fuel *i* production.

(3) Stage energy consumption

Energy uses during a stage include consumption of total process fuels, transportation for feedstock and fuel, and distribution for fuel. The following formula is used to calculate energy use for a particular stage:

$$PEP_{stage} = \sum_i PEP_i \times k_2 \times k_3 + PEP_{tra} + PEP_{dis} \quad \mathbf{(4)}$$

where

PEP_{stage}: energy use for a given stage (feedstock or fuel) (Btu/10^6Btu of fuel throughput);

PEP_{tra}: transportation energy use for a given stage (Btu/10^6Btu of fuel transportation);

PEP_{dis}: distribution energy use for a given stage (Btu/10^6Btu of fuel disttibution);

k_2: loss factor of transportation for a given stage and

k_3: loss factor of distribution for a given stage.

Vehicle Operation Energy Consumption

Vehicle operation energy use is calculated by using following equation:

$$EVO = LHV \,/\, MPG \qquad \textbf{(5)}$$

where

EVO: vehicle operation energy use (Btu/mile)

LHV: low heating value of gasoline (in Btu/gal); and

MPG: gasoline equivalent fuel economies for a given vehicle (mile/per gasoline equivalent gallon)

Well to Wheel Energy Consumption

Well to Wheels energy uses include consumption of operation, fuel, and feedstock. The following formula is used to calculate the energy use from well to wheels:

$$EWTW = EVO \times \left(1 + PEP_{fuel} + k_1 \times PEP_{feedstock}\right) \qquad \textbf{(6)}$$

where

EWTW: energy use from well to wheel for a given vehicle (Btu/mile)

Calculation of GHGs

GHGs include total emissions of CO_2, CH_4, and N_2O in this study. GREET combines these three GHGs with their GWPs for the 100-year time horizon to estimate CO_2-equivalent GHG emissions.

The following formula is used to calculate CO_2 emissions:

$$CO_{2,j,k} = \lfloor Density_j \div LHV_j \times 1000000 \times C_{ratio,j} - (VOC_{j,k} \times 0.5 + CO_{j,k} \times 0.43 + CH_{4,j,k} \times 0.75) \rfloor \times 44 \div 12 \quad \textbf{(7)}$$

where

$CO_{2,\,j,\,k}$: combustion CO_2 emission factor for combustion technology *k* burning process fuel *j* (in g/10^6Btu of fuel *j* burned);

$Density_j$: density of process fuel *j* (in g/gal for liquid fuels, g/scf for gaseous fuels, or g/ton for solid fuels);

LHV_j: low heating value of process fuel *j* (in Btu/gal for liquid fuels, Btu/scf for gaseous, or Btu/ton for solid fuels);

C_{ratio}: carbon ratio by weight for process fuel *j*;

$VOC_{j,\,k}$: VOC emission factor for combustion technology *k* burning process fuel *j* (in g/10^6Btu of fuel *j* burned) and

$CO_{j,\,k:}$ CO emission factor for combustion technology k burning process fuel j (in g/10^6Btu of fuel j burned);

$CH_{4,\,j,\,k:}$ CH_4 emission factor for combustion technology *k* burning process fuel *j* (in g/10^6Btu of fuel *j* burned).

The formula of calculation for CH_4, and N_2O is shown in section 2.5

Calculation of Criteria Pollution

Emissions of Combustion Process

(1) Direct emissions of pollution i

Emissions from combustion of process fuels for a given stage are calculated by using the following formula:

$$DEM_{cm,i} = \left(\sum_j \sum_k EF_{i,j,k} \times PEP_j \times S_{teck} \right) \div 1000000 \quad \textbf{(8)}$$

where

$DEM_{cm,\, i}$: combustion emissions of pollution *i* in g/10^6 Btu of fuel throughput;

$EF_{i,\, j,\, k}$: emission factor of pollution *i* for process fuel *j* with combustion technology k (g/10^6 Btu of fuel burned) and

S_{teck}: share of combustion technology *k* out of all combustion technologies for fuel $j\,(\sum_k S_k = 1)$.

(2) Total emissions of pollution *i*

Total emissions from combustion of process fuels for a given stage should take account of direct emissions and upstream emissions (or called indirect emissions, include feedstock emissions and fuel emissions for process fuels). The following equation was developed to account for total emissions of combustion process:

$$TEM_{cm,i} = DEM_{cm,i} + \sum_j \left(k_{1,j} \times TEM_{feedstock,i,j} + TEM_{fuel,i,j} \right) \div 1000000 \quad \textbf{(9)}$$

where

$TEM_{cm,\, i}$: total combustion emissions of pollution *i* in g/10^6 Btu of fuel throughput;

$TEM_{feedstock,\, i,\, j}$: feedstock stage emissions of pollution *i* for process fuel *j* (g/10^6 Btu of feedstock recovery) and

$TEM_{fuel,\, i}$: fuel stage emissions of pollution *i* for process fuel *j* (g/10^6 Btu of fuel production).

Stage Emissions of Pollution i

Emissions of pollution *i* for a given stage include total combustion emissions, non-combustion emissions, and emissions from transportation and distribution. The stage emissions are calculated by using the following formula:

$$TEM_i = TEM_{cm,i} \times k_2 \times k_3 + TEM_{ncm,i} + TEM_{tra,i} + TEM_{dis,i} \quad (10)$$

where

TEM_i: Emissions of pollution *i* in g/10^6 Btu of fuel throughput for a given stage;

$TEM_{ncm,i}$: non-combustion emissions of pollution *i* for a given stage (g/10^6 Btu of fuel throughput);

$TEM_{tra,i}$: transportation emissions of pollution *i* for a given stage(g/10^6 Btu of fuel transportation) and

$TEM_{dis,i}$: distribution emissions of pollution *i* for a given stage(g/10^6 Btu of fuel distribution).

Emissions of SO_2

In the GREET model, SO_2 emission factors for combustion technologies fueled with all fuel except coal, crude oil, and residual oil are calculated by assuming that all sulfur contained in these process fuels is converted into sulfur dioxide. The following formula is used to calculate the SO_2 emissions of combustion technologies:

$$SO_{2,j} = Density_j \div LHV_j \times 1000000 \times S_{ratio,j} \times 64 \div 32 \quad (11)$$

where

$SO_{2,j}$: SO_2 emission factor for combustion of process fuel *j* (in g/10^6Btu of fuel *j* burned) and

S_{ratio}: sulfur ratio by weight for process fuel *j*.

Urban Emissions of Pollution

In the GREET model, the urban components (shares) of total emission are used to calculate the sum of urban emissions. The formula is composed of the right items of equation (10) times urban shares, respectively.

Uncertainty Analyses

The inputs of GREET have uncertainties. Actually, they have their probability distribution functions. Previously, a sensitivity analysis was conducted to estimate the effects of varying the input assumption. However, this analysis cannot change the other input assumptions at the same time. In fact, it is reasonable that all the input assumptions can change their value on their distribution in random. In this study, we used the key inputs probability distribution functions explored by ANL [7-9]. These distribution functions are incorporated into the GREET model. A commercial software, Crystal Ball™ [11], is used in GREET to conduct Monte Carlo simulations. Instead of the point estimates in previous study[12], the results of this study were probability distributions.

HYDROGEN PATHWAYS

This study focuses on hydrogen based FCV fuel-cycle analysis in Shanghai. The results are used to assist decision makers in selecting the hydrogen supply systems that achieve the greatest energy and emission benefits. Therefore, the hydrogen pathways are chosen on current and anticipated future options (pathways) for Shanghai. Table **2** lists 10 hydrogen pathways in this study.

Table 2: Fuel pathways included in this study.

No.	Feedstock	Fuel	Abbreviation
1	Petroleum	Gasoline	
2	Natural Gas	GH_2 Central Plant	G NG C
3	Natural Gas	GH_2 Refueling Station	G NG R
4	Natural Gas	LH_2 Central Plant	L NG C
5	Natural Gas	LH_2 Refueling Station	L NG R
6	Naphtha	GH_2 Central Plant	G N C
7	Naphtha	LH_2 Central Plant	L N C
8	Coal	GH_2 Central Plant	G C C
9	Coal	LH_2 Central Plant	L C C
10	Electricity	GH_2 Refueling Station	G E R
11	Electricity	LH_2 Refueling Station	L E R

There are hydrogen production plants that rely upon coal, naphtha, and electricity as input. Although there is no current hydrogen production plant from natural gas in Shanghai, the capability exists elsewhere in the world. With the development of pipelines, a steady supply of natural gas would exist and would allow for building plants capable of producing hydrogen from natural gas.

Hydrogen is applied for FCV in gaseous and liquid condition. Gaseous H_2 (GH_2) needs to be compressed to be about 6000 pounds per square inch (psi) in order to store enough energy for a reasonable FCV driving range. FCVs fueled by liquid H_2 (LH_2) can have a longer driving range than those fueled with GH_2. However, LH_2 storage requires strict heat insulation for LH_2 tanks and can suffer from the boil-off loss of LH_2. Besides compression and liquid, some other H_2 storage technologies, such as metal hydrides, are currently under R&D. These other storage technologies are not discussed here.

Hydrogen plants location has two options (central plant or refueling station) also in this study. While H_2 production is in central plants, transportation of massive amounts of H_2 from central plants to refueling stations for FCV applications can require a significant capital investment. To avoid the formidable, costly H_2 transportation infrastructure, H_2 can be produced in refueling stations. The feedstocks in the fueling stations have NG and electricity in this study.

In summary, hydrogen pathways have 10 possible production combinations, including types of feedstock, various production locations, and productions of gaseous or liquid hydrogen. Using petroleum to produce gasoline is used as the baseline fuel pathway for GV application to which hydrogen pathways for FCV application are compared. Figs. **4** to **15** present the 11 fuel pathways respectively.

DATA SOURCE AND KEY ASSUMPTIONS

Key input assumptions regarding both WTP and PTW stages affect WTW energy and emissions results. Due to time constraints, it is impossible to include all of China's data in the GREET model. The study assumes China's emissions factor of combustion technologies and vehicle operation in 2010 equal to those in the United States in 2000. In other words, China lags behind the United States by 10 years in environment emissions. Thus, we can use the current emissions default data in GREET to simulate the projections 10 years later in China. The baseline fuel/technology pathways after 10 years in China were CG (Conventional

Gasoline Vehicle) in near-term in GREET. The following paragraphs summarize the key assumptions used in this study.

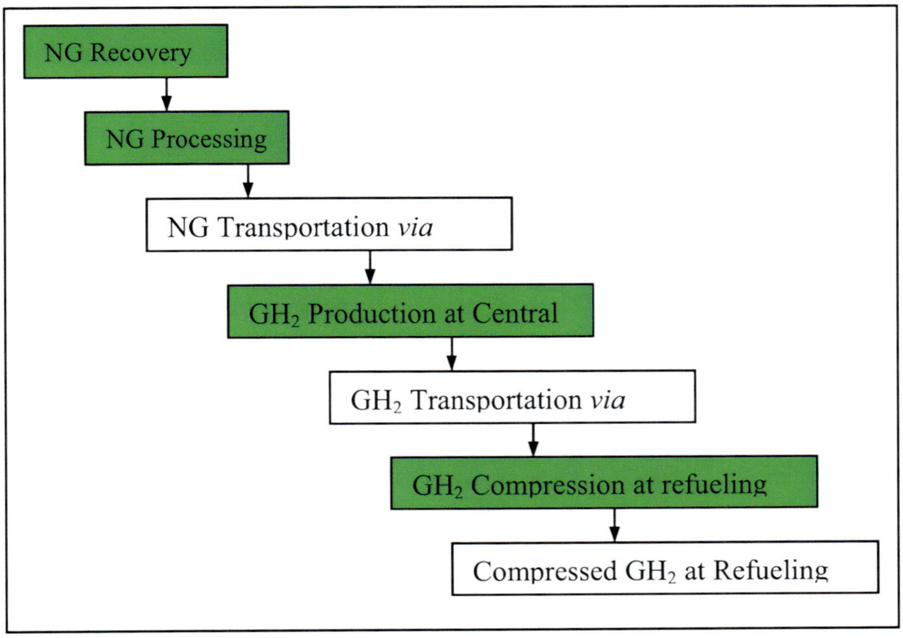

Figure 4: NG-to-GH_2 pathways: central plant production.

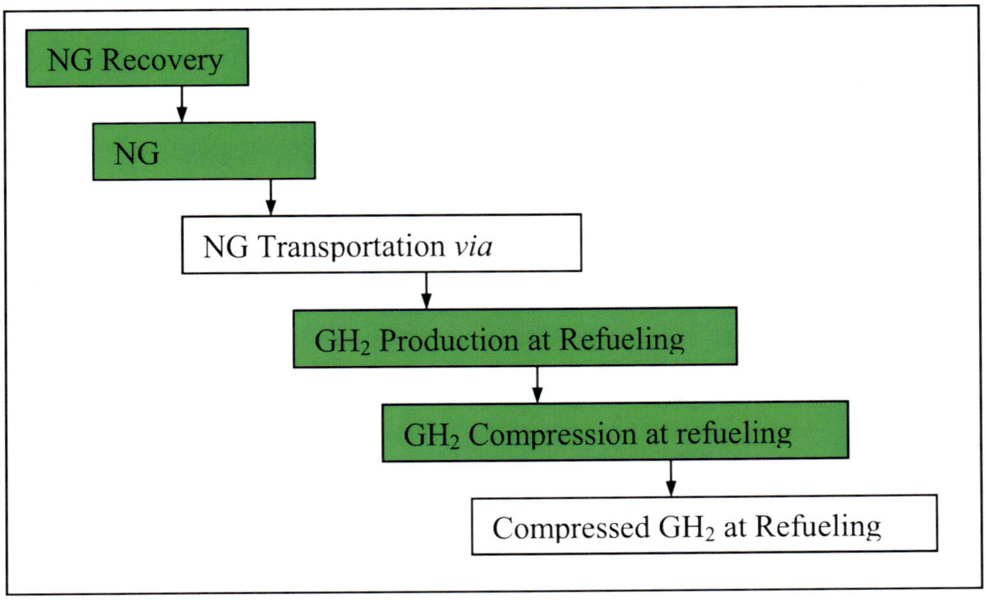

Figure 5: NG-to-GH_2 pathways: refueling station production.

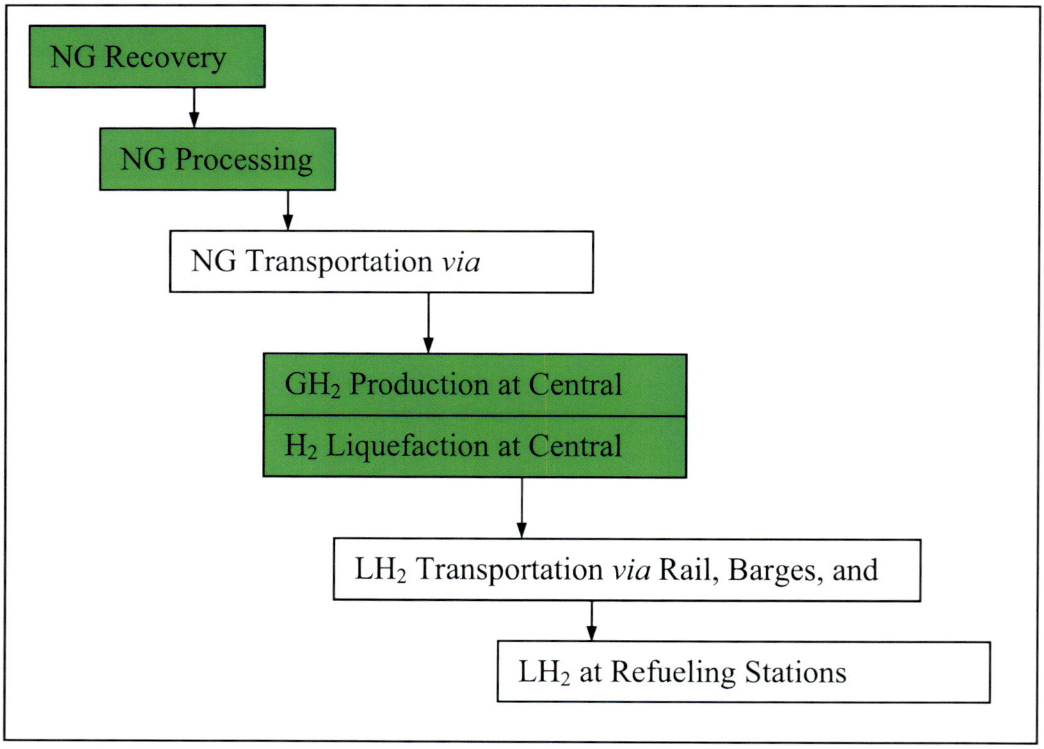

Figure 6: NG-to-LH_2 pathways: central plant production.

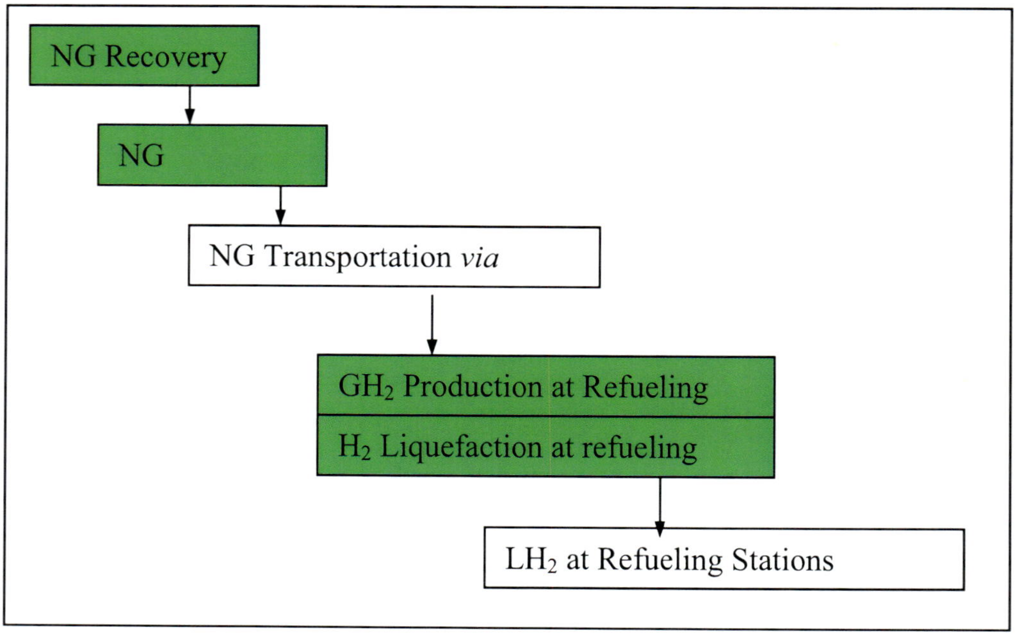

Figure 7: NG-to-LH2 pathways: refueling station production.

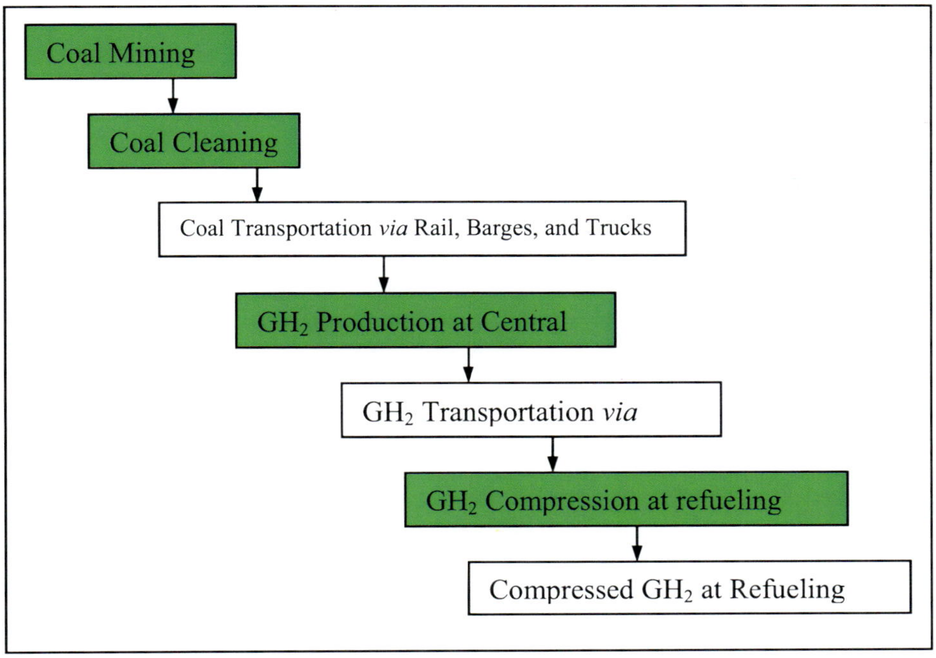

Figure 8: Coal-to-GH_2 pathways: central plant production.

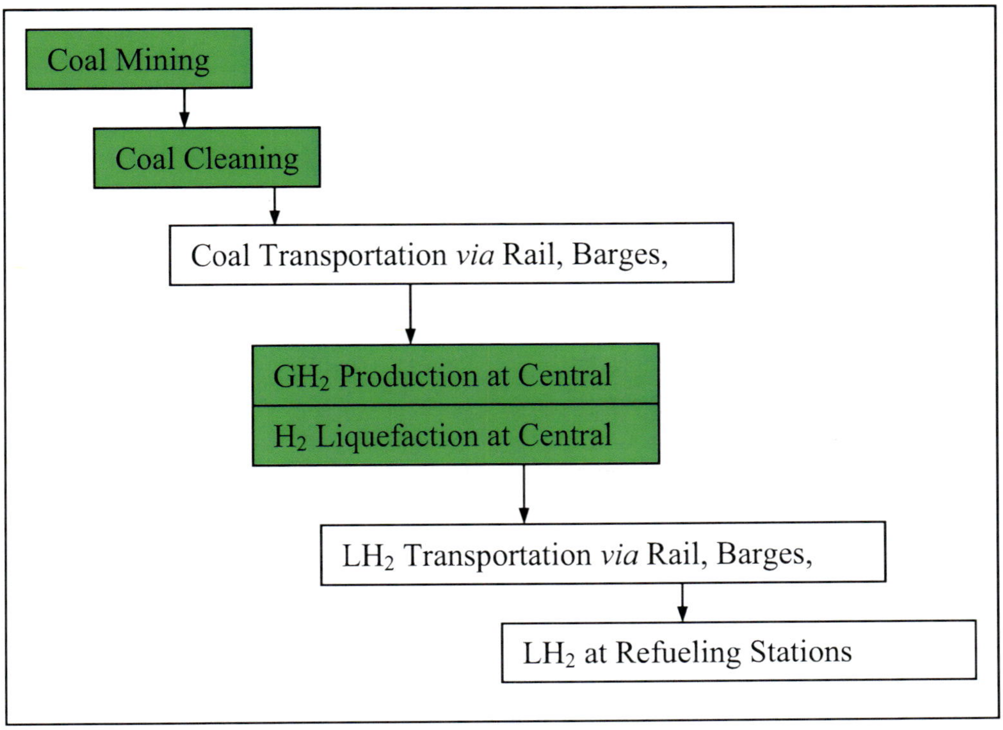

Figure 9: Coal-to-LH_2 pathways: central plant production.

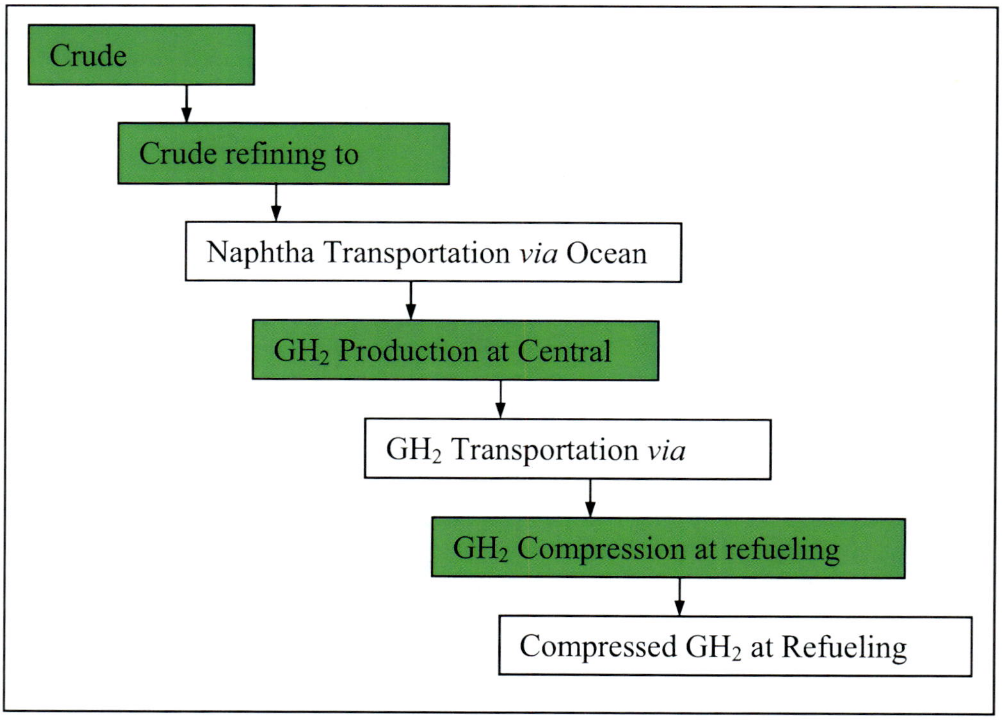

Figure 10: Naphtha-to-GH_2 pathways: central plant production.

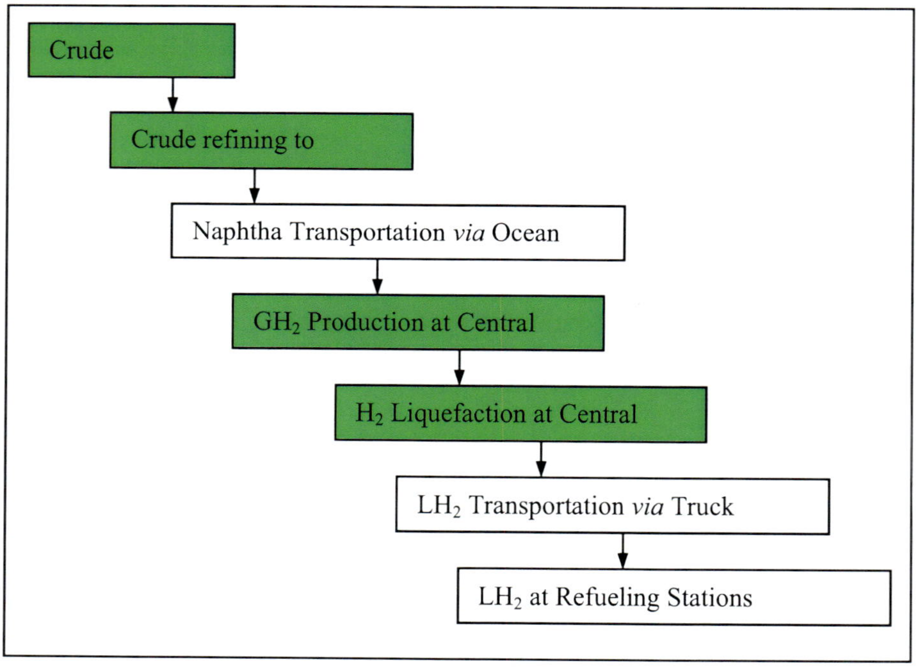

Figure 11: Naphtha-to-LH_2 pathways: central plant production.

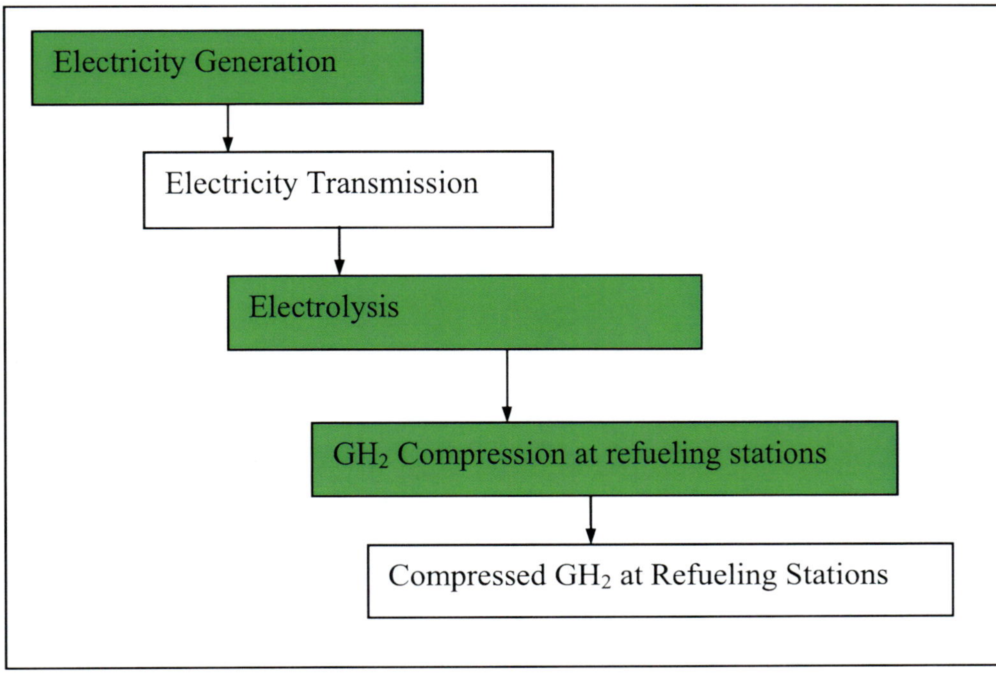

Figure 12: Electrolysis-to-GH_2 pathways: refueling stations production.

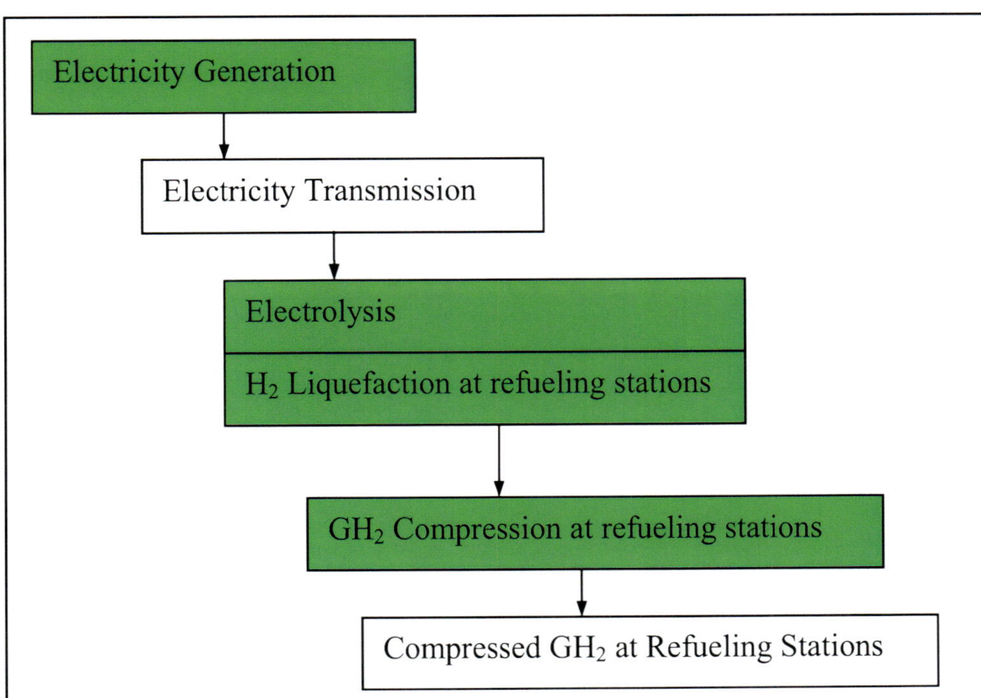

Figure 13: Electrolysis-to-LH_2 pathways: refueling stations production.

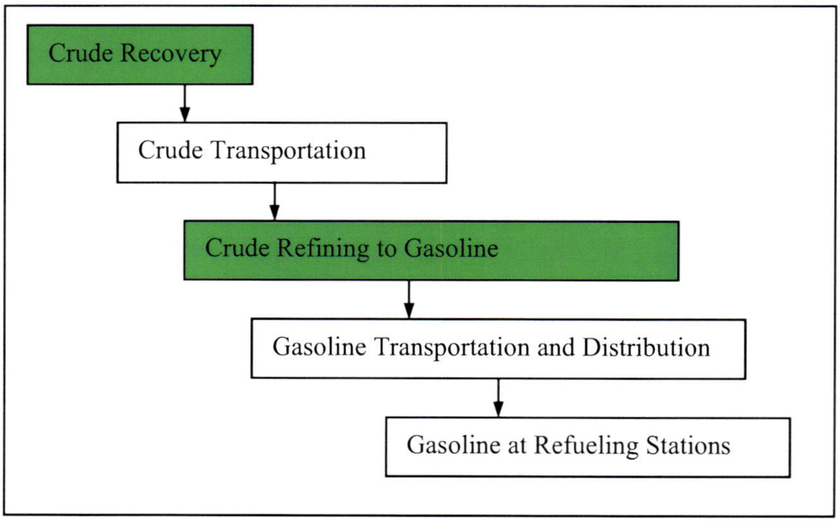

Figure 14: Crude-to-gasoline pathways: baseline pathway.

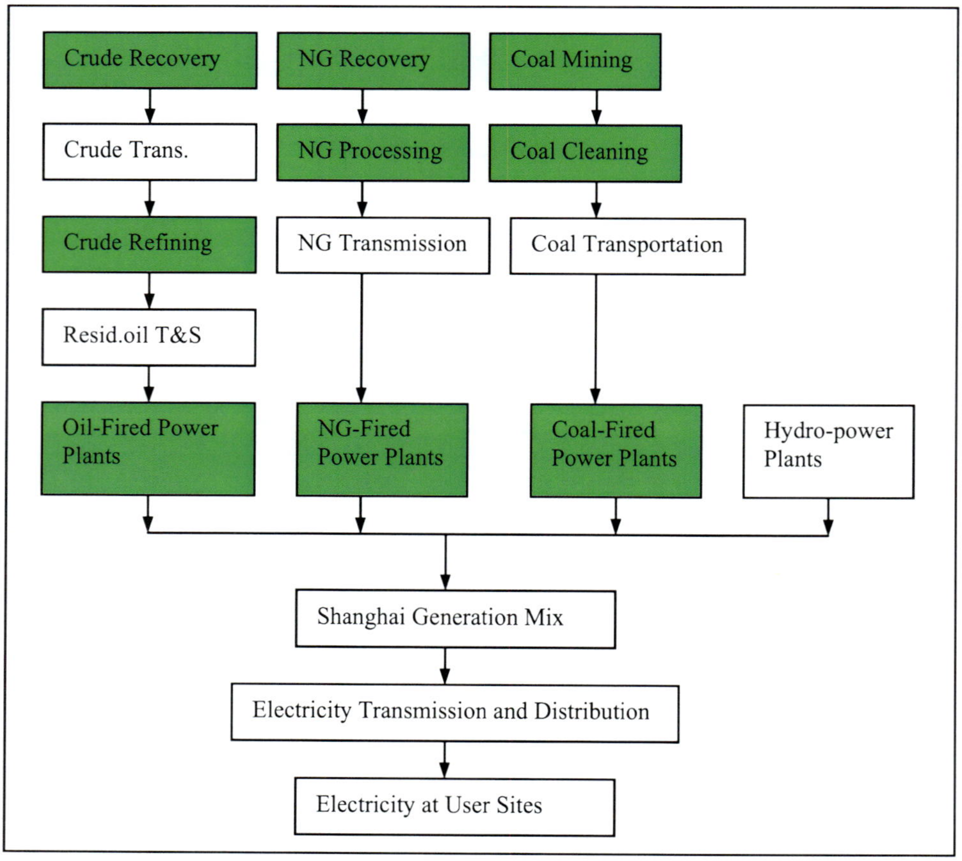

Figure 15: Electricity generation pathways.

In this study, we address uncertainties in fuel-cycle analysis with the GREET model. Using the Crystal Ball^{TM} software, we conducted Monte Carlo simulations for the pathways included in this study. To develop probability distribution functions for key parameters, we first developed a ranking of the uncertainty level involved in each given input parameters. Then, based on the range of values for the parameter from published results and the developed uncertainty level, we determined the values for the input parameter (such as energy efficiency) at probability of 20%, 50% and 80%. Furthermore, we assumed that the input parameter (except as noted) follows the normal distribution curve in order to establish the probability distribution function for the parameter.

Well-to-Pump Assumptions

Natural Gas to Hydrogen

This study includes four NG-based pathways producing Hydrogen (see Table **2**). The Natural Gas will come from Xingjian to Shanghai through pipelines. Table **3** presents efficiency assumptions for these four pathways (from Figs. **4** to **7**). Central hydrogen plant can be designed to produce steam and/or electricity. If steam and/or electricity are co-produced with hydrogen, the overall plant energy efficiency can be improved to some extent. Here, we evaluate plant designs that do not include steam and/or electricity export.

Table 3: Energy efficiencies for NG-H_2 pathways.

Stage	**Energy Efficiency(%)**		
	P20[b]	**P50[b]**	**P80[b]**
NG recovery	96.0	97.5	99.0
NG processing	96.0	97.5	99.0
Central GH_2			
GH_2 production	68.0	71.5	75.0
GH_2 Compression[a]	90.0	92.5	95.0
Central LH_2			
GH_2 production	68.0	71.5	75.0
H_2 liquefaction	65.0	71.0	77.0
Station GH_2			
GH_2 production	62.0	67.0	72.0
GH_2 Compression[a]	91.5	94.0	96.5

Table 3: cont….

Station LH_2			
GH_2 production	62.0	67.0	72.0
H_2 liquefaction	60.0	66.0	72.0

[a] A triangle distribution curve assumed.
[b] P20: probability of 20%; P50: probability of 50%; P80: probability of 80%.

The majority of existing central H_2 plant use SMR technology. Steam is added to the NG feed, and the mixture of NG and steam is preheated before entering the reformer, where CH_4 is converted into the syngas (H_2, CO, and CO_2) in the presence of catalysts. The produced hot syngas, at a temperature of 900-930 °C, exits the SMR reformer and is cooled before entering the shift converter, where shift catalysts convert CO and steam to CO_2 and additional H_2. The gas from the shift converter is further cooled to ambient temperature before entering a pressure swing adsorption (PSA) unit, where high-purity H_2 is produced; the remaining gas mixture, the so-called tail gas, is used in SMR reformer as supplemental fuel for the burners. We assumed energy efficiencies of 68% to 75% [9].

For LH_2 production option, we assumed that GH_2 is produced in central plants first, and liquefied at the same facility. Thus, GH_2 production efficiencies assumed in the previous paragraph are applied to this pathway. In addition, H_2 liquefaction efficiencies are applied to this pathway.

Power requirements for refrigeration compressors during liquefaction consume a large amount of energy. Because the boiling point of LH_2 is much lower than that of LNG (-253°C *vs.* –163°C), the amount of energy required for H_2 liquefaction is much higher than that required for NG liquefaction. In this study, we assumed H_2 liquefaction efficiencies of 65% to 77% for central H_2 plant liquefaction [6].

To avoid long-distance transportation and storage of GH_2 and LH_2, which can be very expensive, GH_2 may be produced from NG in refueling stations with small-scale SMR systems. We expected that efficiencies of H_2 production and liquefaction at refueling stations are lower than that in central plants. We assumed they are 62%-72% and 60%-72%, respectively [6].

GH_2 may need to be stored onboard FCVs at pressures above 5000 psi. So GH_2 may need to be compressed to 6000 psi or greater at refueling station. For GH_2

produced in central plants and transported to station *via* pipelines, we assumed an initial pressure of 250 psi for GH_2 coming out of pipelines at refueling stations. We estimated H_2 compression efficiencies of 90% to 95% for electric compressors. For GH_2 at refueling stations, we assumed the initial pressure was 500 psi, so the compression efficiencies could be higher than that of central plants. We assumed efficiencies of 91.5%-96.5% [6].

Liquid H_2 pathways suffer two major energy losses: GH_2 production and H_2 liquefaction. In addition, LH_2 is subject to boil-off losses. If LH_2 is stored for a long period before use, boil-off losses can be significant. Due to these losses, LH_2 pathways generally consume more energy and produce more emissions than GH_2 pathways do.

In H_2 plants, all the carbon in CH_4 eventually ends up as CO_2. The produced CO_2 could be sequestered into depleted oil and gas wells to limit CO_2 emissions from H_2 plants or to enhance oil or gas recovery in oil and gas fields. However, without economic incentives or regulations, it is uncertain whether CO_2 from H_2 plants will be sequestered. In this study, we did not assume CO_2 sequestration in H_2 plants.

Electrolysis Hydrogen

Hydrogen can be produced from electricity *via* electrolysis of water in refueling stations. Since the electric transmission and distribution system is already extensive, electricity can readily be transmitted to refueling stations. Thus, this pathway helps avoid long-distance transportation and storage of H_2. In particular, if H_2 will eventually become a fuel-cell fuel, this pathway could help overcome inadequate H_2 distribution infrastructures in the early stage of FCV introduction and in areas outside the major metropolitan areas.

Table 4: Average electricity generation mix in Shanghai.

Coal (%)	**Oil (%)**	**NG (%)**	**Nuclear (%)**	**Other[a] (%)**
77.2	3.7	0.6	1.3	17.2

[a] including hydropower, geothermal energy, organic wastes, solar energy and wind power.

The energy sources used for electricity generation embody the most important factor in determining energy use, greenhouse gases emissions, and criteria pollutions emissions of electrolysis H_2, it is assumed that electricity used for electrolysis come from electricity grid which is the mixture of average electricity generation. Thus, we chose Shanghai generation mix (Table **4**) for the simulation [13].

Table 5: Energy efficiencies from electrolysis to H_2 pathways.

Stage	Energy Efficiency (%)		
	P20	**P50**	**P80**
China average electricity generation			
Oil-fired power plant:steam boilers	32.0	35.0	38.0
NG-fired power plant:steam boilers	32.0	35.0	38.0
NG-fired power plant:combined-cycle turbines[a]	50.0	55.0	60.0
Coal-fired power plant:steam boilers	30.0	32.0	34.0
Coal-fired power plant:advanced technologies	32.0	35.0	38.0
Electricity generation from renewable sources	100.0	100.0	100.0
Electrolysis H_2 production at refueling stations			
GH_2 production	67.0	71.5	76.0
H_2 liquefaction (for LH_2)	60.0	66.0	72.0
H_2 Compression (for GH_2): Electric compressors[a]	91.5	94.0	96.5

[a] A triangle distribution curve is assumed for these parameter.

Table **5** presents assumptions regarding electrolysis H_2 production at fueling stations. For electricity delivered to refueling stations, we use an electric transmission and distribution loss of 10%, the China average loss. In our calculations, we use a conversion efficiency of 100% from renewable energy sources to electricity because, for renewable sources, resource consumption is not a concern, and there aren't any process fuel combustion emissions.

Coal to Hydrogen Pathways

From coal to hydrogen, first step is gasification of coal. The advanced large-scale gasification of coal is TEXACO technology. In TEXACO process, coal is ground into powder. Then coal powder becomes slurry with water. After that, coal slurry exhibits incomplete combustion with pure oxygen into syngas (CO, H_2, CO_2). Finally, the mixture is through water to create syngas (CO, H_2, CO_2). After the syngas is cleaned, it enters the shift converter, where shift catalysts convert CO and steam to CO_2 and additional H_2. The gas from shift converter enters a pressure swing adsorption (PSA) unit, where high-purity H_2 is produced. We assumed energy efficiencies of 45% to 55% [14] (see Table **6**).

Table 6: Energy efficiencies for coal to H_2 pathways.

Stage	**Energy Efficiency (%)**		
	P20	**P50**	**P80**
Coal mining and cleaning		99.3	
GH_2 production at central plants	45	50	55
H_2 liquefaction[a] (for LH_2)	65	71	77
H_2 compression[a] (for GH_2)	90	92.5	95

[a] A triangle distribution curve is assumed for these parameter.

Naphtha to Hydrogen Pathways

The large-scale H_2 production from naphtha is also an SMR technology. The feed naphtha is processed through a bed of zinc oxide sorbent to remove the sulfur. Steam is added to the desulfurized naphtha feed in the reformer, where naphtha is converted into CO, H_2, CH_4, and CO_2. These gases enter the shift converter, where shift catalysts convert CO and steam to CO_2 and additional H_2. The gas from the shift converter entered a PSA unit, where high-purity H_2 is produced. The remaining gas mixture reentered the reformer. We assumed energy efficiencies of 60%-70% [15] (see Table **7**).

Table 7: Energy efficiencies of naphtha to H_2 pathways.

Stage	**Energy efficiency (%)**		
	P20	**P50**	**P80**
Crude recovery[a]	96.0	98.0	99.0
Crude refining to naphtha	89.0	91.0	93.0
GH_2 production at central plant	60.0	65.0	70.0
H_2 liquefaction[a] (for LH_2)	65.0	71.0	77.0
H_2 compression[a] (for GH_2)	90.0	92.5	95.0

[a] A triangle distribution curve is assumed for these parameter.

Pump to Wheels Assumptions

The key PTW parameter determining WTW energy use and GHG emissions associated with vehicle/fuel systems is the fuel economy of H_2 FCV. Wang [5] reviewed many researches about the fuel economy of HEVs and FCVs Nevertheless, in this study, the assumptions of H_2 FCVs fuel economy are little lower than Wang's assumptions (see Table **8**). Fig. **16** presents H_2 FCVs fuel economy distribution functions. With these distribution functions and the WTP distribution functions presented in previous section, the GREET model was run to generate energy use and emission results with probability distributions.

Table 8: Distribution functions for H_2 FCVs fuel economy ratios.

Vehicle Type	**P5**	**P50**	**P95**
Baseline GV fuel economya(MPG)	22.0	27.0	33.0
H_2 FCV[a] (Michael)(%)	210.0	235.0	260.0
H_2 FCV[a] (this study)(%)	195.0	200.0	210.0

[a] A Weibull distribution curve is assumed.

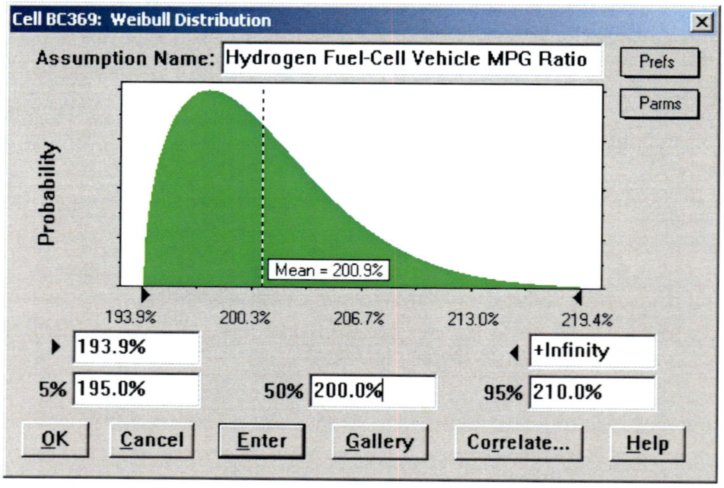

Figure 16: A sample input distribution function for stochastic simulation: H_2 FCV fuel economy ratio.

RESULTS

On the basis of the WTP and PTW distribution assumptions presented in Section 4, we use the GREET model to conduct stochastic simulations of WTW energy use, GHG emissions, and criteria pollutions emissions for hydrogen pathways of FCV included in this study. GREET generates results for a given output item with a probability distribution. Fig. **17** shows a sample output profile from Crystal Ball^{TM} stochastic simulations.

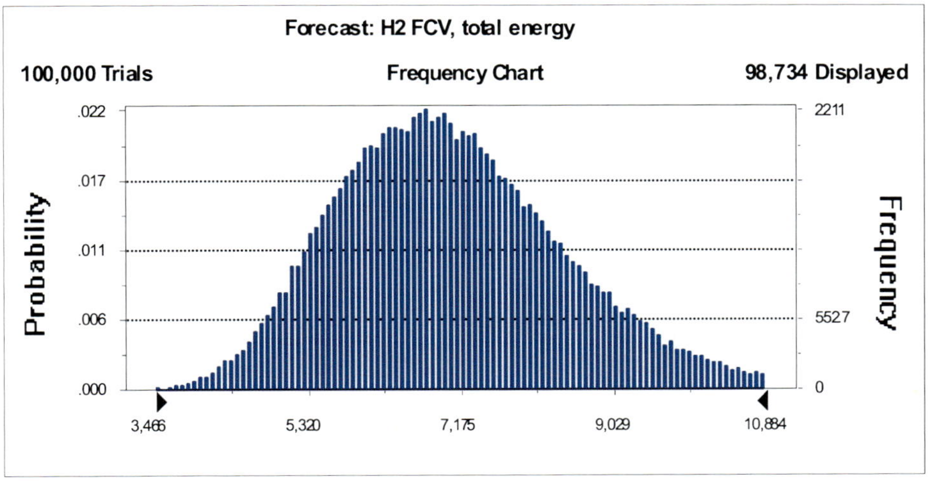

Figure 17: A sample output of crystal ball stochastic simulations: WTW total energy use of H2 FCV.

The forecast chart reveals the range of total energy consumption of H_2 FCV in WTW processes. Each bar on the chart represents the probability of WTW total energy use of H_2 FCV. The cluster of columns near the center indicates that the most likely energy use level is about 7175. There is also a small chance of energy using about 3466 and a small chance of energy about 10884.

WTP Energy Efficiencies

Fig. **18** shows WTP energy efficiencies for 11 fuel pathways. These efficiencies are affected by the energy losses from primary energy feedstock to fuel available at refueling stations. In this chart and following figures, the bars present average values, and the lines on the bars represent ranges from a probability of 20% to a probability of 80%.

The crude-to-gasoline pathway has an efficiency of about 80%. All H_2 pathways have lower WTP energy efficiencies than this baseline pathway. The WTP efficiencies are mainly affected by feedstock, H_2 conditions, and plant location, but the feedstock is the most important factor. H_2 from NG has the highest efficiency. Next is the pathway from naphtha. The third pathway is from coal. H_2 from electrolysis has the least efficiency. Efficiency of G NG C is 1.2 times of G N C, 1.9 times of G C C, and 2.6 times of G E R. The production of LH_2 is subject to more energy losses than GH_2. Finally, due to large-scale production at central plants, the efficiency at central plants is higher than at refueling stations.

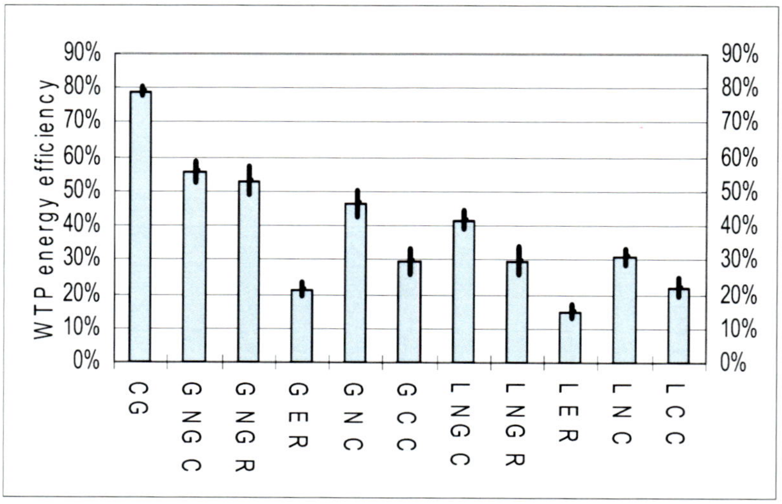

Figure 18: Well-to-pump energy efficiencies.

WTW Total Energy Use Changes

Although H_2 pathways have lower WTP energy efficiencies than GVs, the FCVs have higher fuel economy than the baseline pathway, so the WTW total energy use changes can explain whether these H_2 pathways have higher energy efficiency than GVs. Fig. **19** presents WTW total energy use changes by FCVs relative to baseline GVs fueled with CG.

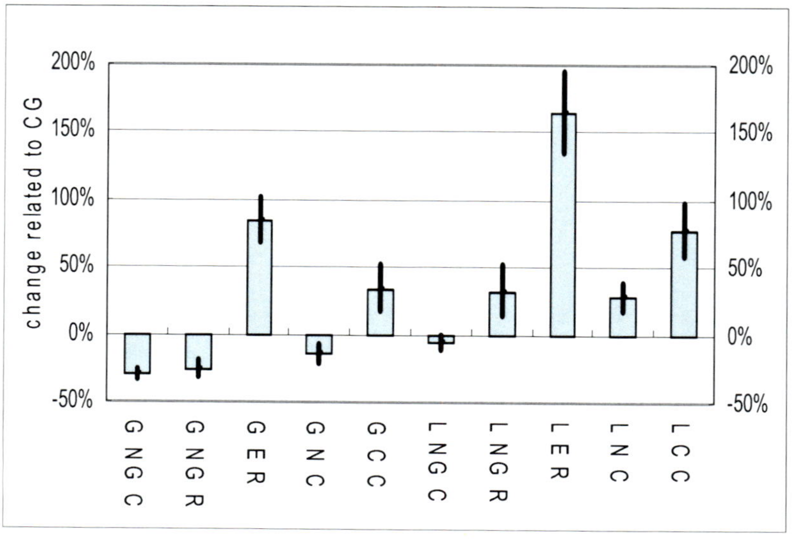

Figure 19: Changes in well-to-wheels total energy use (relative to baseline GVs).

Fig. **19** shows that four pathways (G NG C, G NG R, G N C, and L NG C) achieve 6-30% reductions in total energy use, the other six pathways suffer 30%-160% increases in total energy use. The increases are caused by significant energy losses during the WTP stages. The WTP energy losses of these pathways are so large that even the improved mpg of FCVs is not enough to offset the losses. This chart also tells that even efficient FCVs may not achieve energy benefits if inefficient fuel pathways are used to provide hydrogen for them.

WTW Fossil Energy Use Changes

WTW fossil energy use changes differ significantly from total energy use changes for the fuel based on renewable sources, because the energy sources of H_2 pathways in this study depend mainly on fossil energy, except for electricity generation mix which has a hydropower percentage of 17.2. The changes in WTW fossil energy use are basically the same as the changes in WTW total energy use (see Fig. **20**). There are the same four pathways that have 6-30% reduction in fossil energy use, and the same six pathways that suffer 30-130% increases in fossil energy use.

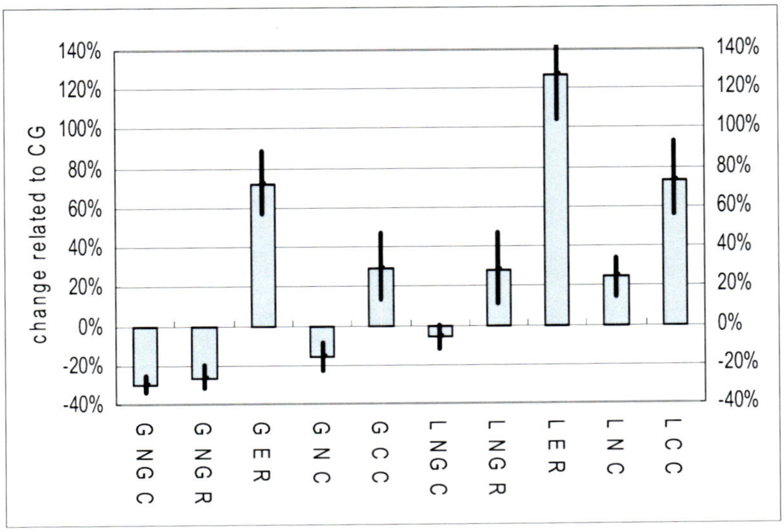

Figure 20: Changes in well-to-wheels fossil energy use (relative to baseline GVs).

WTW Petroleum Use Changes

Fig. **21** shows the changes in WTW petroleum use. Except for two pathways using petroleum feedstock (G N C and L N C), all other pathways achieve over 80%

reduction in petroleum use. The pathway of G N C has about 20% reduction in petroleum, because it has reduction in total energy use and fossil energy. Due to using electricity in H_2 liquefaction, the pathway of L N C has some reduction in petroleum use too.

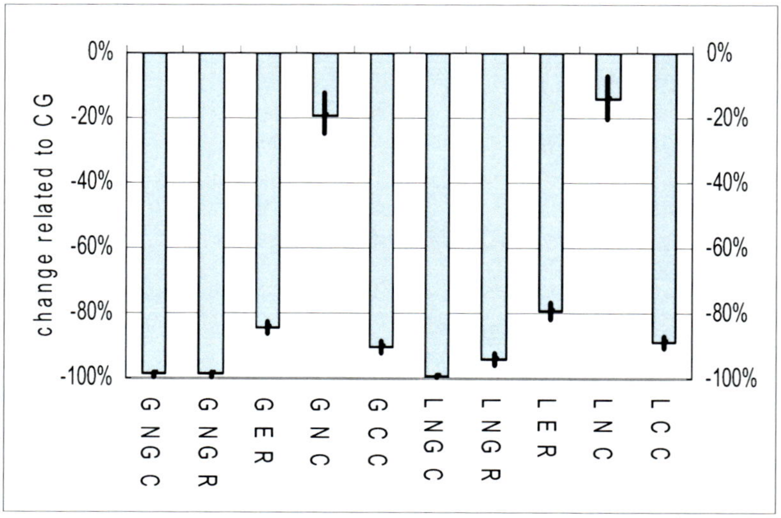

Figure 21: Changes in well-to-wheels petroleum use (relative to baseline GVs).

WTW Greenhouse Gases Emissions Changes

GHG emissions here include CO_2, methane (CH_4) and N_2O. The three gases are combined with global warming potentials (1 for CO_2, 21 for CH_4, and 310 for N2O). Fig. **22** presents the changes in WTW GHG emissions. There are four pathways (G NG C, G NG R, G N C, and L NG C), which achieve 10%-40% reduction in WTW GHG emissions. There are the other six pathways that suffer 40%-200% increases in WTW GHG emissions. This chart also shows that FCVs may not achieve GHG emissions if inefficient H_2 pathways are used to supply hydrogen for FCVs.

WTW Total Criteria Pollutions Emissions Changes

Criteria pollutions emissions in this study include volatile organic compounds (VOCs), carbon monoxide (CO), nitrogen oxides (NO_x), sulfur oxides (SO_x) and particulate matter with diameters of 10 micrometers or less (PM_{10}). We are not only interested in total criteria pollutions emissions, but also concerning urban criteria pollutions emissions. Therefore, we calculated two kinds of criteria pollutions emissions in WTW. The urban criteria pollutions emissions will be explained in the following section. Figs. **23** to **27** show the changes in WTW total criteria pollutions emissions.

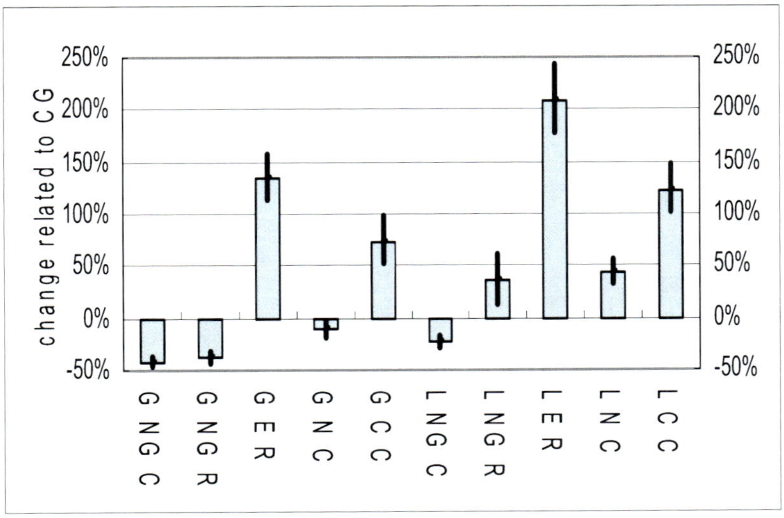

Figure 22: Changes in well-to-wheels greenhouse gases emissions (relative to baseline GVs).

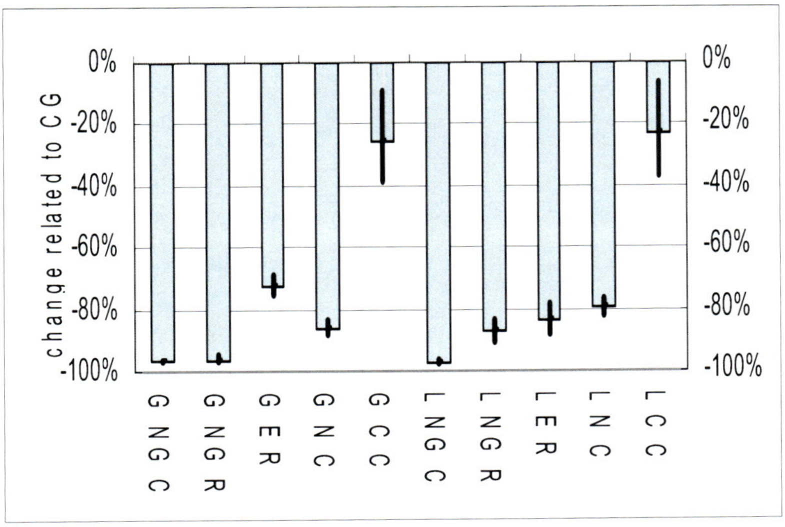

Figure 23: Changes in well-to-wheel total VOCs emissions (relative to baseline GVs).

Fig. **23** shows the changes in WTW total VOCs emissions. This is because total VOCs emissions in baseline GVs operation stage has 75% of WTW total VOCs emissions (see Fig. **28**), and in the FCV operation stage there is a zero VOCs emission. Except two hydrogen pathways from coal, all other pathways achieve over 80% reduction in WTW VOCs emissions. The two H_2 pathways from coal reduce about 20% total VOCs emissions, because these ways have more VOCs emissions in fuel stage than other do.

Fig. **24** presents changes in WTW total CO emissions. From Fig. **28**, we know the total CO emissions in baseline operation stage share 99% of WTW total CO emissions, and FCVs have no CO emissions any more. Therefore, all pathways almost eliminate total CO emissions.

Fig. **25** shows changes in WTW total NO_x emissions. There are four pathways that have total NO_x emissions increases. Because of the mix of electricity generation share over 70% coal electricity, these four pathways include a lot of coal combustion, which emit more NO_x than other fuel combustion. The FCV operation zero NO_x emissions cannot offset the increase in hydrogen production stage in these pathways. These four pathways suffer 25%-50% increases in WTW total NO_x emissions. On the other hand, the other six pathways achieve 20%-80% reduction in WTW total NO_x emissions.

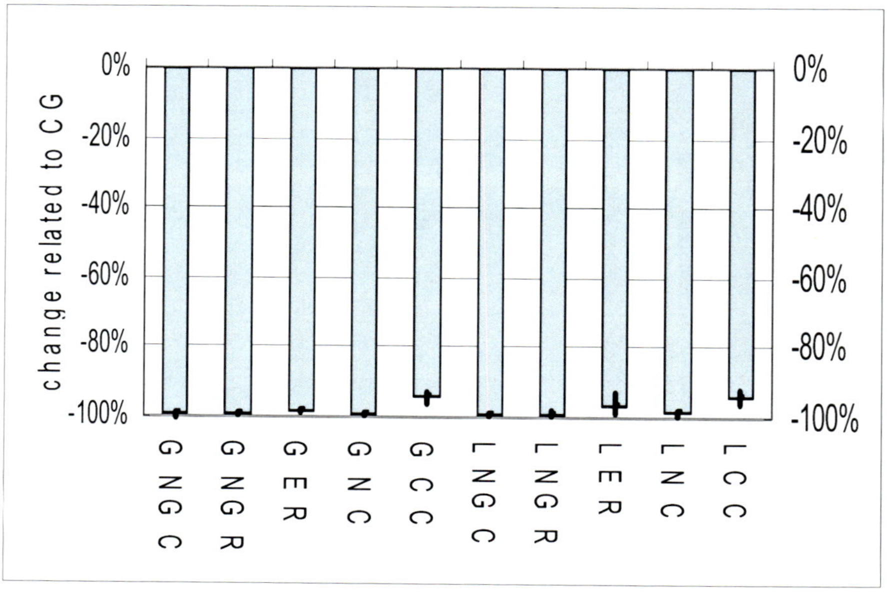

Figure 24: Changes in well-to-wheel total CO emissions (relative to baseline GVs).

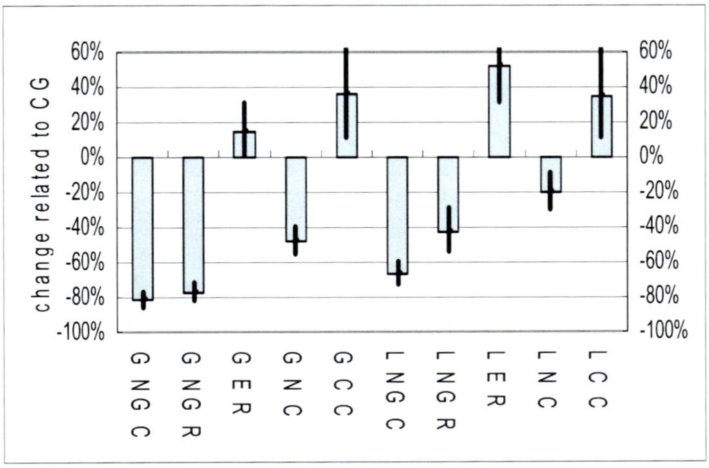

Figure 25: Changes in well-to-wheel total NO_x emissions (relative to baseline GVs).

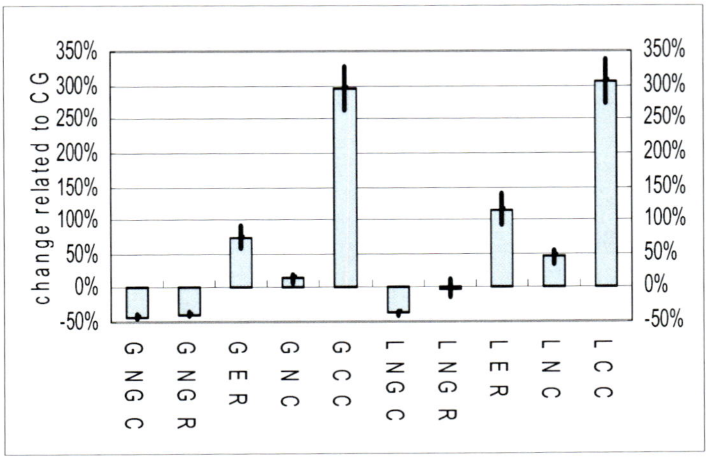

Figure 26: Changes in well-to-wheel total PM_{10} emissions (relative to baseline GVs).

Fig. **26** presents changes in WTW total PM_{10} emissions. Except the four H_2 pathways from NG, all other six pathways suffer increases in WTW total PM_{10} emissions. Especially, the two pathways from coal have four times total PM_{10} emissions than the baseline. The two pathways from electrolysis suffer 75%-125% increases in WTW total PM_{10} emissions. The large increases of total PM_{10} in previous four pathways come from coal combustion. The two pathways from naphtha suffer 13%-45% increases. On the other hand, refueling station liquid hydrogen from natural gas has almost no changes with the baseline. The other three pathways achieve about 40% reduction in WTW total PM_{10} emissions.

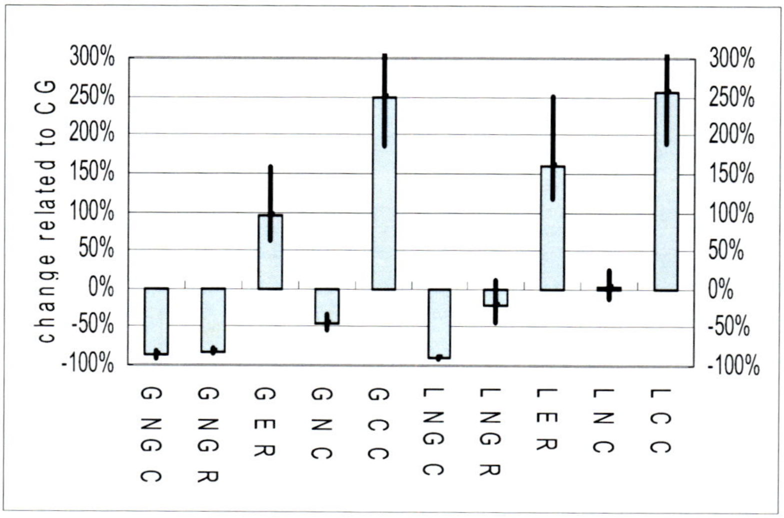

Figure 27: Changes in well-to-wheel total SO_x emissions (relative to baseline GVs).

Fig. **27** shows changes in WTW total SO_X emissions. The operation stage total SO_X emissions share only 30% of WTW total SO_X emissions in baseline GVs. Therefore, WTP stages emissions majorly affected the changes. Because SO_X emission factor of coal combustion is larger than other fuels, the two pathways from coal and two pathways from electrolysis suffer 100%-300% increases in WTW total SO_X emissions. On the other hand, the liquid hydrogen from naphtha has almost no changes in WTW total SO_X emissions and the other five pathways achieve 20%-90% reduction.

WTW Urban Criteria Pollutions Emissions Changes

Urban air quality is mainly affected by vehicles emissions. Therefore, the changes in WTW urban criteria pollutions emissions are important results in this study. Figs. **29** to **33** show changes in WTW urban criteria emissions. From Fig. **28**, except urban SO_X, of which the operation stage shares half WTW emissions, the other WTW urban criteria emissions account for a great half (over 70%) in operation stage. However, FCVs have zero urban criteria emissions. Therefore, WTW urban criteria emissions almost achieve large reductions (see Figs. **29** to **32**), except urban SO_X emissions. Because SO_X emission factor is large, the urban SO_X emissions suffer 230%-250% increases in two H_2 pathways from coal (see Fig. **33**).

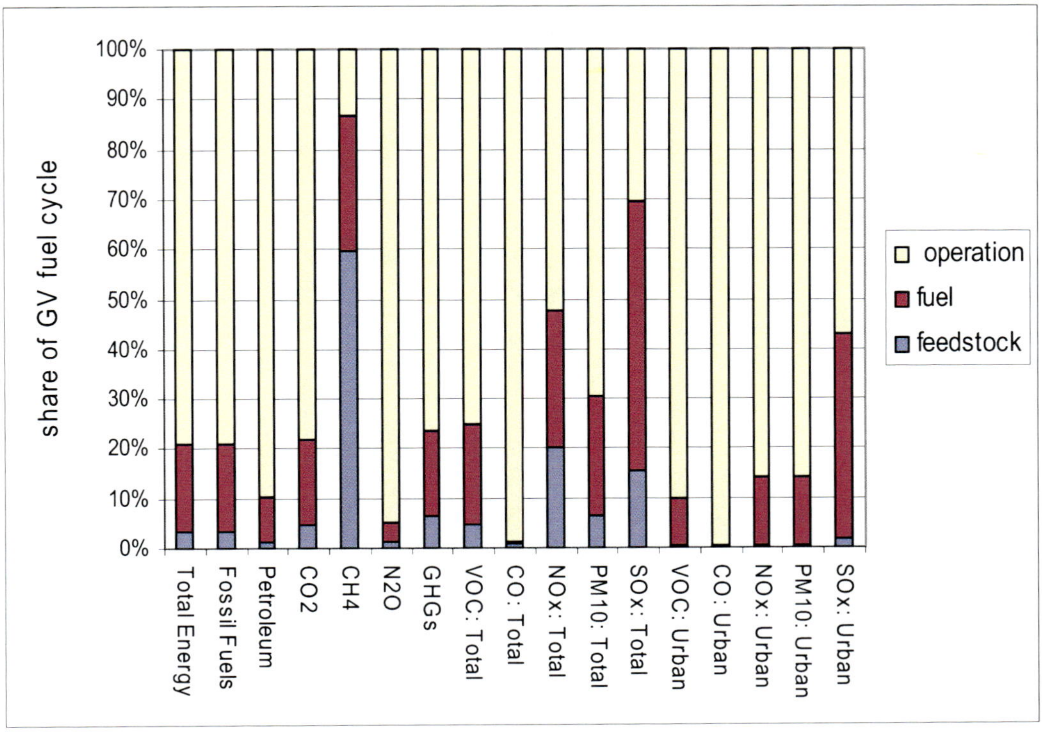

Figure 28: Shares of fuel-cycle energy use and emissions by stages: baseline GVs.

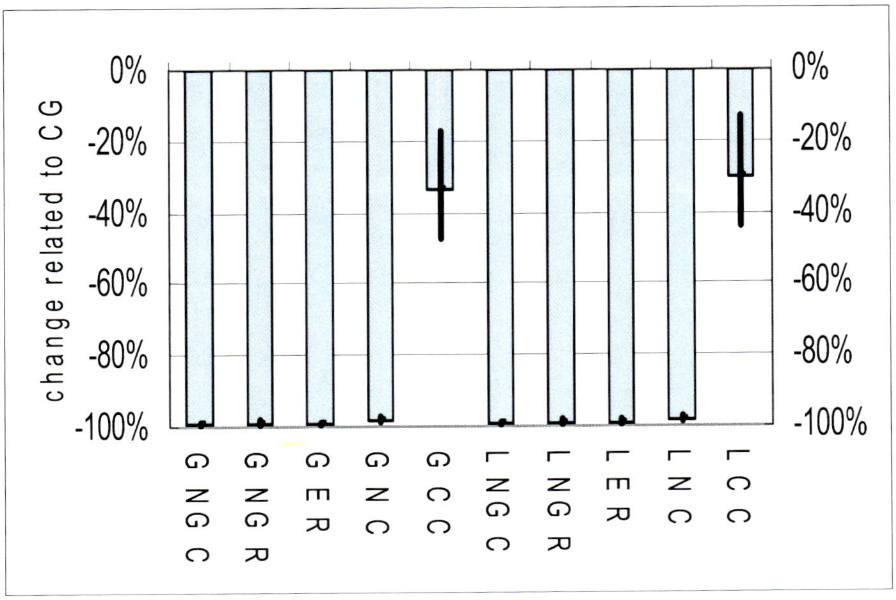

Figure 29: Changes in WTW urban VOCs emissions.

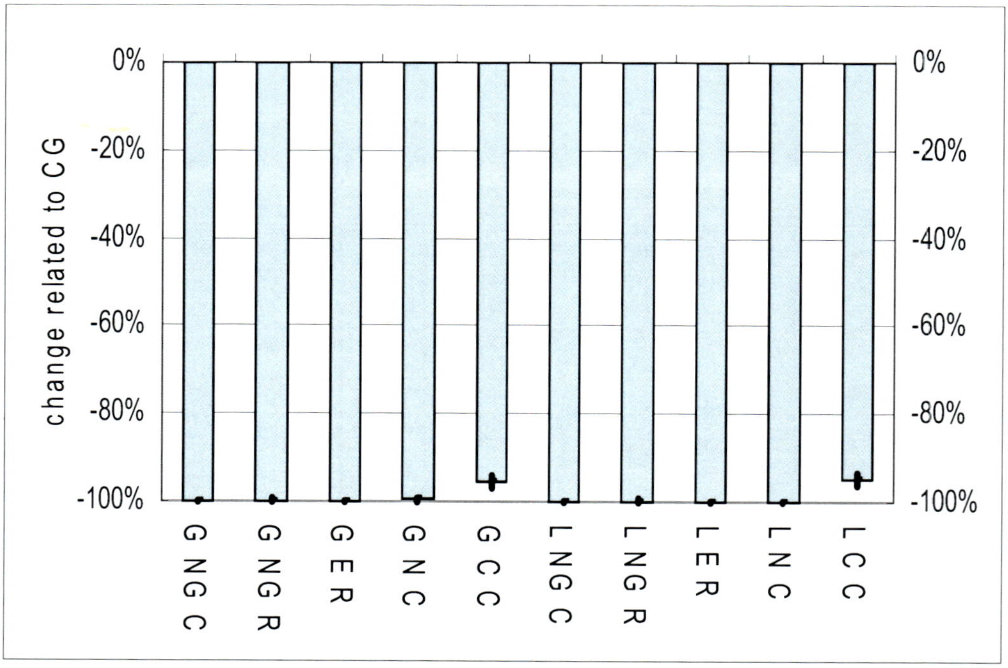

Figure 30: Changes in WTW urban CO emissions.

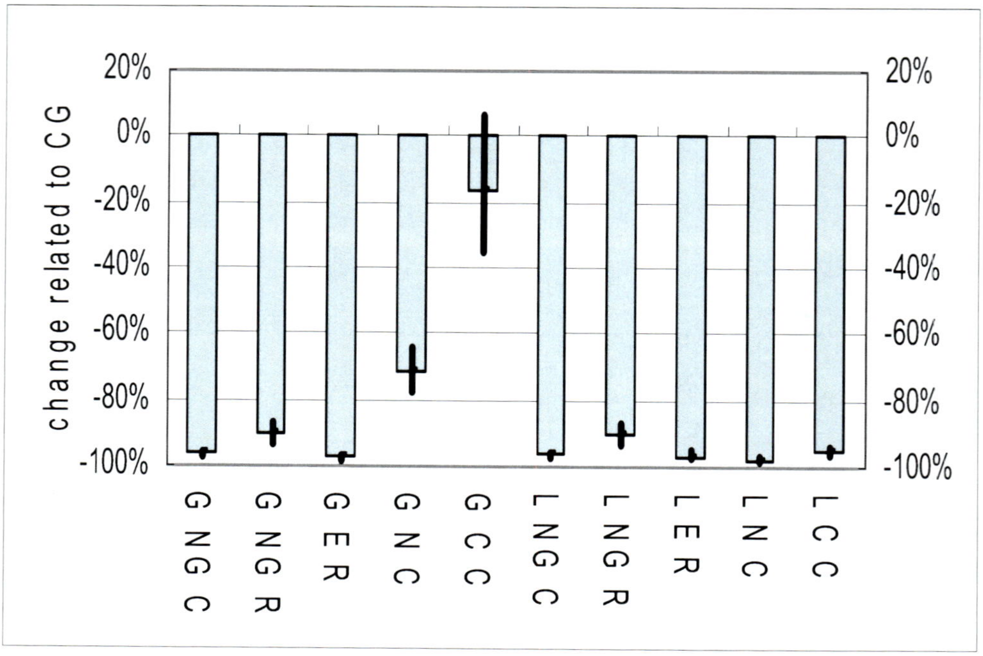

Figure 31: Changes in WTW urban NO_x emissions.

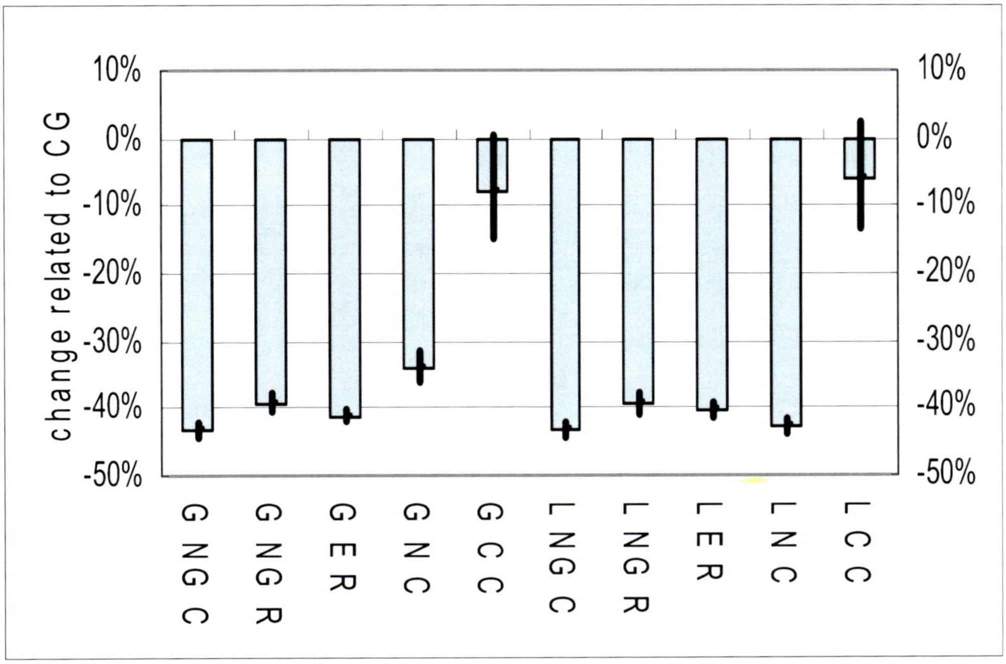

Figure 32: Changes in WTW urban PM_{10} emissions.

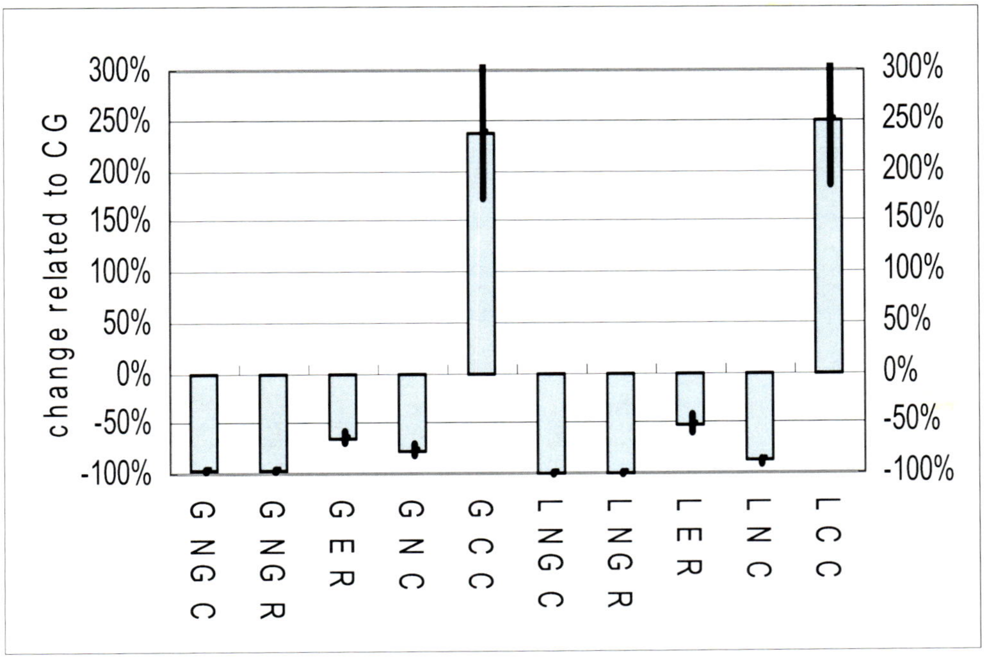

Figure 33: Changes in WTW urban SO_x emissions.

CONCLUSIONS

1) All the pathways have significant reduction in WTW petroleum use, except two H_2 pathways from naphtha, which achieve about 20% reduction in WTW petroleum.

2) All the pathways have significant reduction in WTW urban criteria pollutions emissions, except two H_2 pathways from coal, which offer significant increase in WTW urban SO_x emissions.

3) The NG-based H_2 pathways have best WTW energy efficiencies, and the electrolysis H_2 pathways have worst WTW energy efficiencies. The WTW energy efficiencies of H_2 pathways from naphtha and H_2 pathways from coal are between NG-based pathways and electrolysis pathways. The pathways from naphtha have higher energy efficiencies than the pathways from coal. Only four pathways (G NG C, G NG R, G N C, and L NG C) offer WTW energy benefits, and the other six pathways consume more WTW energy than baseline-conventional gasoline vehicles.

4) Changes in WTW GHGs emissions have nearly identical results with changes in WTW energy use.

5) For WTW total criteria pollutions emissions, all pathways can achieve significant reduction in WTW total VOCs and CO. The other criteria pollutions emissions, NO_x, PM_{10} and SO_2, are certainly reduced in NG and crude oil-based H_2 pathways, but increase significantly in electrolysis and coal-based pathways.

6) Because different H_2 pathways have very different energy and emissions effects, the well-to-wheels analyses are necessary for adequately evaluating fuel/vehicle systems. The pathways used to supply H_2 for FCVs must be carefully examined by their WTW energy use, GHGs emissions, total criteria pollutions emissions, and urban criteria pollutions emissions.

ACKNOWLEDGMENTS

This study was funded by Natural Resource Defense Council, and Michael Wang in Argonne National Laboratory of US Department of Energy offered advisor and research situation. Therefore, I'd like to thank Barbara Finamore, Tina Clegg, Hannah Stutzman, and Weisi Zhu of NRDC for work for me coming to USA. I was grateful to Dr. Michael Wang, Dr. Guokui Liu, Dr. Ira Goldberg, and Dr. Hanjie Lee of Argonne for help in my research and living. Finally, I would also like to thank Prof. Xu Zhang, and Prof. Gan Wan of Tongji University Shanghai for offering me the chance to study in America.

DISCLOSURE

Part of information included in this chapter has been previously published in ***Energy*** Volume 31, Issue 4, March 2006, Pages 471-489 (http://www.sciencedirect.com/science/article/pii/S0360544205000617).

REFERENCES

[1] China Statistic Bureau. China statistic annual. Beijing: China Statistic Publishing Company, 2001.

[2] Z. Huang, X. Zhang, G. Wan, "Development stratagem of alternative fuel of automobile", *China Energy*, vol. 8, pp. 30-33, 2001.

[3] P.L. Spath, M. Mann, "Life cycle assessment of hydrogen production *via* natural gas steam reform" in NREL/TP-570-27637, Colorado: National Renewable Energy Laboratory, Golden, 2001.

[4] M.Q. Wang, "GREET1.5 – Transportation Fuel-Cycle Model, Volume 1: Methodology, Development, Use, and Results" in ANL/ESD-39, Center for Transportation Research Argonne National Laboratory, August 1999.

[5] M.Q. Wang, "GREET1.5 – Transportation Fuel-Cycle Model, Volume 2: Appendices of Data and Results" in ANL/ESD-39, Center for Transportation Research Argonne National Laboratory, Argonne, August 1999.

[6] M.Q. Wang, H.S. Huang, "A Full Fuel-Cycle Analysis of Energy and Emissions Impacts of Transportation Fuels Produced from Natural Gas" in ANL/ESD-40, Center for Transportation Research Argonne National Laboratory, Argonne, December 1999.

[7] General Motors Corporation, Argonne National Laboratory, BP, ExxonMobil, and Shell, "Well-to-Wheel Energy Use and Greenhouse Gas Emissions of Advanced Fuel/Vehicle Systems" in North American Analysis, Vol. 1, June 2001.

[8] General Motors Corporation, Argonne National Laboratory, BP, ExxonMobil, and Shell, "Well-to-Wheel Energy Use and Greenhouse Gas Emissions of Advanced Fuel/Vehicle Systems" in North American Analysis, Vol. 2, June 2001.

[9] General Motors Corporation, Argonne National Laboratory, BP, ExxonMobil, and Shell, "Well-to-Tank Energy Use and Greenhouse Gas Emissions of Transportation Fuel" in North American Analysis, Vol. 3, June 2001.

[10] M. Wang, "Fuel choice for fuel-cell vehicles: well-to-wheels energy and emissions impacts", *J. of Power Source* vol. 112, pp. 307-321, 2002.
[11] Decisioneering, Inc. Crystal Ball 2000.2 User Manual, 2002
[12] R. Bharvirkar, "Quantification of Variability and Uncertainty in Emission Factors and Emissions Inventories", M.S. thesis, North Carolina State University, NC, 1999.
[13] China Statistic Bureau. China energy statistic annual. Beijing: China Statistic Publishing Company, 2000.
[14] W. Ni, X. Wang, "Hydrogen infrastructure research of fuel cell bus" in AA501986, Tsinghua University, 2002.
[15] G. Wang, X. Zhang, Z. Huang, "Life cycle assessment of hydrogen pathways for fuel cell vehicle" in 2001AA501987, Tongji University, 2002.

APPENDICES

Units of Measure

Due to English units of measure in GREET model, English units of measure are used in this report. Therefore, material consumption is reported in units based the pound, energy consumption based on the Btu, and distance based on the mile. When it can contribute to the understanding of the analysis, the metric system equivalent is stated in parenthesis. The English units used for each parameter are given below, with the corresponding to Metric units.

Mass:	pound = 0.4535 kilogram (kg)
	ton = 0.9074 metric tonne (T)
Distance:	mile = 1.6129 kilometer (km)
Volume:	gallon = 3.785 litres (l)
Pressure:	pound per square inch (psi) = 6.8965 kilopasicals (kpa)
Energy:	Btu = 1.0539 kilojoule (kJ)
Temperature:	$^{\circ}F = 32 + {}^{\circ}C \times 5/9$

Abbreviation and Terms

ANL	Argonne National Laboratory
Btu	British thermal units
CG	Conventional gasoline

CH_4	Methane
CO	Carbon monoxide
CO_2	Carbon dioxide
FCVs	Fuel-Cell Vehicles
GWP	Global warming potential
GHGs	Greenhouse gases
GREET	Greenhouse gases, Regulated Emissions, and Energy use in Transportation
GH2	Gaseous hydrogen
GVs	Gasoline Vehicles
G NG C	Pathway of gaseous hydrogen from natural gas at central plants
G NG R	Pathway of gaseous hydrogen from natural gas at refueling stations
G E R	Pathway of gaseous hydrogen from electrolysis at refueling stations
G N C	Pathway of gaseous hydrogen from naphtha at central plants
G C C	Pathway of gaseous hydrogen from coal at central plants
IPCC	Intergovernmental Panel on Climate Change
LH_2	Liquid hydrogen
LCA	Life-cycle assessment
L NG C	Pathway of liquid hydrogen from natural gas at central plants
L NG R	Pathway of liquid hydrogen from natural gas at refueling stations
L E R	Pathway of liquid hydrogen from electrolysis at refueling stations
L N C	Pathway of liquid hydrogen from naphtha at central plants
L C C	Pathway of liquid hydrogen from coal at central plants
Mpg	Mile per gallon of gasoline equivalent

NG	Natural gas
NRDC	Natural Resource Defense Council
N_2O	Nitrous oxide
NO_x	Nitrogen oxide
PM_{10}	Particulate matter with diameter of 10 μm or less
P20	20% probability
P50	50% probability
P80	80% probability
PSA	Pressure swing adsorption
PTW	Pump-to-Wheels
VOCs	Volatile organic compounds
SMR	Steam-methane reforming
SO_x	Sulfur oxide
U.S. DOE	United States Department of Energy
WTW	Well-to-Wheels
WTP	Well-to-Pump

Data used to Prepare Figs. 18 to 33

Table 9: WTP energy efficiencies (data of Fig. **18**).

	CG	G NG C	G NG R	G E R	G N C	G C C	L NG C	L NG R	L E R	L N C	L C C
P20	77.8%	53.0%	48.9%	19.5%	42.6%	26.0%	39.6%	25.7%	13.4%	28.5%	19.8%
P50	78.8%	55.7%	52.9%	21.3%	46.3%	29.5%	41.8%	29.7%	14.9%	30.6%	22.1%
P80	80.0%	58.6%	57.0%	23.1%	49.8%	33.3%	44.2%	34.1%	16.6%	32.9%	24.6%

Table 10: WTW total energy use changes (data of Fig. **19**).

	G NG C	G NG R	G E R	G N C	G C C	L NG C	L NG R	L E R	L N C	L C C
P20	-33.3%	-31.2%	69.4%	-21.5%	17.6%	-11.4%	15.0%	135.9%	18.8%	59.5%
P50	-29.6%	-25.9%	84.0%	-15.1%	33.1%	-6.1%	32.2%	164.1%	28.1%	77.4%
P80	-25.7%	-19.7%	101.3%	-7.7%	51.5%	-0.4%	52.7%	194.1%	37.9%	97.8%

Table 11: WTW fossil energy use changes (data of Fig. **20**).

	G NG C	G NG R	G E R	G N C	G C C	L NG C	L NG R	L E R	L N C	L C C
P20	-33.8%	-31.7%	58.1%	-22.5%	13.8%	-11.4%	11.7%	104.8%	15.3%	56.1%
P50	-30.1%	-26.5%	72.1%	-16.1%	28.7%	-6.1%	28.0%	127.1%	24.3%	73.4%
P80	-26.2%	-20.4%	88.3%	-8.8%	46.2%	-0.4%	47.2%	152.2%	33.8%	93.1%

Table 12: WTW petroleum use changes (data of Fig. **21**).

	G NG C	G NG R	G E R	G N C	G C C	L NG C	L NG R	L E R	L N C	L C C
P20	-99.1%	-99.0%	-85.9%	-24.7%	-91.6%	-99.6%	-95.4%	-81.7%	-20.3%	-90.0%
P50	-98.9%	-98.8%	-84.5%	-19.1%	-90.5%	-99.3%	-94.0%	-79.6%	-14.3%	-88.8%
P80	-98.6%	-98.6%	-83.0%	-12.4%	-89.1%	-99.1%	-92.4%	-77.3%	-7.5%	-87.4%

Table 13: WTW GHGs emissions changes (data of Fig. **22**).

	G NG C	G NG R	G E R	G N C	G C C	L NG C	L NG R	L E R	L N C	L C C
P20	-45.0%	-42.5%	114.8%	-18.6%	53.8%	-26.8%	14.8%	177.9%	33.0%	101.7%
P50	-42.0%	-38.2%	133.7%	-11.6%	74.0%	-22.8%	35.3%	208.3%	44.0%	122.4%
P80	-38.8%	-33.5%	155.9%	-3.7%	97.4%	-18.5%	60.0%	242.5%	55.6%	147.7%

Table 14: WTW total VOCs emissions changes (data of Fig. **23**).

	G NG C	G NG R	G E R	G N C	G C C	L NG C	L NG R	L E R	L N C	L C C
P20	-97.0%	-96.7%	-75.3%	-88.0%	-38.9%	-97.5%	-90.2%	-88.0%	-81.7%	-37.0%
P50	-96.6%	-96.2%	-72.4%	-86.3%	-25.7%	-97.1%	-87.6%	-84.0%	-79.5%	-23.5%
P80	-96.2%	-95.3%	-69.1%	-84.2%	-9.6%	-96.6%	-84.0%	-78.6%	-76.9%	-6.8%

Table 15: WTW total CO emissions changes (data of Fig. **24**).

	G NG C	G NG R	G E R	G N C	G C C	L NG C	L NG R	L E R	L N C	L C C
P20	-99.7%	-99.5%	-98.7%	-99.2%	-95.6%	-99.4%	-99.1%	-97.9%	-98.8%	-95.5%
P50	-99.5%	-99.3%	-98.5%	-99.0%	-94.4%	-99.2%	-98.9%	-96.2%	-98.6%	-94.1%
P80	-99.3%	-99.0%	-98.3%	-98.8%	-92.8%	-98.9%	-98.5%	-93.2%	-98.4%	-92.6%

Table 16: WTW total NO_x emissions changes (data of Fig. **25**).

	G NG C	G NG R	G E R	G N C	G C C	L NG C	L NG R	L E R	L N C	L C C
P20	-84.8%	-81.0%	1.3%	-54.5%	11.5%	-71.9%	-53.1%	32.1%	-28.8%	12.0%
P50	-81.7%	-77.1%	15.3%	-47.6%	35.9%	-66.5%	-42.1%	52.5%	-19.4%	35.3%
P80	-77.2%	-72.4%	30.9%	-39.3%	63.1%	-59.4%	-28.7%	75.1%	-8.7%	63.7%

Table 17: WTW total PM_{10} emissions changes (data of Fig. **26**).

	G NG C	G NG R	G E R	G N C	G C C	L NG C	L NG R	L E R	L N C	L C C
P20	-44.4%	-40.9%	61.1%	8.0%	266.2%	-38.8%	-12.3%	95.3%	36.0%	273.9%
P50	-42.9%	-39.0%	74.3%	13.6%	296.1%	-36.9%	-1.2%	115.0%	44.5%	304.0%
P80	-41.4%	-37.1%	89.1%	19.7%	327.4%	-34.8%	12.5%	137.0%	53.7%	335.4%

Table 18: WTW total SO_x emissions changes (data of Fig. **27**).

	G NG C	G NG R	G E R	G N C	G C C	L NG C	L NG R	L E R	L N C	L C C
P20	-89.2%	-84.5%	65.4%	-52.2%	188.2%	-90.3%	-41.7%	117.6%	-12.3%	189.9%
P50	-87.2%	-82.4%	93.3%	-44.3%	249.4%	-89.6%	-21.6%	159.4%	1.4%	254.7%
P80	-83.2%	-78.4%	156.0%	-33.5%	327.3%	-88.8%	11.0%	249.1%	24.2%	333.8%

Table 19: Shares of fuel-cycle energy use and emissions by stages (data of Fig. **28**).

	Feedstock	**Fuel**	**Operation**
Total Energy	3.5%	17.4%	79.0%
Fossil Fuels	3.5%	17.4%	79.2%
Petroleum	1.3%	8.9%	89.7%
CO2	4.7%	17.2%	78.1%
CH4	59.5%	27.1%	13.3%
N2O	1.2%	3.8%	95.0%
GHGs	6.6%	17.2%	76.2%
VOC: Total	4.8%	20.1%	75.0%
CO: Total	0.7%	0.7%	98.6%
NOx: Total	20.1%	27.4%	52.5%
PM10: Total	6.4%	24.1%	69.5%
SOx: Total	15.6%	53.9%	30.4%
VOC: Urban	0.4%	9.3%	90.3%
CO: Urban	0.0%	0.2%	99.7%
NOx: Urban	0.6%	13.6%	85.8%
PM10: Urban	0.3%	14.0%	85.7%
SOx: Urban	1.5%	41.4%	57.1%

Table 20: WTW urban VOCs emissions changes (data of Fig. **29**).

	G NG C	G NG R	G E R	G N C	G C C	L NG C	L NG R	L E R	L N C	L C C
P20	-99.4%	-99.3%	-99.4%	-98.5%	-46.7%	-99.4%	-99.4%	-99.2%	-98.6%	-43.8%
P50	-99.4%	-99.2%	-99.2%	-98.0%	-33.4%	-99.3%	-99.2%	-99.0%	-98.1%	-30.2%
P80	-99.3%	-98.2%	-99.0%	-97.4%	-17.3%	-99.3%	-98.3%	-98.7%	-97.3%	-13.2%

Table 21: WTW urban CO emissions changes (data of Fig. **30**).

	G NG C	G NG R	G E R	G N C	G C C	L NG C	L NG R	L E R	L N C	L C C
P20	-99.9%	-99.9%	-100.0%	-99.7%	-96.5%	-99.9%	-99.9%	-100.0%	-100.0%	-96.3%
P50	-99.9%	-99.8%	-100.0%	-99.7%	-95.3%	-99.9%	-99.8%	-100.0%	-100.0%	-95.1%
P80	-99.9%	-99.6%	-100.0%	-99.6%	-93.9%	-99.9%	-99.6%	-99.9%	-100.0%	-93.6%

Table 22: WTW urban NO_X emissions changes (data of Fig. **31**).

	G NG C	G NG R	G E R	G N C	G C C	L NG C	L NG R	L E R	L N C	L C C
P20	-96.8%	-93.0%	-97.9%	-77.7%	-35.1%	-96.6%	-93.0%	-97.3%	-98.1%	-96.3%
P50	-96.4%	-90.6%	-97.4%	-71.7%	-16.7%	-96.2%	-90.6%	-96.6%	-97.8%	-95.1%
P80	-95.7%	-87.5%	-96.6%	-64.1%	6.0%	-95.8%	-87.4%	-95.5%	-97.5%	-93.6%

Table 23: WTW urban PM_{10} emissions changes (data of Fig. **32**).

	G NG C	G NG R	G E R	G N C	G C C	L NG C	L NG R	L E R	L N C	L C C
P20	-44.0%	-40.5%	-41.9%	-36.1%	-14.8%	-44.1%	-40.7%	-41.1%	-43.6%	-13.4%
P50	-43.2%	-39.2%	-41.1%	-34.1%	-7.9%	-43.2%	-39.3%	-40.3%	-42.8%	-6.2%
P80	-42.4%	-37.8%	-40.3%	-31.6%	0.5%	-42.4%	-37.8%	-39.5%	-42.0%	2.5%

Table 24: WTW urban SO_X emissions changes (data of Fig. **33**).

	G NG C	G NG R	G E R	G N C	G C C	L NG C	L NG R	L E R	L N C	L C C
P20	-98.1%	-98.0%	-68.1%	-80.6%	173.4%	-99.7%	-99.4%	-58.2%	-89.8%	185.0%
P50	-97.8%	-97.7%	-63.8%	-76.4%	235.0%	-99.7%	-99.2%	-52.0%	-88.2%	248.8%
P80	-97.4%	-97.2%	-57.8%	-71.1%	307.6%	-99.6%	-99.0%	-43.4%	-86.2%	324.6%

CHAPTER 3

Contribution to the Valorization of Moroccan Oil Shales

A.K. Abourriche[a,*], M. Oumam[b], H. Hannache[b], A.M. Abourriche[b], M. Birot[c], R. Pailler[d] and R. Naslain[d]

[a]Laboratoire Matériaux, Procédés, Environnement et Qualité, École Nationale des Sciences Appliquées, B.P. 63, 46000 Safi, Morocco; [b]Faculté des Sciences Ben M'sik, B.P. 7955 Casablanca, Morocco; [c]Université Hassan II, Faculté des Sciences Ben M'sik, B.P. 7955 Casablanca, Morocco; and [d]Université Bordeaux 1/CNRS/CEA/SAFRAN, Laboratoire des Composites Thermostructuraux (LCTS), 3 allée de la Boétie, F-33600 Pessac, France

Abstract: Oil shale has constituted for a long time an economical hope for countries that possess important reserves of these rocks and that view to use them as an energy source substitute for petroleum.

Morocco, with estimated reserves of 93 billion tons, is increasingly looking at oil shale as an alternative energy source. A lot of studies have concentrated on oil shale located in Timahdit and Tarfaya, because of their high percentage of organic matter. Most of the studies focus either on the effect of various parameters on the yield and the quality of the oil obtained by conventional pyrolysis, or on the characterization of these oils by different physical and chemical techniques.

This paper explores the possibility to produce new materials, starting from the Moroccan oil shale, for different applications. More specifically, we aimed to demonstrate that the organic fraction of the oil shale could be used as precursors of low cost carbon fibres or graphitizable carbon, after appropriate chemical treatments resulting in a "maturation" of this organic phase. We also showed that this organic fraction of the Moroccan oil shale has interesting bioactive properties and that it could be used as a source of compounds with pharmaceutical interests.

Keywords: Oil, shale, supercritical, extraction, carbon, fibers, activation, pitch, graphitizable carbon, phenol, Raman, spectroscopy, bioactive properties, antibacterial, cytotoxicity, DNA, valorization.

INTRODUCTION

Oil shales have constituted for a long time an economical hope for countries that

*Address correspondence to A.K. Abourriche: Laboratoire Matériaux, Procédés, Environnement et Qualité, École Nationale des Sciences Appliquées, B.P. 63, 46000 Safi, Morocco; Tel: (+212)6 64456721; Fax: (+212)5 24668012; E-mail: krimabou@hotmail.com

possess important reserves of these rocks and that view to use them as an energy source substitute for petroleum. Morocco, with estimated reserves of 93 billion tons [1], is increasingly looking at oil shales as an alternative energy source. Many studies have concentrated on oil shales located in Timahdit and Tarfaya, because of their high proportion of organic matter [2]. Most of the studies focused either on the yield and the quality of the oils obtained by conventional pyrolysis or on the characterization of these oils by different physical and chemical techniques [3-6].

In addition, oil shales have a certain potential for the production of several synthetic products such as cement, sulfur, ammonia, adsorbent carbons, carbon fibers and other chemicals [7-9].

This chapter evaluates the possibilities to produce new materials, starting from Moroccan oil shales, for different applications. More specifically, we aim at demonstrating that the organic fraction of the oil shales could be used as precursor of low-cost carbon fibers or graphitizable carbon, after appropriate chemical treatments resulting in a "maturation" of this organic phase. We also show that this organic fraction has interesting bioactive properties and that it could be used as a source of compounds of pharmaceutical interest.

ACTIVATED CARBON FIBERS FROM MOROCCAN OIL SHALES

This section gives an overview on the preparation of the first carbon fibers activated with phosphoric acid from Moroccan oil shale [10, 11].

Preparation of the Precursor of Carbon Fibers (Pitches)

The oil shale used in this work was from the Tarfaya deposit located in the South of Morocco. This deposit consists of several layers that are in turn subdivided in sub-layers, each having a different amount of organic matter. The samples were obtained from the so-called **R_3** sub-layer characterized by its high content of organic matter [12]. Its chemical composition is given in Table **1**.

Preparation of the Raw Material

The carbonate-free oil shale was obtained by dissolution of carbonates with HCl [13, 14]. The powdered **R_3** shale (20 g, grain size 0.063–0.08 mm) and 80 mL of concentrated HCl (7 M) were introduced in an Erlenmeyer. The mixture was then subjected to magnetic stirring for 4 h. The formed CO_2 was trapped by bubbling in a solution of barium hydroxide. After filtration, the solid residue (referred to as **RH**) was washed carefully with distilled water, dried at 100 °C and stocked in a sealed plastic bag.

Table 1: Chemical composition of the R_3 sub-layer [12].

Composition	**(wt %)**
Carbonates	70.0
Kerogen	20.0
Silicates	7.1
Pyrite	1.0
Bitumen	0.9
Metals traces	1.0

Procedures

Supercritical extraction of **RH** (10 g) with toluene was conducted in a 120 mL stainless steel autoclave equipped with a pressure gauge and heated in a tubular furnace whose temperature, as well as the heating rate, was controlled (Fig. **1**). The temperature of extraction was 390 °C with a heating rate of 16 $°C \cdot min^{-1}$ [15]. The maximum pressure reached during this treatment of 120 min was 5.3 MPa and 7.5 MPa for a volume of 50 mL and 70 mL respectively, beyond the critical point of toluene (320 °C, 4.2 MPa), After being cooled to room temperature, the mixture was extracted in a Soxhlet apparatus with chloroform for 12 h. The solvent was removed under reduced pressure, and then the organic material was dried for 12 h at 40 °C and weighed. The recovered oil was treated with *n*–hexane in a 1/10 oil to solvent mass ratio [16]. After stirring for 12 h, the two fractions, soluble (maltenes) and insoluble (asphaltenes), were separated by filtration through Whatman paper, and then dried for 12 h at 40 °C and 80 °C respectively. The maltenes were fractionated in a silica gel column (70–230 mesh, 1 m long and 1.5 cm diameter). Elution of paraffinic, aromatic and polar compounds was performed with hexane, toluene and methanol, respectively [16]. The high–molecular fraction (asphaltenes) constitutes the pitch.

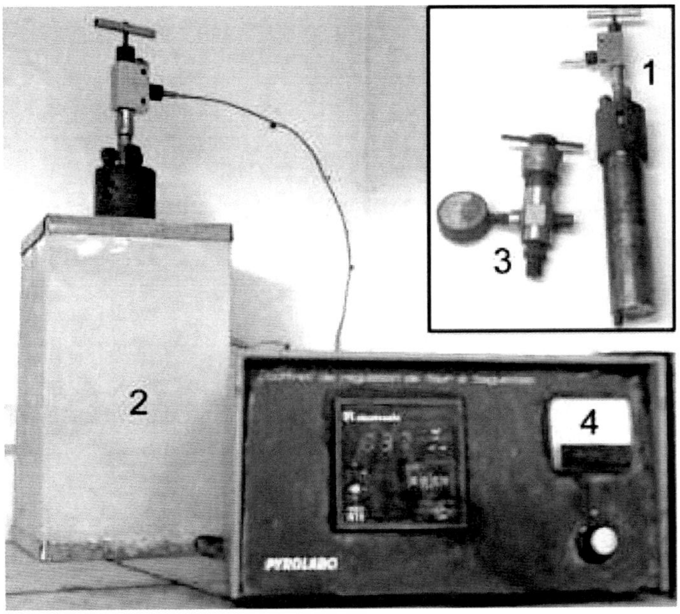

Figure 1: Photograph of the apparatus used for supercritical extraction: (1) reactor; (2) furnace; (3) pressure gauge; (4) regulator.

Preparation of Carbon Fibers

The black, solid precursor is soluble in toluene and its softening point is about 120 °C. It was melted at 150 °C in a laboratory-made spinning device with a single spinneret, equipped with a filter to remove solid particles from the melt. After extrusion and mechanical stretching, the fiber was thermoset by oxidation in dry air by being slowly heated at atmospheric pressure to the temperature in the range 60–280 °C. The thermoset fibers were pyrolyzed under nitrogen flow. The temperature was progressively raised up to 950 °C with a rate of 5 $°C \cdot min^{-1}$.

Activation of the Carbon Fibers

The pyrolyzed carbon fibers were activated with phosphoric acid [17] at different temperatures. The samples were impregnated in H_3PO_4 (10 %) for 1 h and then dried in air at 120 °C for 2 h. The activation reactions were performed in a thermolyne silica electric oven with a temperature control device linked to a thermocouple. The steam generator was placed at the entrance of the reactor.

The activation was carried out either under nitrogen or under steam flow. In both cases, the carbon fibers were placed in the reactor and the temperature was progressively raised at a rate of 5 °C·min^{-1} up to 500 °C, 700 °C and 850 °C, under nitrogen flow, and then the system was set at the maximum temperature that was maintained for 1 h. The first series (FHN) was treated under nitrogen atmosphere and the second series (FHV) under water vapor (Table **2**). At the end of the activation, the oven was cooled to ambient, and then the activated carbon fibers were retrieved and boiled for 2 h in distilled water, then washed, dried and retained for subsequent analyses.

Table 2: Experimental conditions.

Sample	**Temp (°C)**	**Burn off (%)**
FHN500	500	4
FHN700	700	12
FHN850	850	19
FHV500	500	10
FHV700	700	18
FHV850	850	24

Characterization of the Filaments

At 950 °C the carbonization yield was of about 49 %. Elemental analysis of the pyrolyzed filaments (C: 98.83; S: 1.08; Fe: 0.03 and Si: 0.06 in at. %) showed that the method used was very effective, since it was likely to extract the organic matter with a minimum of metal impurities.

Mechanical tensile tests performed on monofilaments of about 20 μm in diameter, revealed a tensile strength of 360 MPa and a Young's modulus of 40 GPa. These values are modest when compared to those found for fibers elaborated from precursors with oriented textures. They are however comparable to those of fibers obtained from coal tar or petroleum isotropic pitches [18, 19].

Characterization of the Activated Carbon Fibers

These activated carbon fibers are essentially microporous. The adsorption isotherm is of type I of the BDDT classification [20] and the *t*-plot [21], which compares the nitrogen adsorption isotherm obtained for the studied porous solid to a reference isotherm determined on a nonporous solid, presents a negative deviation from straight line, which is characteristic of micropores. This observation is confirmed by the B hysteresis characteristic of pores, according to the de Boer classification [22]. The surface area and the pore diameter calculated by the BET method [23] are given in Table **3**.

Table 3: Surface area and pore diameter of prepared samples.

Sample	S_{BET} **(m²·g⁻¹)**	R_p **(nm)**
FHN500	39	2.3
FHN700	481	1.2
FHN850	542	1.9
FHV500	516	1.8
FHV700	698	2.1
FHV850	846	2.4

Scanning Electron Microscopy (SEM)

Samples FHN850 and FHV850 were analyzed by SEM. The micrographs in Fig. **2** show little pores at the outer surface of the two activated fibers, but micrographs of cross section surfaces present a great difference between the samples. They reveal that activation under steam flow produced fibers with a homogeneous repartition of porosity. In the case of activation under nitrogen, the experimental conditions used were insufficient to create a large porosity, which is observed only at the surface of the fiber.

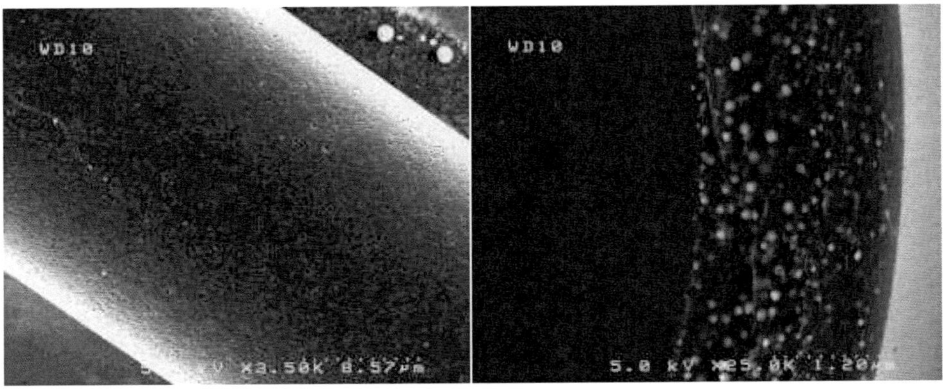

Figure 2: SEM micrographs of the activated carbon fibers FHV (left) and FHN (right).

Chemical Composition of the Fibers

Elementary analysis of the activated fibers (FHV850 and FHN850) provided information on the concentration profile of the elements along the fibers radius *R*, which are inhomogeneous as shown in the SEM micrographs. The analyzed elements were carbon, oxygen, sulfur and phosphorus. The composition of the fiber was locally determined in the core of the fiber ($r_f = 0$), at half of the radius ($r_f = R/2$) and finally at the surface of the fiber ($r_f = R$). The results are given in Table **4**.

Table 4: Atomic composition of the fibers.

	Element	$r_f = 0$	$r_f = R/2$	$r_f = R$
	C	89.23	88.87	88.56
FHV850	O	8.00	8.01	8.13
(at %)	S	0.37	0.35	0.53
	P	2.40	2.77	2.78
	C	99.74	99.69	84.21
FHN850	O	0	0	12.13
(at %)	S	0.26	0.25	0.65
	P	0	0.06	3.01

These results confirm that activation under nitrogen flow took place only at the surface of the fibers. When activation was carried out under steam flow, the phosphorus content was stable and regular along the fiber radius. Activation under nitrogen under these experimental conditions was not efficient. The penetration of phosphoric acid was limited at $r = R - 4$ µm, which in that case is less than $R/2$. This limitation is obviously responsible for the poor surface area that can be increased by increasing the time of immersion of the fibers in phosphoric acid, or by increasing the concentration of phosphoric acid.

NEW GRAPHITIZABLE CARBON OBTAINED FROM MOROCCAN OIL SHALES

Our objective was to study the possibility to produce new pitches by using the insoluble fraction resulting from the separation by hexane of the oils recovered from Moroccan oil shales. The evolution of the degree of graphitization of carbons resulting from the pyrolysis of oil shale pitches in the temperature range 1100–2800 °C was studied by Raman spectroscopy in order to determine the operating conditions that were likely to give a pitch leading to a graphitizable carbon [24, 25].

Two pitches obtained either by supercritical extraction with toluene (BHT) or by subcritical extraction with phenol (BHPH) are considered in this work.

Pitches BHT and BHPH were subjected to pyrolysis at 1100 °C and subsequent heat treatments between 1400 and 2800 °C, in order to obtained carbon materials with different degrees of graphitization. After each treatment, the residue was characterized by Raman spectroscopy. Moreover, the different samples were analyzed by X-Ray diffraction (XRD) after treatment at 2800 °C. The results are presented in Figs. **3** and **4**.

The Raman spectra of pitch BHT treated between 1100 and 2800 °C (Fig. **4**) show the D and G bands at 1342 and 1580 cm^{-1}, respectively [26]. They are consistent with those of non-graphitizable carbons treated at high temperatures [27]. It can be seen that the intensity of the D band, characteristic of poorly organized sp^2 carbon, increases with the temperature treatment. This result could be explained by the presence of significant amounts of oxygen in pitch BHT, which inhibits the graphitization process [24, 25]. Other studies [28] have shown that pitches containing heteroatoms led to materials having poor graphitization ability. On the contrary, in the Raman spectra of pitch BHPH (Fig. **3**), a decrease of the intensity of the D band was observed when the treatment temperature increased. In parallel, the intensity of the G band characteristic of organized sp^2 carbon increased.

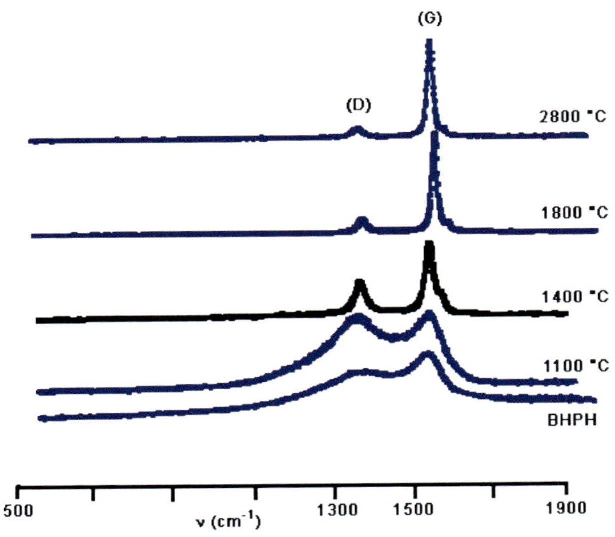

Figure 3: Raman spectra of pitch BHPH treated at different temperatures.

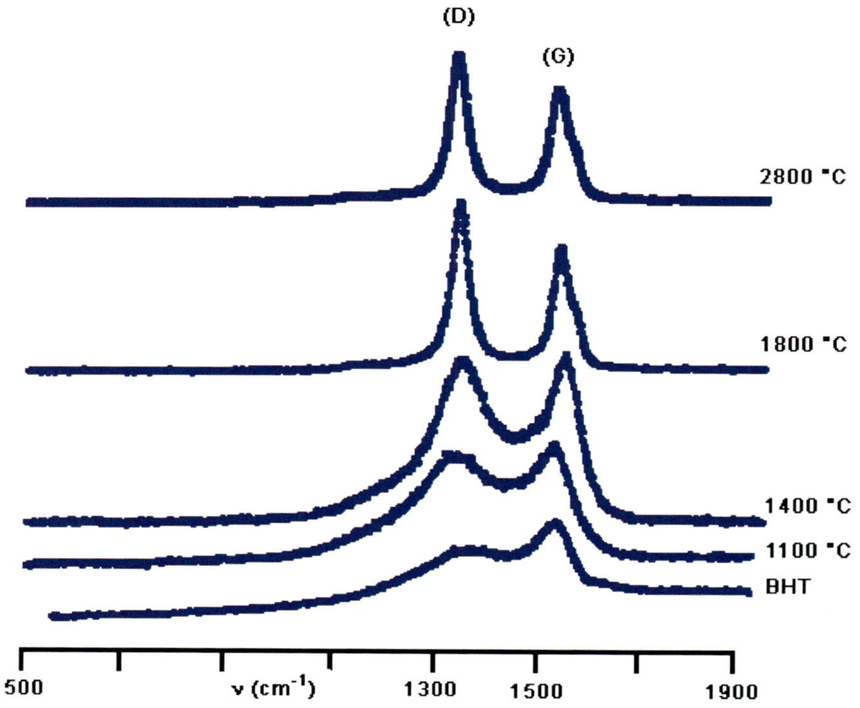

Figure 4: Raman spectra of pitch BHT treated at different temperatures.

The Raman spectroscopy investigation showed that organization of the carbon obtained by pyrolysis of the oil shale pitches began before 1800 °C. Indeed, a marked evolution above 1400 °C was observed in all samples, showing the formation of a more or less ordered material, according to the composition of the pitch.

To confirm these results, we examined by XRD the 002 carbon line for the pitches treated at 2800 °C. This line was used to determine L_c, the coherence length, from the equation:

$$L_c = 0.9\lambda / B\cos\theta$$

where λ is the X-ray wavelength, B the half-height width and θ the Bragg angle of the 002 band.

The values are L_c = 0.6 nm (BHT) and 3.7 nm (BHPH). By comparing these values, it can be concluded that carbon resulting from pitch BHPH has the highest structural organization. This corroborates with the conclusions drawn from the Raman data: the pitch obtained by extraction with phenol leads to carbon that presents the best aptitude for the graphitization.

BIOACTIVE PROPERTIES OF EXTRACTS FROM MOROCCAN OIL SHALES

In this section, we present a new way to increase the standing of oil shales: those of natural products with pharmaceutical interest. Indeed, our investigations about extracts of Moroccan oil shale allow us to study their antibacterial activities against *Agrobacterium tumefaciens*, *Escherichia coli*, *Pseudomonas aerogunosa*, *Staphylococcus aureus* and *Enterococcus faecalis*; their cytotoxicity against larva of *Artemia salina* and to search metabolites that bind DNA in procedure using high-pressure liquid chromatography (HPLC) system for detection of antitumor agents [29].

Extraction Procedure

Oil shale was exhaustively extracted with a Soxhlet using $CHCl_3$/EtOH (v/v) until the refluxing solvent mixture became clear. We obtained a black solution that was concentrated by evaporation under reduced pressure. The residual extract was then extracted successively with hexane, diethyl ether, dichloromethane, ethyl acetate and methanol [29].

Bioassays

Each extract was tested for antibacterial activity, cytotoxicity against larva of *Artemia salina*, and interaction with DNA.

Antibacterial tests

Antibacterial assays were conducted using the standard disc-diffusion assay [30]. Extracts were applied to 6 mm sterile discs in aliquots of 10 μL of solvent, allowed to dry at room temperature, and placed on agar plates and incubated at 37 °C for 24 h. Zones of growth inhibition, if any, were measured following incubation. Product was assayed twice at a concentration of 10 $\mu g \cdot disc^{-1}$.

Antibacterial activity of oil shale extracts was measured as radius of zone of inhibition around the disc Table **5**.

Table 5: Antibacterial activity of oil shale extracts.

	Zone of inhibition (mm) for each extract (10 μg)				
Bacteria	**Hexane**	**Diethyl ether**	**Dichloromethane**	**Ethyl acetate**	**Methanol**
A. tumefaciens	13	0	11	10	12
E. coli	0	0	0	10	10
S. aureus	10	0	8	10	12
P. aerogunosa	9	0	0	8	12
E. faecalis	0	0	10	0	13

In the same way, we have evaluated tetracycline activity against these bacteria. Inhibition zones are, 21, 13, 20, 17 and 15 mm, respectively against *A. tumefaciens, E. coli, P. aerogunosa, S. aureus* and *E. faecalis.*

Except diethyl ether extract, all extracts of oil shale are active against bacteria. *Agrobacterium tumefaciens* is more sensitive to extracts than other bacteria. Activity of ethyl acetate extract against *Escherichia coli* is comparable to tetracycline activity.

Cytotoxicity Tests

Samples were prepared by dissolving extracts in dimethylsulfoxide (DMSO). Brine shrimp eggs were hatched in a shallow rectangular dish filled with artificial seawater that was prepared with a commercial salt mixture and double-distilled water. After 24 h the phototropic nauplii were collected by pipette from the lighter side, having been separated by the divider from their shells. Ten shrimps were transferred to each sample vial. The nauplii can be counted macroscopically in the stem of the pipette against a lighted background.

The vials were maintained under illumination. Survivors were counted, after 24 h, and the percent deaths at each dose and control (solvent) were determined [31]. LD_{50}'s was determined from the 24 h counts. The LD_{50} was derived from the best-fit line obtained by linear regression analysis, after transforming dose-response data into a straight line by means of a logarithmic transformation.

Oil shale extracts have been tested at 20, 40, 60 and 100 $\mu g \cdot mL^{-1}$. LD_{50} values are given in Table **6**.

Table 6: LD50 values of oil shale extracts.

Extract or reference	LD_{50} ($\mu g \cdot mL^{-1}$)
Hexane	32.07
Diethyl ether	41.75
Ethyl acetate	80.67
Dichloromethane	117.33
Methanol	192.21
Potassium dichromate	35.70
Podophyllotoxin	2.30

Oil shale extracts are toxic against *Artemia salina* larva, hexane extract was the most cytotoxic, being comparable to potassium dichromate but significantly less active than podophyllotoxin.

A comparison of toxicity of oil shale extracts with toxicity of some products reported in literature [32] shows that all extracts (except methanol extract) are more toxic than digitalin (LD_{50} = 151 $\mu g \cdot mL^{-1}$) and caffeine (LD_{50} = 306 $\mu g \cdot mL^{-1}$).

DNA Test

Calf thymus DNA solutions were prepared in H_2O (0.1 $mg \cdot mL^{-1}$), and test substances were dissolved in MeOH (1.0 $mg \cdot mL^{-1}$). The HPLC ODS column was equilibrated with a H_2O-MeOH (80:20) solution. Test samples and DNA solutions were then applied to the HPLC column by pre-mixing the DNA solution with the test sample solution (1:1, v/v), incubating the mixture at ambient temperature for 30 min, and injecting the mixture (40 μL). Inhibition % was evaluated by comparing the DNA peak sizes (DNA + sample versus DNA alone). Before the following experiment, the column was rinsed 30 min with MeOH. In order to be sure of good performance of this procedure, we use in reference Adriamycin, an antitumor agent. Interaction activity with DNA is expressed by percentage of loss interaction with regard to control (DNA alone). Inhibition ratio (%) = $[(A-B)/A]$ x 100, where A is the area of DNA and B is that of DNA with the extract.

Hexane and ethyl acetate extracts show negative responses, diethyl ether, dichloromethane and methanol extracts elicit positive responses. Their percentages of binding DNA are respectively 17.61, 51.62 and 10.07 %.

In the same way, we have evaluated interaction of Adriamycin with DNA, it was found to reduce the DNA peak by 76 % Fig. **5**.

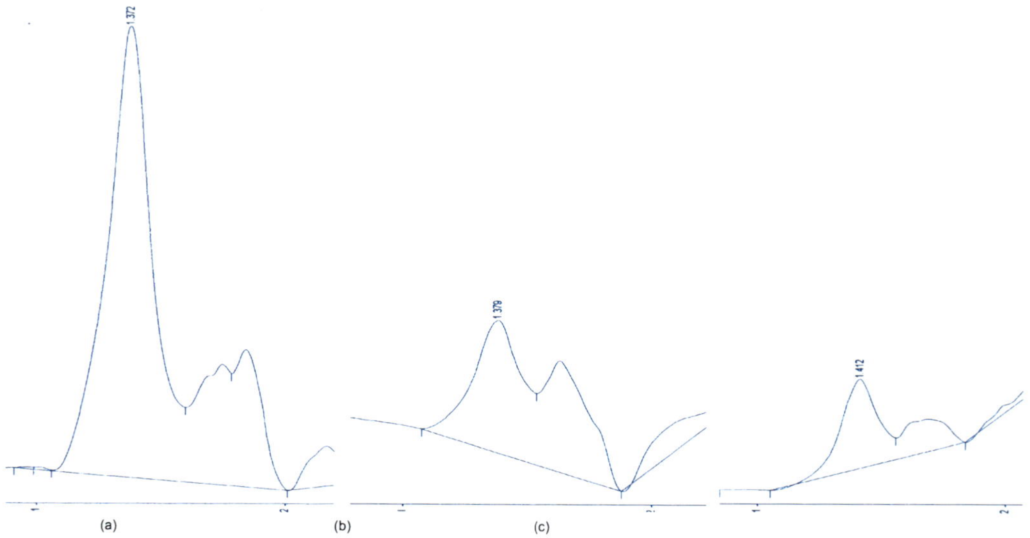

Figure 5: DNA Affinity of oil shale extracts, detection U.V. 254 nm. (a) DNA alone, (b) DNA + diethyl ether extract: a 17.61 % inhibition, (c) DNA + dichloromethane extract: a 51.62 % inhibition.

Dichloromethane extract produced an interesting interaction with DNA but less than Adriamycin. This result suggests that oil shale is able to produce compounds that bind DNA. The results of this test were found in excellent agreement with P388 and KB *in vitro* cytotoxic activity, and have allowed, for example, to isolation of budmunchiamines from *Albizia amarce*.

Recently, it has been described that compounds that interact with nucleic acids are among the most useful cancer chemotherapeutic agents.

In conclusion, we can say that Moroccan oil shale extracts have interesting bioactive properties. It is envisaged that this initial work will lead to a proposal for the use of Moroccan oil shale as source of compounds of pharmaceutical interest.

CONCLUSIONS

The present work is part of a more general program for the exploitation of natural resources, and more particularly the Moroccan oil shale deposit of Tarfaya. We have presented a new process using supercritical extraction to produce activated carbon fibers or graphitizable carbon. The fibers exhibit low mechanical properties owing to the isotropic character of the precursor and the absence of mechanical stretching during pyrolysis. The activation of the fibers with phosphoric acid at 850 °C yields a carbon fiber with a specific area of 800 $m^2 \cdot g^{-1}$. These fibers can be used as adsorbent or filter in environmental applications.

We also prepared new pitches starting from the extraction of Moroccan oil shales. Aptitude for graphitization of carbon resulting from their pyrolysis at increasing temperatures was studied by Raman spectrometry and XRD. The degree of graphitization was strongly dependent on the composition of the pitch. Interestingly, it was found that the organization of carbon began at relatively low temperatures. Thus, beyond 1400 °C, a marked structural evolution occurred, leading to a more or less ordered material, according to the composition of the pitch and its preparation conditions.

The search for bioactive properties of extracts from Moroccan oil shales was carried out on the basis of a certain number of biological tests, in particular antibacterial tests, cycotoxicity and interaction with DNA. This study showed that Moroccan oil shales could be the subject of several pharmaceutical applications and this by the bio-directed study of different fractions resulting from the rough oil shale extract in order to discover molecules responsible for antibacterial, antitumor and cytotoxicity activities.

REFERENCES

[1] Alpern, B. Bituminous shale: composition, reserves and upgrading. *Bull. Centre Rech. Explor. Prod. Elf-Aquitaine*, **1981**, *5*, 319-352.

[2] Tahiri, M.; Sliepcevich, C. M.; Mallinson, R. G. In: *Preprints of the Symposium on Advances in Oil Shale Technology,* Denver, CO, April 5, 1987; ACS Division of Petroleum Chemistry, **1987**; *32*, p. 71.

[3] El Harfi, K.; Mokhlisse, A.; Ben Chanâa, A. Effect of water vapor on the pyrolysis of the Moroccan (Tarfaya) oil shale. *J. Anal. Appl. Pyrolysis*, **1999**, *48*, 65-72.

[4] Thakur, D. S.; Nuttal, H. E. Kinetics of pyrolysis of Moroccan oil shale by thermogravimetry. *Ind. Eng. Chem. Res.*, **1987**, *26*, 1351-1356.

[5] Vantelon, J. P.; Breillat, C.; Gaboriaud, F.; Alaoui Sosse, A. Thermal degradation of Timahdit oil shales: behavior in inert and oxidizing environments. *Fuel*, **1990**, *69*, 211-215.

[6] Abourriche, A.; Oumam, M.; Hannache, H.; Pailler, R.; Naslain, R.; Birot, M.; Pillot, J.-P. Recuperation of the organic matter from oil shales with phenol in subcritical conditions. *Ann. Chim.*, **2005**, *30*, 1-17.

[7] Bunger, J. W.; Devineni, P. A. V.; Cogswell, D. E.; DuBow, J. B.; Huai, H.; Dworzanski, J. P. Market enhancement of shale oil by selective separation. *Preprints of ACS Division of Fuel Chemistry Spring Meeting, San Francisco CA*, **1992**, *37*, pp. 581-589.

[8] Fei, Y. Q.; Derbishire, F.; Jagtoyen, M.; Mochida, I. In: Proceedings of the 1993 Eastern Oil Shale Symposium, Lexington, KY, November 16-19, 1993; Institute for Mining and Minerals Research; Lexington, KY; **1993**; pp. 38-45.

[9] Akar, A.; Ekinci, E. Production of chemicals from oil shale. *Fuel*, **1995**, 74, 1113-1117.

[10] Oumam, M.; Hannache, H.; Pailler, R.; Bourrat, X.; Naslain, R. In: Proceedings of the 2nd ARQUIMACOM Conference, Bordeaux, France, October 7-9, 1998; Martineau, P.; Olivares M., Eds.; Bordeaux, **1998**; pp. 113-117.

[11] Oumam, M.; Hannache, H.; Pailler, R.; Bourrat, X.; Naslain. R. In: Proceedings of the 1st ARQUIMACOM Conference, Seville, Spain, October 24-26, 1996; Laffarga Osteret, J.; Martineau, P.; Olivares M., Eds.; Mairena del Aljarafe, Editan, Seville, Spain, **1996**; pp. 101-108.

[12] Bekri, O.; Ziyad, M. In: Proceedings of the 1991 Eastern Oil Shale Symposium, Lexington, KY, November 13-15, 1991; Institute for Mining and Minerals Research; Lexington, KY; 1991; pp. 437-443.

[13] Yürüm, Y.; Kramer, R.; Levy, M. Interaction of kerogen and mineral matrix of an oil shale in an oxidative atmosphere. *Thermochim. Acta*, **1985**, *94*, 285-293.

[14] Rose, H. R.; Smith, D. R.; Vassallo, A. M. An investigation of thermal transformations of the products of oil shale demineralization using infrared emission spectroscopy. *Energy & Fuels*, **1994**, *7*, 319-325.

[15] Abourriche, A.; Oumam, M.; Hannache, H.; Pailler, R.; Naslain, R.; Birot, M.; Pillot, J.-P. Effect of phenol, temperature and treatment time on the supercritical extraction of oil shale by toluene. *Phys. Chem. News*, **2003**, *12*, 93-101.

[16] Suatoni, J. C.; Swab, R. E. Preparation of hydrocarbon type compounds analysis by high performance liquid chromatography. *J. Chromatogr. Sci.*, **1976**, *14*, 535-537.

[17] Laine, J.; Calafat, A.; Labady, M. Preparation and characterization of activated carbons from coconut shell impregnated with phosphoric acid. *Carbon*, **1989**, *27*, 191-195.

[18] Donnet, J. B.; Bausal, R. C. *Carbon Fibers*, 2nd ed.; Marcel Dekker: New York, **1990**.

[19] Fitzer, E.; Heine, M. In: *Composites Materials Series*; Bunsell A. R. Ed., Elsevier, Amsterdam, **1988**; Vol. *2*, pp. 73-148.

[20] Brunauer, S.; Deming, L. S.; Deming, W. E.; Teller, E. On a theory of the van der Waals adsorption of gases. *J. Am. Chem. Soc.*, **1940**, *62*, 1723-1732.

[21] Lippens, B. C.; de Boer, J. H. Studies on pore systems in catalysts: V. the *t* method. *J. Catal.*, **1965**, *4*, 319-323.

[22] de Boer, J. H. In: *The Structure and Properties of Porous Materials;* Everett, D. H.; Stone, F. S. Eds.; Butterworth: London, UK, **1958**; p. 68.

[23] Brunauer, S.; Emmett, P. H.; Teller, E. Adsorption of gases in multimolecular layers. *J. Am. Chem. Soc.*, **1938**, *60*, 309-319.

[24] Abourriche, A.; Oumam, M.; Adil, A.; Hannache, H.; Pailler, R.; Naslain, R.; Birot, M.; Pillot, J.-P. In: *Supercritical Fluids: Chemical Reactivity and Material Processing in Supercritical Fluids*, Proceedings of the 8th Meeting on Supercritical Fluids, Bordeaux, France, April 14-17, 2002, Besnard, M.; Cansell, F. Eds.; Institut National Polytechnique de Lorraine: Vandoeuvre lès Nancy, France, **2002**; p. 445.

[25] Abourriche, A.; Oumam, M.; Mouhssim, A.; Dahiri, M.; Hannache, H.; Chollon, G.; Pailler, R.; Naslain, R.; Birot, M.; Pillot, J.-P. New pitches enhanced graphitization ability obtained from Moroccan oil shales. *J. Anal. Appl. Pyrolysis*, **2004**, *71*, 935-944.
[26] Tuinstra, F.; Koenig, J. L. Raman spectrum of graphite. *J. Chem. Phys.*, **1970**, *53*, 1126-1130.
[27] Ergun, S.; Schehl, R. R. Analysis of the structure of glassy carbon using the Fourier transform technique. *Carbon*, **1973**, *11*, 127-138.
[28] Hossain, T.; Podder, J. The role of heteroatoms on the carbonization and graphitization of polynuclear aromatic compounds. *Thermochim. Acta*, **1989**, *137*, 225-233.
[29] Oumam, M.; Abourriche, A. K.; Hannache, H.; Abourriche, A. M.; Bennamara, A.; Charrouf, M.; Pailler, R.; Naslain, R. Bioactives properties of extracts from Moroccan oil shale. *Sci. Pharm.*, **2002**, *70*, 271-276.
[30] McCaffrey, E. J.; Endean, R. Antimicrobial activity of tropical and subtropical sponges. *Mar. Biol.*, **1985**, *89*, 1-8.
[31] Meyer, B. N.; Ferrigni, N. R.; Putnam, J. E.; Jacobsen, L. B.; Nichols, D. E.; McLaughlin, J. L. Brine shrimp: a convenient general bioassay for active plant constituents. *Planta Med.*, **1982**, *45*, 31-34.
[32] Mar, W.; Tan, G. T.; Cordell, G. A.; Pezzuto, J. M.; Jurcic, K.; Offermann, F.; Redl, K.; Steinke, B.; Wagner, H. Biological activity of novel macrocyclic alkaloids (budmunchiamines) from Albizia amara detected on the basis of interaction with DNA. *J. Nat. Prod.*, **1991**, *54*, 1531-1542.

APPENDIX: TERMS

BDDT:	Brunauer, Deming, Deming, and Teller theory
BET:	Brunauer, Emmett and Teller method
BHPH:	Pitch obtained by extraction with phenol
BHT:	Pitch obtained by extraction with toluene
DMSO:	Dimethylsulfoxide
FHN:	Fibers treated with phosphoric acid under nitrogen
FHV:	Fibers activated with phosphoric acid under water vapor
HPLC:	High-pressure liquid chromatography
LD50:	Median lethal dose
R3:	raw shale
RH:	carbonate-free shale
SEM:	Scanning electron microscopy
XRD:	X-Ray diffraction

CHAPTER 4

Biofuels – The Optimal Second Best Solution?

G. Stoeglehner[a,*] and M. Narodoslawsky[b]

[a]University of Natural Resources and Applied Life Sciences Department of Spatial, Landscape and Infrastructure Sciences Institute of Spatial Planning and Rural Development, Vienna, Austria and [b]Institute of Process Engineering, Technical University Graz, Austria

Abstract: The aim of this paper is to discuss potentials of biofuels to contribute to environmentally sustainable energy supplies based on the Sustainable Process Index (SPI), an alternative calculation method for ecological footprints. The paper focuses on energy demand for transport. Comparing biofuels with fossil fuels it can be seen that "conventional" first generation biofuels reduce the overall ecological pressure about 30% compared to fossil fuel. If second generation biofuels are used footprint reductions of factors 13 to 16 compared to fossil fuel can be achieved depending on the production methods and the scale of the plant. On the other hand, production limits occur due to overall environmental capacity limits and resource constraints. If we look into technological options for transport, the long term development seems to be electricity based. Yet, we still lack sufficient sustainable electricity production besides wind and biomass. Therefore, we suggest that biofuels may be an advantageous option in a transition period from fossil based to electricity based transport systems, where minor effort is needed to retrofit vehicles until new electricity based technologies for transport means and the corresponding electricity production are available.

Keywords: Ecological footprint, sustainable process, process index, energy footprint, biofuels, renewable, energy, sustainable, development, environmental sustainability, environmental capacity limits, impact on environment, systems approach, life cycle assessment, fossil fuel substitution, optimal solution.

INTRODUCTION

Biofuels are discussed in a very controversial way. On the one hand they offer the possibility to produce fuels for non-stationary applications on the base of renewable resources. It is often argued that they are environmentally friendly and neutral in terms of climate change. Furthermore they can offer regional incomes

***Address correspondence to G. Stoeglehner:** University of Natural Resources and Applied Life Sciences Department of Spatial, Landscape and Infrastructure Sciences Institute of Spatial Planning and Rural Development, Vienna, Austria; Tel: +43 +1 47 654/53 67; Fax: +43 +1 47654/53 53; E-mail: gernot.stoeglehner@boku.ac.at and create regional jobs. On the other hand it is argued that biofuels are competing against food production [1, 2] leading to increased food prices and higher environmental pressures in agricultural areas. The concept of "human appropriated net primary production (HANPP)" [3] argues that in many regions more than 80% of the annual biomass production are already used by humans. This fact suggests that further increase of biomass use for biofuel production might very likely not be sustainable.

This article aims at discussing sustainability issues in-depth from the perspective of environmental capacity limits utilizing an alternative calculation method of ecological footprints, the Sustainable Process Index (SPI) [4]. The SPI calculates the area necessary to embed human activities sustainably into the biosphere by comparing natural flows with flows induced by the activity in question. Calculation of the SPI is based on two principles, namely that a sustainable human society must neither change the long term storage elements of global material cycles (*e.g.,* the carbon cycle) nor the quality of local environmental compartments. This broad definition of ecological sustainability allows the SPI to compare widely different approaches to provide societal services such as energy provision. In particular it is able to compare technologies based on different raw materials such as nuclear, fossil or renewable sources.

At least on the global scale sustainability can be granted if the sum of areas to embed all processes taking place is smaller or equal than the available global surface area. On the local and regional level an SPI bigger than the available area suggests that trade is necessary to utilize a resource hinterland whereas a smaller footprint indicates that land can be "exported" as traded products or services. Materials and energy embedded in traded goods and services are also made visible. The SPI of a product or service comprises subareas for material resources, energy, personnel, process installation (like the area used for producing the machines, but also storage area and the area directly utilized by the production site), product as well as emissions dissipation (taking qualities and quantities of waste materials into account).

The article is structured in the following way: we will show the results of a comparison between different biofuels and fossils fuels. Then we will put the question, how sustainable are biofuels, in a regional context. Taking this discussion one step further, we will assess biofuels compared to other renewable energies. Then, we will enhance the debate in a wider perspective addressing the position of biofuels in the transition process from a fossil-based to a renewable society. Finally, some conclusions are drawn.

This contribution does not aim at explaining different footprint approaches to evaluate energy supplies. We did this extensively in [5, 6]. Here shall be mentioned that the original footprint calculation [7-9] cannot fulfill the requirements to evaluate the sustainability of certain technology options like biofuels, but that some methodological developments like the SPI make the footprint a valuable evaluation tool at this level of detail.

THE ENVIRONMENTAL SUSTAINABILITY OF BIOFUELS *VS.* FOSSILS

The comparison of certain biofuels and fossil fuels concerning their sustainability is based on a life cycle assessment utilizing the SPI [10]. Judging the sustainability of biofuels is complex as biofuel production might involve a wide variety of raw materials and production processes. Furthermore, the size of the production plant is an important factor. A comparison of different options considering the process and size of facilities for ethanol production reveals the following patterns concerning environmental pressures: On the one hand, the environmental pressures per kWh energy content of the fuel decrease with the size of the production plant due to enhanced technological efficiencies. The SPI of all options converges for a 10.000 t/a bioethanol capacity. On the other hand, large scale productions over a certain threshold (60.000 t/a bioethanol in this case study) have a footprint 4 times higher than the 10.000 t/a capacity. This is due to the fact that process energy at large scales may not be supplied by renewable resources but by fossil fuel like natural gas. Transport effort to bring renewable energy resources to a central large plant will increase dramatically, finally eradicating their environmental advantage. Using renewable resources not only an "economy of scale" – bigger plants lead to increased efficiency – but also an "ecology of scale" is emerging. Therefore, new optima have to be found in process engineering taking regional features into account [11].

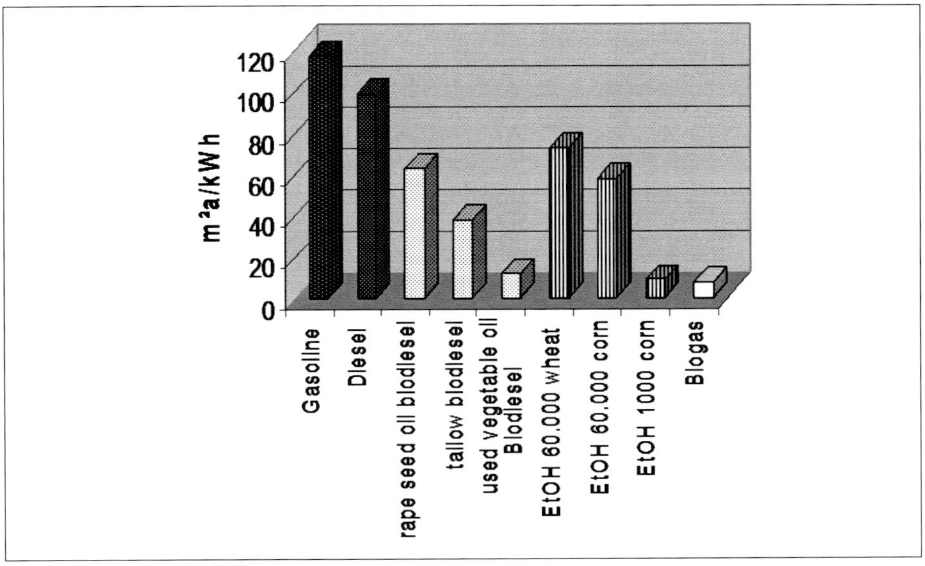

Figure 1: Ecological footprint (SPI-method) of different biofuels compared to fossil fuels [10].

Fig. **1** shows the SPI of biofuels in connection to the raw material and the size of the production plant. The numbers in the ethanol entries in this figure denote the production capacity (in tons ethanol per year) of the respective plants. The entry for 1000 t/a (which shows a considerably lower ecological footprint) uses exclusively renewable energy resources. In this particular case the process energy is heat from a biogas unit that is operated with the crops from a sustainable crop rotation that renders the corn necessary to run the bioethanol plant. Also the SPI of fossil fuels is included. It can be clearly seen that "conventional" biofuels like biodiesel from rapeseed or large scale bioethanol reduce the environmental pressures only to a relatively small extent of approximately 30% compared to fossil fuels. Especially in the case of biodiesel the ecological footprint can be considerably reduced if by-products of other sectors (like tallow) or even waste materials (like used vegetable oils) are used as raw material. As bioethanol from corn has a higher yield per hectare than wheat the SPI of the latter is bigger. The example of corn based bioethanol shows that small scale productions may have a much smaller footprint than large scale productions even though large scale productions are technically more efficient. Footprint reductions up to a factor 8 compared to fossils can be achieved if small scale production takes advantage of renewable resources for process energy.

Compared to bioethanol biogas offers the biggest reduction of environmental pressures due to three reasons. First, the yield per hectare is much higher as the whole plant can be used, whereas in bioethanol production only the grain is processed. Second, less fertilizer is required and no drying is needed in the agricultural production. Third, the technical conversion process has a low impact because of low heat and moderate electricity requirements. With biogas the footprint per kWh can be dramatically reduced by a factor 13 compared to fossil diesel and a factor 16 compared to gasoline.

Summarizing these results, it can be concluded that depending on the raw material, the production process and its size biofuels can be considerably more environmentally sustainable than fossil fuels. Yet it remains open, whether they are sustainable enough to be broadly used because this question has to be answered in a wider context of an overall vision and might differ considerably from region to region, as for instance, the conditions for agricultural production of the raw material(s) might differ considerably [6]. In the following section we discuss the environmental sustainability of biofuels in comparison to other renewables.

THE ENVIRONMENTAL SUSTAINABILITY OF BIOFUELS *VS.* OTHER RENEWABLES

Considering the environmental pressures related to the energy production of different renewable energy sources, we can distinguish the biomass related energy sources and the technical-oriented energy sources like wind, water power and solar energy. Although the direct solar input and useable energy share from solar radiation is considerably higher with technical-oriented energy sources than biomass, the SPI does not necessarily differ considerably taking all related environmental effects into account. Yet, we have to consider that especially qualitative environmental effects can only partly be included into the footprint (and SPI) concept [12].

Table **1** shows the different SPI footprints of different renewable energy sources for heat and electricity in comparison to biofuels and fossils.

Table 1: Different SPI footprints[a].

Energy Source	SPI (m^2/kWh)
Fossils	
Gasoline	117, 3
Diesel	99, 0
Brown coal	437, 2
Hard coal	160, 02
Natural gas	54, 20
Biofuels-Mobility	
Biodiesel from rape seed oil	62, 7
Biodiesel from tallow	38, 2
Biodiesel from used vegetable oils	12, 0
Bioethanol 60.000 t/a from corn	64, 8
Bioethanol 1000t/a from corn	8, 9
Biogas (96%-CH4)	16, 76
Renewables-Electricity	
Hydropower	1, 4
Wind energy	11, 3
Photovoltaics thin layer	60, 9
Biomass ORC	9, 97
Renewables-Heat	
Fuelwood	3, 4
Woodchips	3, 74
Pellets (dried)	17, 35
Solar thermal	38, 9

[a]SPI of different energy sources [13].

We can derive the following conclusions considering the state of the art of energy production: First, no matter if it is for heating, power generation or mobility, fossil energy carriers always have a footprint considerably higher than the "worst" renewable solution.

Second, the direct use of the sun for electricity and heat production causes relatively high environmental pressures compared to other renewable energy sources. This can be reasoned by the relatively high material and energy inputs for the production of the devices which is mainly covered by fossil energy under current production processes. In a completely fossil-free society the environmental impacts would have to be re-evaluated.

Third, hydropower has a small footprint, but this is mainly true because many environmental effects of hydropower cannot be quantified in the footprint concept, like biodiversity losses due to damages to the river systems. Therefore, we argue that for most industrialized nations, except for some small scale hydropower stations, hydropower capacities cannot be further increased because of their strong ecological impacts not quantifiable by the footprint concept.

Fourth, the “best” biomass based renewable energy sources are competitive to wind energy in terms of ecological footprints for electricity production. This is especially important for the substitution of fossils for the transport sector. Wenzel [14] states that biofuels will not solve future energy problems of the transport sector inter alia because of their limited availability. He states that electricity might be the best solution. This implies the question at which environmental constraints the electricity might be produced. In the following section we will apply a systems approach to discuss the role of biofuels for the transformation process in renewable energy supplies with a focus on mobility.

THE ROLE OF BIOFUELS IN A SYSTEMS APPROACH

There is no doubt about the fact that biofuels are no “drop-in” substitution to conventional fossil fuels. Wenzel [14] as well as Hedegaard *et al.* [15] argued convincingly that biomass may not reach the necessary substitution rates to support human society in a non-fossil energy scenario. Depending on the scenarios a substitution level between 18 and 62% by 2030 seems possible. This means that a system change in energy provision as well as utilisation is in any case necessary to achieve sustainable development.

In accordance with Wenzel [14] Table **1** clearly shows that the “biggest ecological bang for the buck” *i.e.* the biggest reduction of ecological pressure per kWh is possible when biomass substitutes fossil resources for heat generation and electricity provision. It is clear that this trajectory is the most preferable way to put biomass to use in energy terms, at least in the long run.

Table **1** however also shows another fact: besides hydropower (which is restricted due to other environmental concerns not covered by the SPI), wind and biomass there is still no environmentally benign technology available to provide electricity. Therefore if the future of transportation is seen as electricity based we have still to find ways to provide it in the necessary quantity with low environmental pressures. A large potential in this respect lays in advancing photovoltaic electricity provision as well as in thermo-electrical technologies based on solar radiation. Only if we can achieve sustainable electricity production it makes sense to change the transport sector to electricity. The available low impact technologies hydropower, wind and biomass alone however will not do the trick.

If we now look at the role biofuels can play in the medium and long term, Table **1** can again be a guide. First generation technologies like large scale bioethanol production and biodiesel from rape seed as well as palm oil are clearly alternatives with limited merits. On the one hand their ecological impact is by far too large to offer a sustainable way to support transport (although it is lower than conventional fossil fuels). On the other hand they may take away valuable acreage from more sustainable use, be it for food provision or for the provision of energy in more sustainable ways than as biofuels. In the long run only biodiesel from by products and waste materials (tallow, used vegetable oil) will become viable sustainable transportation fuels.

A particularly different situation however arises with biogas. As can be seen from Table **1** biogas production exerts a low ecological pressure and offers large mileage per hectare. On top of that it is readily applicable to conventional Otto engines (as well as to diesel engines after substantial re-engineering). The cost of re-fitting cars for using methane is relatively low. On top of that there is an extensive distribution grid for gas available that might be used for distributing biogas along side natural gas. At least for a medium term transition period it may therefore be advantageous to convert a large part of cars, buses and trucks to gas motors. This makes all the more sense as natural gas itself has a SPI value comparable to conventional first generation biofuels. The rule may therefore be to use biogas in CHP whenever the by product heat can be utilised and to inject it into the grid (for use in transportation as well as other gas applications) whenever no recipient for heat is available at the biogas plant. The same logic may be applied to Synthetic Natural Gas (SNG) production, based on lingo-cellulose material. In both cases the technology utilises "lower end" material in terms of quality and are therefore outside the competition with food production.

CONCLUSIONS

All these considerations lead to the conclusion that with present technologies energy savings are the most important pillar and precondition to design sustainable energy supplies. In order to reach energy efficiency in the transport sector, not only the improvement of technical systems is important, because past experiences show that especially in the transport sector efficiency gains were overcompensated by increases of transport volumes. Therefore, efficiency improvements also have to address the way settlement structures are organized in order to avoid mobility. Such energy efficient principles are the minimization of urban sprawl, keeping and/or re-establishing nearness between different spatial functions like housing, working, supply, recreation *etc.* as well as optimizing urban density in order to allow for efficient infrastructures on the one hand, and to provide for a high quality of living on the other hand [16].

In terms of the energy base of future transport systems, electricity may become the most important energy form used. This however depends on the build-up of a sustainable electricity provision system. Wind and biomass CHP are already currently available elements of such a system, advances in photovoltaic technology as well as thermo-electric systems are however necessary in the long run.

The role of biomass in a sustainable energy system will certainly be dominated by its application to heat and power generation in stationary systems. Biofuels produced by first generation technologies do not offer particular advantages in terms of reduction of environmental pressures and may indeed be a costly diversion. Biodiesel from by-products and waste materials however may prove to be sustainable and contribute in a small segment to a future transportation system.

Biogas, as well as SNG may play a major role in a transition period as the conversion of the existing fleet as well as technology adaptations to new vehicles are easy to achieve. Both offer high mileage per hectare at a low environmental impact. As natural gas may also be a medium term alternative to conventional liquid fuels and a well established grid available for distribution these technologies offer the best transition trajectory.

Whatever the product, if it is liquid biofuels, bio commodities or biogas itself, conventional economy of scale has to be augmented by "ecology of scale" considerations. As biomass (and the residues of transforming it into energy) are transport intensive, more de-central industrial structures will become necessary to achieve overall sustainability.

REFERENCES

[1] I. Catic and M. Rujnic-Sokele, "Agriculture products – Food for living beings or for machinery", *Gummi, Fasern, Kunststoffe*, vol. 61, no.11, pp. 701-708, 2008.

[2] P. Hellegers, D. Zilbermann, P. Stedto and P. McCornick, "Interactions between water, energy, food and environment: evolving perspectives and policy issues", *Water Policy*, vol. 10, pp. 1-10, 2008.

[3] H. Haberl, "The global socioeconomic energetic metabolism as a sustainability problem", *Energy*, vol. 31, pp. 87-99, 2006.

[4] C. Krotscheck and M. Narodoslawsky, "The Sustainable Process Index. A new dimension in ecological evaluation", *Ecological Engineering*, vol. 6, pp. 241-258, 2006.

[5] G. Stoeglehner and M. Narodoslawsky, "Implementing ecological footprinting in decision-making processes", *Land Use Policy*, vol. 25, pp. 421-431, 2008.

[6] G. Stoeglehner and M. Narodoslawsky, "How sustainable are biofuels? Answers and further questions arising from an ecological footprint perspective", *Bioressource Technology*, vol. 100, pp. 3825-3830, 2009.

[7] M. Wackernagel and W. Rees, "How big is our ecological footprint – a handbook for estimating a community's appropriated carrying capacity", Vancouver: UBC Department of Family Practice, 1993.

[8] C. Monfreda, M. Wackernagel and D. Deumling, "Establishing national natural capital accounts based on detailed Ecological Footprint and biological capacity assessments", *Land Use Policy*, vol. 21, pp. 231-246, 2004.

[9] J. Kitzes, A. Galli, S.M. Rizk, A. Reed and M. Wackernagel, "Guidebook to the National Footprint Accounts: 2008 Edition", Oakland: Global Footprint Network, 2008.

[10] G. Gwehenberger, M. Narodoslawsky, B. Liebmann and A. Friedl, "Ecology of scale versus economy of scale of bioethanol production", *Biofuels, Bioproducts and Biorefining,* vol. 1, no. 4, pp. 264-269, 2007.

[11] M. Narodoslawsky, "Structural Chances and Challenges for Bio Commodity Processes", *Food Technology and Biotechnology,* in press.

[12] G. Stöglehner, "Ecological Footprint – a tool for assessing sustainable energy supplies", *Journal of Cleaner Production*, vol. 11, pp. 267-277, 2003.

[13] Homepage of the SPI calculation, [Online] Available: www.SPIonEXCEL.at. [Accessed Feb. 22, 2010]

[14] H. Wenzel, "Biofuels: the good, the bad, the ugly – and the unwise policy", *Clean Technologies and Environmental Policy*, DOI 10.1007/s10098-009-0231-6, 2009.

[15] K. Hedegaard, K.A. Thyø and H. Wenzel, "Life Cycle Assessment of an Advanced Bioethanol Technology in the Perspective of Constrained Biomass Availability", *Environmental Science and Technology*, vol. 42, pp. 7992-7999, 2008.

[16] G.Stoeglehner, N. Niemetz, and K.-H.Kettl, "Spatial dimensions of sustainable energy systems: new visions forintegrated spatial and energy planning", *Energy, Society and Sustainability*, vol.1:2, pp. 1-9, 2011.

APPENDIX: TERMS

Biofuels: The term biofuels in general covers a wide range of solid, liquid or gaseous fuels made from biomass or organic waste products. Liquid biofuels in particular are undergoing a rapid innovation process: first generation biofuels are mainly produced from agricultural products like wheat, corn or sugarcane. Second generation biofuels are based on non-food agricultural products such as wood or harvest residues. Third generation liquid biofuels are made from algae.

Ecological Footprint: Ecological footprint is an indicator for environmental sustainability to transfer all human activities into land use. On the global level, the area used by mankind shall not increase the actually available bioproductive land area of our planet. The ecological footprint methodology is changing over time and some alternative calculation methods were added to the original calculation method. Now, a "footprint family of indicators" exists that offers different calculation methods for different assessment problems.

Sustainable Process Index (SPI): The SPI is an alternative concept to measure the ecological footprint based on a strict set of ecological sustainability criteria comparing natural material flows with flows induced by human activities. The SPI is in particular able to compare technologies and their products that are based on different raw materials such as nuclear, fossil and renewable resources. The SPI uses a life cycle approach to measure the ecological impact including raw material generation, technological processes as well as wastes and emissions generated over the whole life cycle of a product.

CHAPTER 5

Design of an Optimal Standalone Wind Power Generation System

A. Roy and S. Bandyopadhyay*

Department of Energy Science and Engineering, Indian Institute of Technology, Bombay, Powai, Mumbai, India

Abstract: Generation of electrical energy from the wind can be a suitable proposition for off grid power supply at locations having a favorable wind regime. Proper design of wind power generation system is of utmost importance to assure maximum benefit to the consumer in terms of economic competitiveness as well as power supply reliability. Designing a wind power system involves appropriate sizing of different components based on the availability of wind speeds and the energy demand. Since, wind as a resource is intermittent and variable by nature, the mismatch between the generation and demand can be leveled by provision of a battery bank as a storage medium. A methodology for designing an efficient wind-battery power system is presented in this chapter. The major system design parameters are identified to be the wind rotor diameter, the generator rating and the storage capacity. By considering the energy interactions between the generator, the storage system and the load over a given time horizon, a number of feasible design solutions can be generated. A diagrammatic representation of all feasible solutions enables a system designer to understand the tradeoffs between different system design variables, corresponding maximum and minimum limits, and arrive at an optimum solution system subject to an appropriate design objective as well.

Keywords: Wind, power, generation system, performance analysis, rotor, model, sizing wind-based system, standalone, power system, wind-battery power system, wind speed, physical design, rotor diameter, generator rating, design, optimal.

INTRODUCTION

Wind energy is an economically viable renewable source of energy today. The approximate cost of grid connected wind power lies in the range of € 0.05 - 0.12/kWh [1]. It is also one of the fastest growing energy markets today with a growth rate of 32% per year [2]. One of the niche areas for wind power generation today is decentralized or off-grid power supply. This is a particularly attractive option

***Address correspondence to S. Bandyopadhyay:** Department of Energy Science and Engineering, Indian Institute of Technology, Bombay, Powai, Mumbai, India; Tel: +91-22-25767894; Fax: +91-22-25726875, 25723480; E-mail: santanu@me.iitb.ac.in

Hatim Machrafi (Ed)

where the cost of extending transmission and distribution network is very high. The key concern in making any standalone power project economically competitive is to size different components of the system so as to obtain an optimum match between the load to be served and the resource available. A major constraint in the case of a standalone power system is that the site of installation of the wind generator has to be always in the vicinity of the load to be served. Moreover, ensuring reliability of power supply from standalone systems is difficult as there is no grid back-up when there is no wind available. Therefore it is essential to store excess energy when available and supply it when required in a storage device such as a battery bank. This chapter addresses the issues of sizing a standalone wind-battery power system supplying a specified load at a particular location. The major parameters affecting wind turbine design are discussed first and subsequently a methodology for optimum sizing of a standalone system comprising of wind generator with battery storage is described.

HISTORICAL ACCOUNT OF WIND POWER DEVELOPMENT

The first historical evidence of the use of wind as a source of mechanized power dates back to the 7th century AD when the Persians used vertical axis machines with a number of radially-mounted sails for grain grinding [3]. Acquaintance of wind power spread from Asia to Europe with the first account available from England in the 12th century [4]. The most revered design in Europe was the giant corn grinding wind mills built by the Dutch. The Dutch wind mill in its refined design had four blades each of which were twisted and tapered similar to the modern wind turbines. Dutch settlers brought this concept to America. A severe requirement of pumping ground water stimulated the development of a much smaller multi-bladed (12-20 blades) American wind machine which had a high torque and adequate efficiency for pumping applications. The first electricity producing wind machine was developed in 1888. This machine was rated at 12 kW with 144 numbers of 17m diameter blades [5]. However, the design was rather inefficient due to a number of slow moving blades. Subsequently, the Danes pioneered the design of electricity generating wind machines and postulated that generation of electricity from the wind requires a small number of aerodynamically shaped blades which could rotate at a higher speed. Based on this concept, a number of wind turbines in the range of 5-30 kW were built in

Denmark [5]. Conceptual understanding of aerodynamic shape and position of center of forces of zero moment reduced the structural problem of supporting the blade. This was in the second decade of the twentieth century and is an important mile-stone in the history of wind machines. Thereafter, longer blades of aerodynamic shape could be designed and used. A major landmark in wind turbine development in the year 1940 was the experimental 1.25 MW Smith-Putnam wind turbine [6] which was the largest wind turbine built till the 1970's. This machine was directly connected to the electric grid. Nevertheless, the higher cost/kW of such machines compared to a grid connected thermal power plant led to a declined interest in wind power. Later in 1957, the 200 kW Gedser wind turbine of Denmark and the 100 kW Hütter turbine of Germany were designed and successfully operated [4]. Since then, wind power technology has advanced by leap and bounds to the state of art multi-megawatt propeller wind turbines of the 21st century. Fig. **1** shows the schematic of a modern three bladed propeller type wind machine.

In order to make power generated by wind turbines a suitable option; it becomes important that they should be cost effective. As a prerequisite to analyze the performance of wind machines, it becomes necessary to first identify the parameters affecting the behavior of wind machines. The following section describes the primary parameters which would enable the evaluation of the performance of a wind turbine system.

PERFORMANCE ANALYSIS OF WIND MACHINES

The blades of the wind turbine intercept the kinetic energy of the flowing wind and extract power there from. It is not possible to extract all the energy contained in a unit volume of air. Otherwise, the wind velocity behind the rotor would be practically negligible which is impossible according to the law of conservation of mass (equation of continuity). Therefore, it is necessary to ascertain what fraction of the incident energy is extracted by the rotor. Besides, the power captured by a wind rotor is specific to the design and orientation of the wind machine *e.g.*, horizontal *vs.* vertical axis, propeller *vs.* drag translator, upwind *vs.* downwind, *etc.* This suggests the need of a common parameter to assess the performance of all classes of wind machines. Following parameters have evolved from the dimensional analysis of wind machines for the purpose of characterizing their performance.

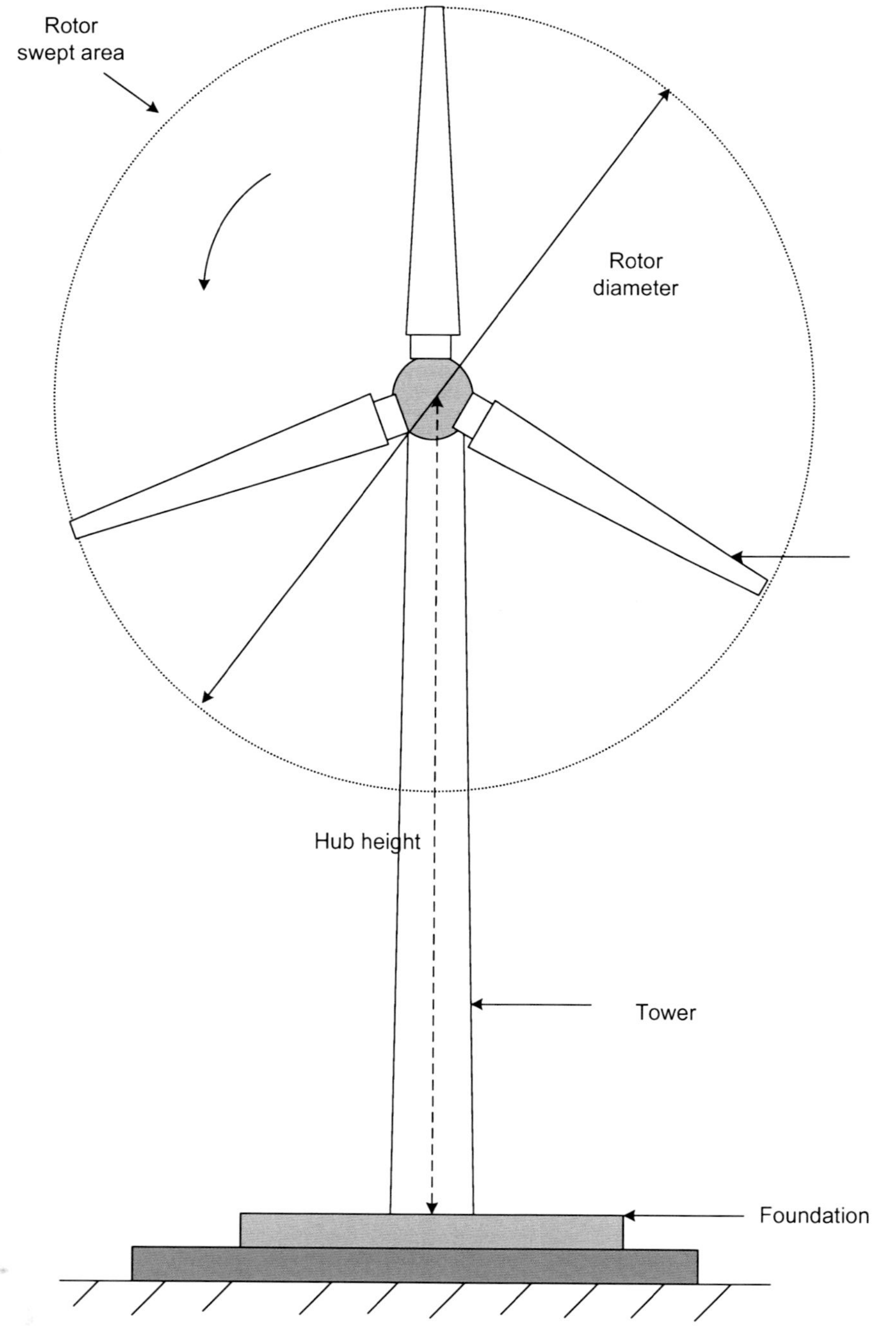

Figure 1: A horizontal axis three bladed propeller type wind machine.

Wind Machine Parameters

Coefficient of Performance (C_p)

Coefficient of performance (C_p) of a wind machine having a rotor with swept area A_s is defined as the ratio of the mechanical power output at the rotor shaft (P_m) to the power in the un-obstructed wind seen by the rotor.

$$C_p = \frac{P_m}{\frac{1}{2}\rho A_s V^3} \qquad \textbf{(1)}$$

where, ρ is the density of air and V is speed of air far upstream. C_p is a non-dimensional parameter which allows for comparison across machines of same type with different blade geometry as well as varied types of wind machines. The maximum value of C_p for horizontal axis propeller type machine is limited to 0.593 and is known as the Lanchester - Betz limit [7]. As seen from the above expression, for a constant C_p, the extractable power increases in proportion to the swept area of the rotor (A_s) and as cube of incident wind speed (V). Additionally, the rotational speed of the machine, Ω, number of blades, B, blade geometry (chord and twist distribution along the blade radius), lift and drag characteristics of the rotor blade are among other parameters which affect power extraction and C_p. The difference in air pressure between the upper and lower surfaces of an aerofoil (cross-section of the blade) generates the lift and drag forces. Power extraction can be maximized by increasing the lift force and decreasing the drag force simultaneously. This can be done by a proper choice of aerofoil section [8]. The role of these parameters in the performance of a wind machine is discussed in the following sections.

Tip Speed Ratio (λ)

The tip speed ratio (λ), defined as the ratio of the linear speed of the blade at its tip (ΩR) to the speed of the un-obstructed wind upstream of the rotor, signifies how fast the blade moves with respect to the oncoming wind and is mathematically expressed as follows:

$$\lambda = \frac{\Omega R}{V} \qquad \textbf{(2)}$$

For a given shaft power output, a key requirement for electricity generating wind machines is to have an increased Ω with relatively lower torque, whereas for mechanized forms of power utilization such as pumping and grinding, high torque is the primary requirement. Thus, electricity generating wind machines are designed for higher values of λ (typically 5 to 10) so as to reduce gearing requirements and hence the cost of a gear box. The C_p *vs.* λ characteristic is the fundamental measure used to compare the performance of different types of wind machines. Fig. **2** shows the typical C_p *vs.* λ characteristic of different designs of wind rotors. The C_p of an ideal propeller approaches the value of 0.593 with an increase in λ. It can be seen from Fig. **2**, that for all practical machines, as λ increases, C_p increases to a maximum value and then decreases. Hence, there exists an optimum value of tip speed ratio for power extraction. Further, propeller type machines can be used to achieve higher values of C_p compared to any other type of machine. This has led to the widespread utilization of propeller type of wind machines.

Solidity of the Rotor (σ)

Solidity of a rotor is defined as the ratio of projected blade area to the swept area of the rotor. For a horizontal axis propeller type wind machine, solidity is expressed as follows:

$$\sigma = \frac{B\,c}{\pi R} \qquad \textbf{(3)}$$

Where, c is the mean chord of the airfoil and B is the number of blades of the rotor. The higher the solidity, the higher the driving torque is. This is especially important to ensure that the turbine is self starting. Applying Newton's third law, if the air imparts a cranking torque on the rotor in a particular direction (say anti-clockwise), an opposite (clockwise) torque would be imposed on the air by the rotor. The torque thus imparted to the air on the downstream side of the rotor sets it into a swirling motion, known as the rotor wake. Therefore higher solidity leads to higher wake losses and lower power extraction. However, strength and rigidity of the blade increases with higher solidity, so that it can withstand larger forces and moments. Savonius turbines have a solidity of 1; multi bladed wind machines have solidities approximately in the range of 0.8 while 2 or 3 bladed propeller machines would have solidities less than 0.1. It may be observed from Fig. **2** that high solidity machines like the Savonius and the American multi-bladed wind machines operate at relatively low tip speed ratios as compared to propeller type of machines. This is so because if the rotor angular speed is increased (for a constant radius), the aerodynamic losses in addition to the losses in the rotor wake become large. This puts a practical limit on the use of higher tip speed ratio for these machines.

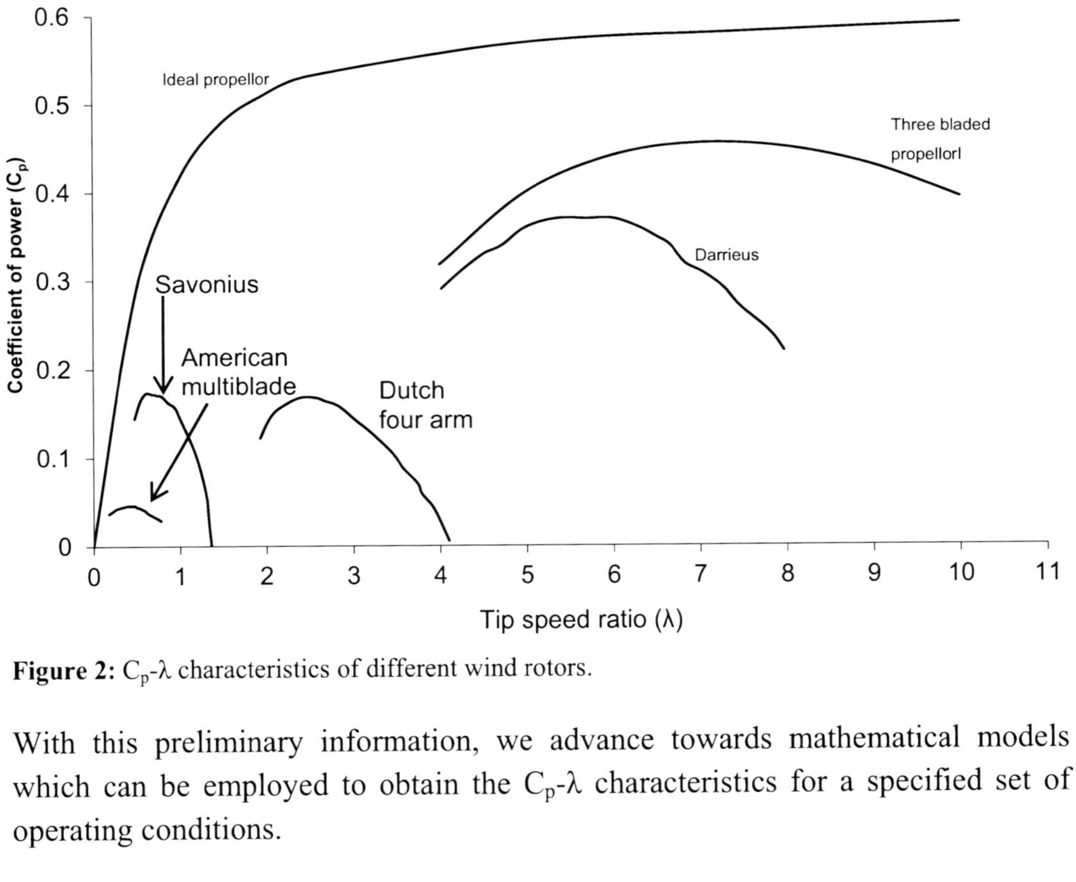

Figure 2: C_p-λ characteristics of different wind rotors.

With this preliminary information, we advance towards mathematical models which can be employed to obtain the C_p-λ characteristics for a specified set of operating conditions.

Rotor Model

In order to determine the performance of a wind turbine rotor, a number of theories have been proposed. In general, understanding of rotor behavior can follow two pathways:

1. By replacing the real rotor with a hypothetical disc (also known as actuator disc) of equivalent area placed in a control volume which

forms the basis of the 'axial momentum theory'. The contribution of this method lies in the fact that, from the axial momentum analysis of a one dimensional flow, it was shown that the velocity at the disc (rotor) is different from the free stream wind velocity upstream of the rotor. Lanchester and Betz used the concept to determine the ideal maximum power extraction by a wind rotor which is 0.593 times the power in the undisturbed upstream wind. This line of thought assumes that the forces on the rotor are uniformly distributed along its radius.

2. By considering the forces on individual rotor blade segments which form the 'blade element theory'. A real rotor, is never uniformly loaded as assumed by the actuator disc model. In order to analyze the radial load variation along the blades, the entire rotor blade is divided into finite number of elements and flow at each element is computed independently. From this it was possible to link the rotor performance with the rotor blade geometry.

The two methods are used in combination to determine the rotor characteristic, called the 'blade element momentum (BEM) theory'. It is reported that BEM methodology can produce reasonably accurate results within an error band of ± 2.5%, against the field test performance [9].

To illustrate the principle of the BEM method, let us consider a cross-section of a blade element i of thickness dr positioned at a radial distance r_i from the hub of the turbine as shown in Fig. **3**. This element has a chord length of c_i and moves with a local tip speed ratio of λr_i. Due to rotation of the blade element with a speed Ωr_i and the air with a speed V, the element at a station i will see a relative velocity (V_{rel}). The blade pitch angle (β_i) is the angle between the chord line of the blade and the rotor plane. It is a static angle and represents the orientation of the blade with respect to the rotor plane. The geometrical parameters of the blade include the blade pitch angle (β_i), the twist angle (β_{ti}) and the chord (c_i) which are known as a function of the elemental radius (r_i).

By considering the deceleration of the air at the rotor plane (a) and the rotational acceleration (a′) imparted to the wake behind the rotor, Vrel can be further corrected to account for the local flow variations caused by the turbine. The symbols a and a′ are referred as the axial induction factor and rotational induction factor respectively. As a and a′ are a functions of the elemental radius r_i, they are designated as ai and ai′ respectively and are obtained by considering the balance of axial and angular momentum on the upstream and downstream of the rotor. The relative flow angle (φ) is defined as the angle between the relative wind velocity (V_r) and the rotor plane and can be derived in terms of the axial and rotational induction factors (Fig. **3**):

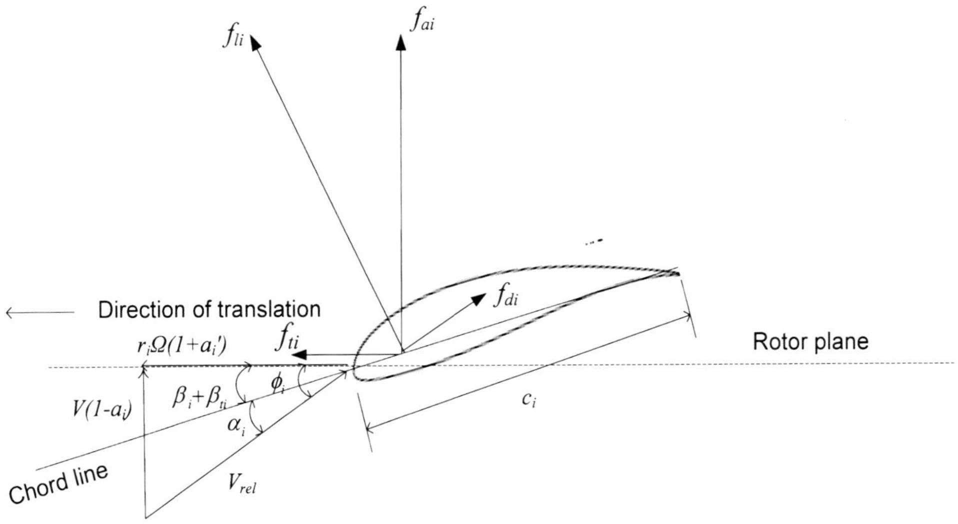

Figure 3: Forces and velocities at blade elements.

$$\tan\varphi_i = \frac{(1-a_i)}{(1+a_i')\lambda_{r_i}} \quad \textbf{(4)}$$

Flow of the air around the aerofoil establishes a pressure differential across the upper and lower surface of the blade element giving rise to a force in a direction perpendicular to flow called the lift force f_l. Moreover, a drag force f_d is also generated in the direction of flow. The lift and drag force per unit length are expressed in terms of the lift and drag coefficient C_l and C_d as:

$$f_l = \frac{1}{2}\rho c V_{rel}^2 C_l(\alpha) \quad \textbf{(5)}$$

$$f_d = \frac{1}{2}\rho c V_{rel}^2 C_d(\alpha) \quad \textbf{(6)}$$

The lift and drag forces are a function of the angle of attack (α) defined as the angle between the relative velocity and the chord line.

$$\alpha_i = \varphi_i - \beta_i - \beta_{ti} \quad \textbf{(7)}$$

In order to maximize the ratio of lift to drag force at every blade element, the angle of attack should be maintained constant at a specific value. Due to the variation of Ωr along the blade length, it becomes necessary to twist the blade by an angle β_t so as to maintain a constant value of α. The lift (f_l) and drag (f_d) can further be resolved into a tangential force (f_t) in the direction of blade motion and an axial thrust force (f_a) in the direction of the wind flow. In normalized form, the coefficient of the normal force (C_n) and coefficient of tangential force (C_t) can be expressed as a function of the relative flow angle (φ), the lift coefficient (C_l) and drag coefficient (C_d):

$$C_{ni} = C_{li}\cos\varphi_i + C_{di}\sin\varphi_i \quad \textbf{(8)}$$

$$C_{ti} = C_{li}\sin\varphi_i - C_{di}\cos\varphi_i \quad \textbf{(9)}$$

In order to account for the tip losses due to a finite number of blades and the thrust factor for values of a greater than 0.4, a Prandtl's tip loss correction factor (F_i) and a Glauert's correction ($a_c \approx 0.2$) are applied to increase the accuracy of the model [10]. The expressions for a_i and a_i' can be given by:

$$a_i = \frac{1}{2}\left[2 + K_i(1-2a_c) - \sqrt{(K_i(1-2a_c)+2)^2 + 4(K_i a_c{}^2 - 1)}\right] \quad a > a_c$$
$$= \left[K_i - 1\right]^{-1} \quad \text{otherwise} \quad \textbf{(10)}$$

$$a_i' = \left[\frac{8F_i \pi r_i \sin\varphi_i \cos\varphi_i}{Bc_i C_{ti}} - 1\right]^{-1} \quad \textbf{(11)}$$

where

$$F_i = \frac{2}{\pi}\cos^{-1}\left[\exp\left\{-\frac{B}{2}\frac{1}{\sin\varphi_i}\left(\frac{R}{r_i}-1\right)\right\}\right] \tag{12}$$

$$K_i = \frac{8\pi r_i F_i \sin^2\varphi_i}{B c_i C_{ni}} \tag{13}$$

Equations (4) to (13) are solved simultaneously to determine the axial as well as the rotational induction factors for each blade element. The coefficient of performance is calculated by integrating the incremental contribution to the power from each blade element.

$$C_p = \frac{B}{\pi R^2}\int_{r_h}^{r_t} c_i \lambda_{ri}(1-a_i)^2 \frac{C_{li}}{\sin\varphi_i}\left(1-\frac{C_{di}}{C_{li}}\right)dr \tag{14}$$

The total number of rotor blades is B while r_h and r_t are the radii at the hub and tip of the blade respectively. For illustration purpose, the C_p-λ characteristics for a rotor of diameter 6 m at different blade pitch angles are shown in Fig. **4**. It may be observed that the maximum power coefficient of the rotor is a function of the tip-speed ratio and the blade pitch angle. A variable speed configuration of the wind turbine enables its operation in the proximity of the maximum power coefficient. Typically, the turbine rotor speed is varied in proportion to the wind speed to maintain a constant tip-speed ratio. This ensures the maximum power coefficient for the wind rotor and hence, the maximum power extraction for different wind speed [5]. As the variable speed configurations are more efficient to the fixed speed configurations, a variable speed configuration is assumed with the design tip speed ratio being equal to the optimum tip-speed ratio (λ_{opt}). Mechanical power at the wind rotor shaft (P_m) at any wind speed with respect to the cut-in (V_c), rated (V_r) and cutoff (V_f) velocities can be expressed as:

$$\begin{aligned} P_m &= P_{mr} && \text{whenever } V_r \le V \le V_f \\ &= \frac{1}{2}\rho A V^3 C_{p\max} && \text{whenever } V_c \le V \le V_r \\ &= 0 && \text{otherwise} \end{aligned} \tag{15}$$

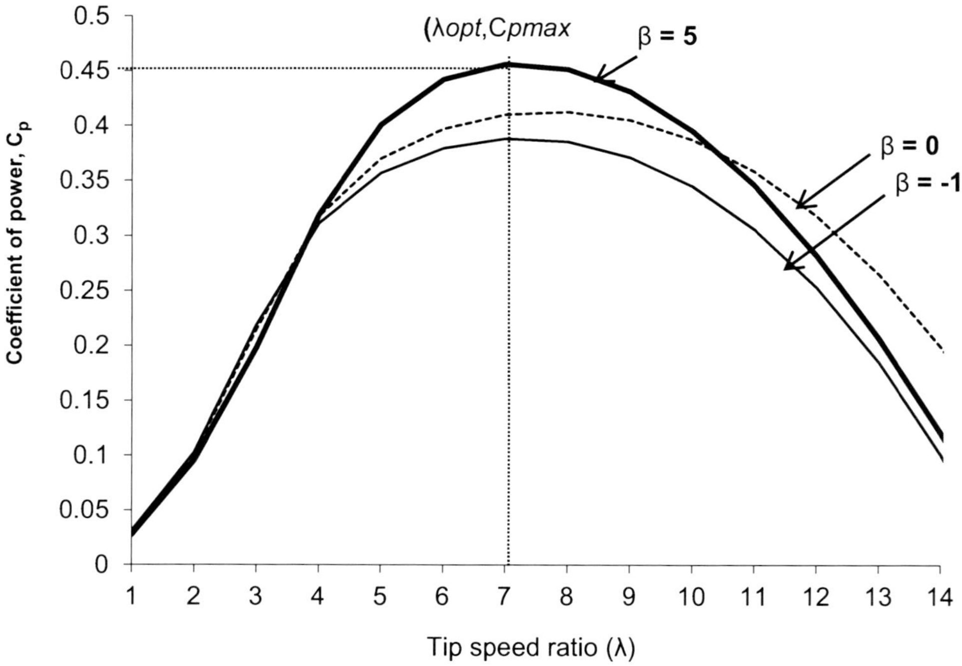

Figure 4: C_p-λ characteristics for a 6m diameter rotor for various pitch angles (β).

P_{mr} is the mechanical power available at the rotor at rated wind speed (V_r), designated as the low speed shaft rated power. It is assumed that power control above rated wind speed is achieved by progressively yawing the turbine rotor out of the wind in such a manner that $P_m = P_{mr}$.

In order to predict the power output from the wind turbine it is necessary to incorporate the characteristics of the electrical generator and transmission system. The models for these sub-systems are discussed later on.

Transmission System

Mechanical efficiency (η_m) of the drive train can be modeled by assuming that the loss in gear transmission per stage is a constant fraction (k) of the mechanical rated power, P_{mr} [4]:

$$\eta_m = \frac{P_t}{P_m} = 1 - \frac{qkP_{mr}}{P_m} \tag{16}$$

Where P_t is the power transmitted through the gear box shaft and q is the number of stages. The power (P_t) and torque (T_t) available at the gear box output shaft is then obtained as a function of the angular speed of the wind rotor shaft (low speed shaft), Ω and the gear ratio, GR.

$$P_t = P_m \eta_m \tag{17}$$

$$T_t = \frac{P_t}{\Omega GR} \tag{18}$$

Electrical Generator

A self excited induction generator is considered in the present study. Such a generator draws the reactive power which is supplied by charged capacitor banks connected across the stator terminals. The capacitor current in such a scheme can be controlled so as to allow a variable frequency generation. With known values of generator circuit parameters and given input wind torque, the output power from the induction machine can be determined by accounting for various losses in the generator circuit from its equivalent circuit model [11]. In order to supply a load that requires a constant frequency, the variable voltage variable frequency output of the generator can be converted into constant voltage constant frequency by means of a power electronic device consisting of an AC-DC and DC-AC converter. It is assumed that at rated condition the generator three phase voltage is 415V and the frequency of current is 50Hz for a 4-pole electrical machine.

By integrating the rotor, the transmission and the generator models, the power output characteristics of the entire wind machine can be obtained. For illustration, power curve for a rotor diameter of 10m and a rated electrical power of 10 kW, at a hub height of 30m is shown in Fig. **5**.

The cut-in wind speed is a function of the no-load system losses which in turn depends on the rotor diameter and the rating of the electrical generator. The cut-off wind speed is assumed to be 15 m/s. Beyond 15 m/s of wind velocity, the turbine is shut down by yawing it in a direction perpendicular to that of the predominant wind and no power is produced.

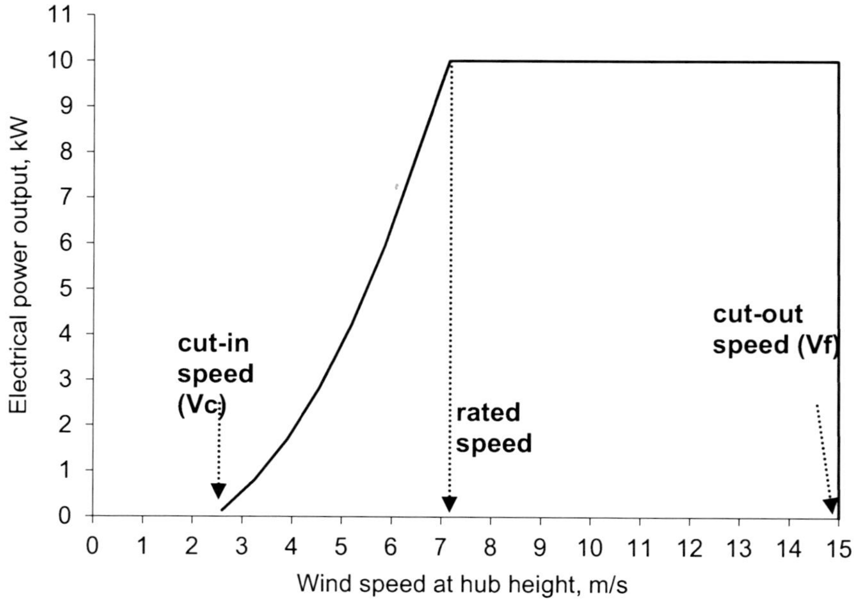

Figure 5: Power curve for a rotor diameter of 10m, rated electrical power of 10 kW, and at a hub height of 30m.

SIZING A WIND-BASED STANDALONE POWER SYSTEM

A standalone power system model comprises of the models predicting the behavior of various system sub-components, the physical interconnection between them and the role of power supply reliability therein. The input to the model is the time series data of the demand load and resource recorded over a specific interval of time (say, one day, one month, entire year). The output from such a model is the size of different system components while abiding by different system constraints. In this study we restrict to a deterministic parameter modeling approach wherein, it is assumed that the demand and wind speed at every time step is constant. This is a simplifying assumption but sufficient to investigate the system behavior with respect to it major design variables.

A generalized methodology is presented to relate the system variables viz., generator rating, rotor diameter and the storage capacity given an energy demand and wind resource profile. A set of feasible combinations of the generator capacity, rotor diameter and the corresponding minimum storage requirement can be determined by performing a time series simulation of the system energy balance over a specified planning horizon subject to various system level design constraints. When these feasible combinations are plotted on a two dimensional space comprising the rotor diameter and generator rating with battery bank size as a parameter, a limiting curve identifying the minimum rotor diameter requirement corresponding to a specific wind generator rating can be obtained. The limiting curve divides the entire space into a feasible and an infeasible design region of which the former embodies the system's design space. The concept of design space was originally applied by Poddar and Polley [12] for design and optimization of heat exchanger networks. Kulkarni *et al.* [13-15], have explored the efficacy of design space representation in the case of solar water heating systems; Arun *et al.* [16-18] and Roy *et al.* [11, 19] have studied specific cases of isolated power systems utilizing the said approach. The proposed methodology further serves as a tool for optimizing and selecting an appropriate system configuration based on the desired objective. The methodology is simple, flexible, does not require significant computational setup and can be applied to a wide range of problems.

Mathematical Model of a Standalone Wind-Battery Power System

Fig. **6** shows a general representation of an isolated power system consisting of a wind turbine, battery bank, bi-directional converter and charge controller. It is observed that these components are arranged on either sides of an AC and DC bus in a way that the AC loads and sources are connected to the AC bus while the DC devices are on the DC side of the bus. Such an interconnection between system sub-components is termed as a parallel configuration which allows the power demand to be met simultaneously through various sources by synchronizing with the converter. Further, it also offers an opportunity to minimize the size of the sources so as to serve the load in an optimal way.

The operation of the system is as follows: when generated power exceeds the demand, the surplus charges the battery or if the battery is fully charged it is sent to dump loads. When generation is insufficient to meet the demand, energy is drawn from the battery, provided the battery has not reached its depth of discharge (DOD). A loss of load occurs if at any instant there is no or less generation from the wind turbine and the battery bank is fully drained [16]. Power transfer across the battery bank (dQ_B/dt) can be calculated by considering the balance between the power supplied by the wind turbine ($P_w(t)$) and the load ($L(t)$) at a given instant t as:

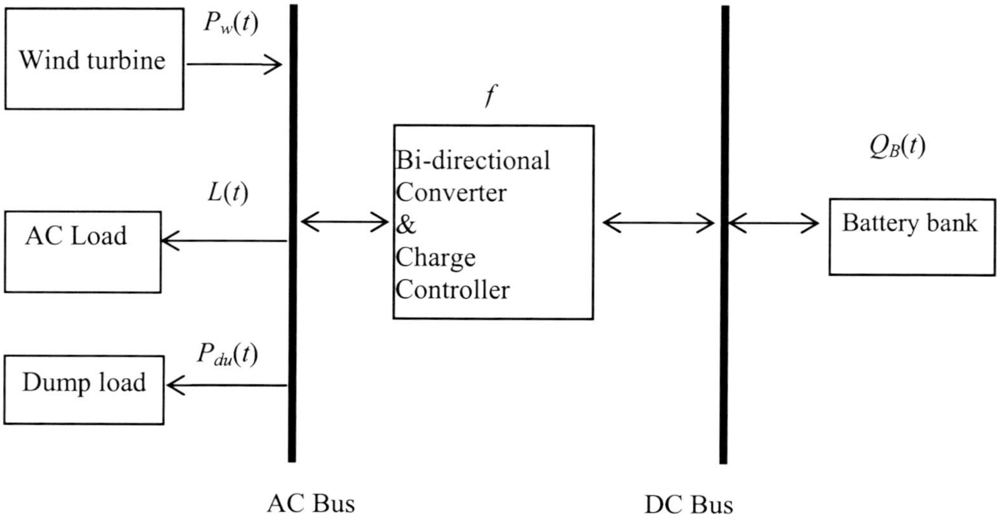

Figure 6: Schematic of a wind-battery isolated power system.

$$\frac{dQ_B}{dt} = [P_w(t) - L(t) - P_{du}(t)]f \tag{19}$$

$$\text{where } f = \eta_c \text{ whenever } P_w(t) \geq L(t) \\ = 1/\eta_d \text{ whenever } P_w(t) < L(t) \tag{20}$$

$P_{du}(t)$ represents the power which is dumped whenever there is a net surplus while η_c and η_d represent the fractional energy loss during the charging and discharging of the combined system consisting of the battery bank and bi-directional converter. It is assumed that η_c and η_d and therefore f do not vary within the time step and the battery self discharge loss is negligible. It may be noted that there is an upper limit to the value of (dQ_B/dt) imposed due to the maximum allowable charging current of the battery. However, the charging current can be limited by considering appropriate voltage of the battery bank and parallel combination of batteries. If a small time step Δt is considered, the system energy balance (19) may be integrated and a relation for the battery energy at a time step $t+\Delta t$ be obtained as follows:

$$Q_B(t+\Delta t) = Q_B(t) + [P_w(t) - L(t) - P_{du}(t)] f \Delta t \quad \textbf{(21)}$$

For each time step in the time horizon, the battery energy is obtained by applying Equation (21). The battery storage capacity (B_r) may now be determined as:

$$B_r = \frac{\max\{Q_B(t)\}}{DOD} \quad \textbf{(22)}$$

By treating the initial battery energy ($Q_B(t = 0)$) and the power dumped at each time step ($P_{du}(t)$) as optimization variables, the battery capacity (Equation 22) can be minimized subject to the constraints imposed by the non-negativity of the battery energy and the power dumped at every time step ($Q_B(t) \geq 0$, $Pdu(t) > 0$) in addition to the equality of the battery energy state at the beginning and end of the time horizon ($Q_B(t=0) = Q_B(t=T)$). Therefore, following the above procedure the minimum battery size for a specific combination of generator rating and rotor diameter can be obtained. The non-negativity constraint of the battery energy affirms that the battery can never be discharged below the level set by the DOD, while the equality constraint suggests that the battery bank is employed only as an energy storage device and no net energy is supplied to or drawn from it. In other words, if $Q_B(t=T)$ is less than $Q_B(t=0)$, there will always be a net energy deficit over the time horizon leading to a loss of load and an infeasible system design. A little insight may reveal that if the power generation at each time step $P_w(t)$ is such that the power dumped is almost zero, then such a system design would correspond to the minimum possible generator rating required to satisfy the load. Hence for specified values of generator rating and rotor diameter, the proposed methodology is able to demarcate between a feasible and infeasible design option. For further understanding, the methodology is applied to the case of a remote village in central India as illustrated in the following section.

Illustrative Example

Fig. **7** shows recorded hourly average time series data for power demand at a remote village hamlet in the state of Madhya Pradesh located in the central part of India. Fig. **8** shows the hourly mean wind speed on a representative day at the location [20]. The daily energy demand of the community is 39.5 kWh, with a peak of 3.8 kW and average demand 1.8 kW respectively. The average wind speed at the site at a hub height of 10 m is 4.4 m/s and the highest wind speed recorded is 7.6 m/s. By comparing the wind speed and load profile, it is observed that during hours 5-8 when there is a significant load, the wind availability is the least whereas for hours 11 to 16 when the wind speed is at its highest the power demand is at its lowest indicating a mismatch between the load and resource. It is further assumed that the demand remains constant over the time step of one hour, the charging and discharging efficiency of the battery bank is constant and the self discharge loss of the battery bank is negligible. The assumptions of constant charging and discharging rates and negligible self discharge rate are justified by Arun *et al.* [16] for the case of an isolated diesel generator power system. The time step of simulation is 1h and the time horizon considered is 1 day. The battery – converter system net charging and discharging efficiency is taken to be 85%, while the depth of discharge of the battery is 70% [21]. The turbine has a hub-height of 30m and the power law index for the site is 0.31.

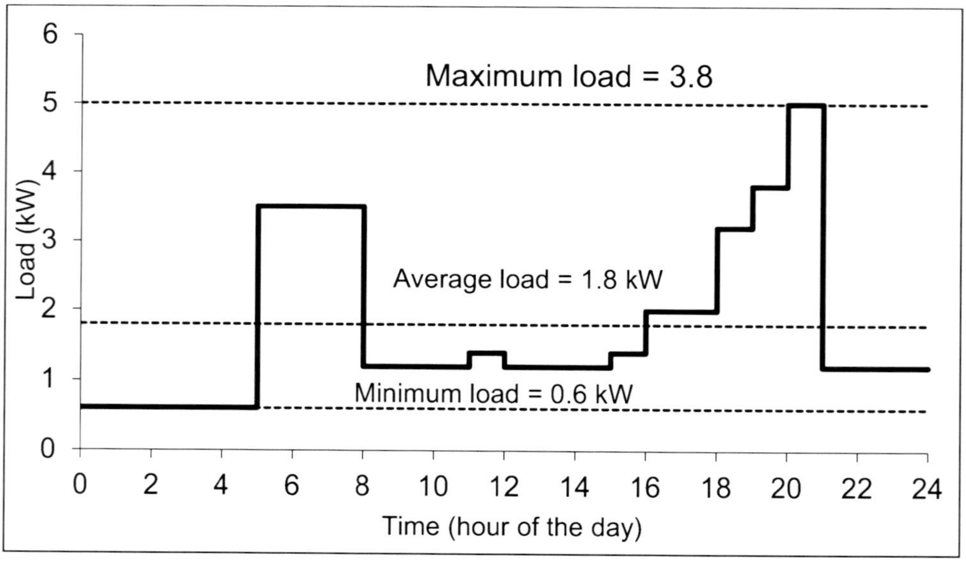

Figure 7: Load variations on a representative day at a remote village in Central India.

Following the procedure illustrated in the previous section, for a specified combination of the wind generator rating (P_r) and rotor diameter (D), the power output at each time step $P_w(t)$ can be predicted and the minimum battery capacity B_r be determined. System simulations show that by maintaining the rotor diameter constant at a particular value and varying the generator rating a feasible system may be designed only between a minimum and a maximum value of P_r. A plot of battery capacity *vs.* generator rating for different constant values of rotor diameter is shown in Fig. **9**. It is noted from Fig. **9** that a minimum generator rating of 1.8 kW is necessary to supply the load. The limit to the minimum possible generator rating is explained on the premise that in order to meet a definite load, the power generated by the turbine has to be of a magnitude which would serve the load during the entire time horizon. The existence of a minimum limit to P_r is further found to be true for all different values of rotor diameter. Further, there is an upper limit to the value of utilizable rotor diameter which is dictated by the hub height of the turbine. More specifically, if the turbine hub height is H, then the maximum rotor diameter (D_{max}) may be such that it leaves a clearance of h_0 from the ground as given by the following relation:

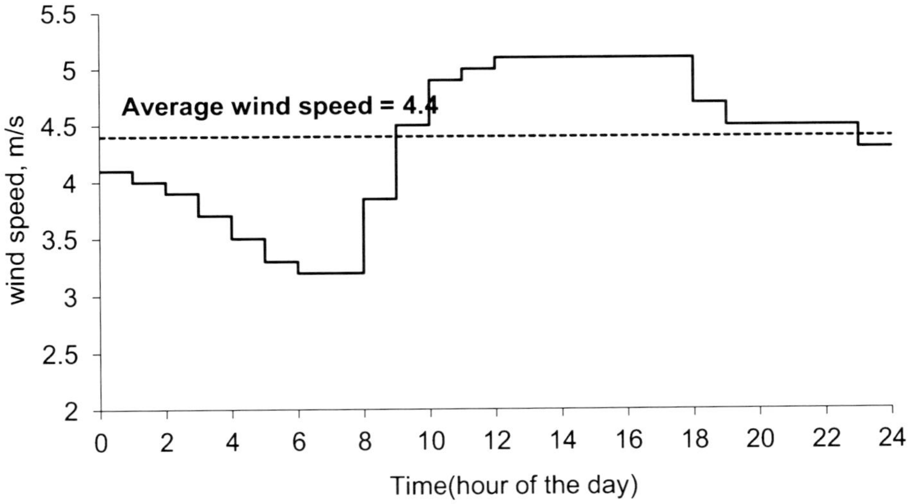

Figure 8: Hourly average wind speed variation at 10m above ground level.

$$D_{max} = 2(H - h_o) \quad \textbf{(23)}$$

In the present example h_0 is taken to be 10 m and the hub height of the turbine is 30m which restricts the maximum possible rotor diameter to 40 m.

Fig. **9** also points out that the global minimum rotor diameter (D_{min}) which would satisfy the load is 8.6 m. A system designed with a rotor diameter less than D_{min}

will fail to capture adequate energy to meet the demand load and the internal losses of the system. This is the reason for the existence of a limit to the minimum possible diameter. Further observation of different constant diameter lines brings to light that there is a variation between the behavior of different constant diameter lines. For example, constant diameter lines from 8.6 to 9.1m exhibit an ever increasing storage requirement with increase in generator rating; while constant diameter lines from 9.2 to 40 m show an initial decline in the battery capacity until a minimum is reached and further continuous increase in storage capacity.

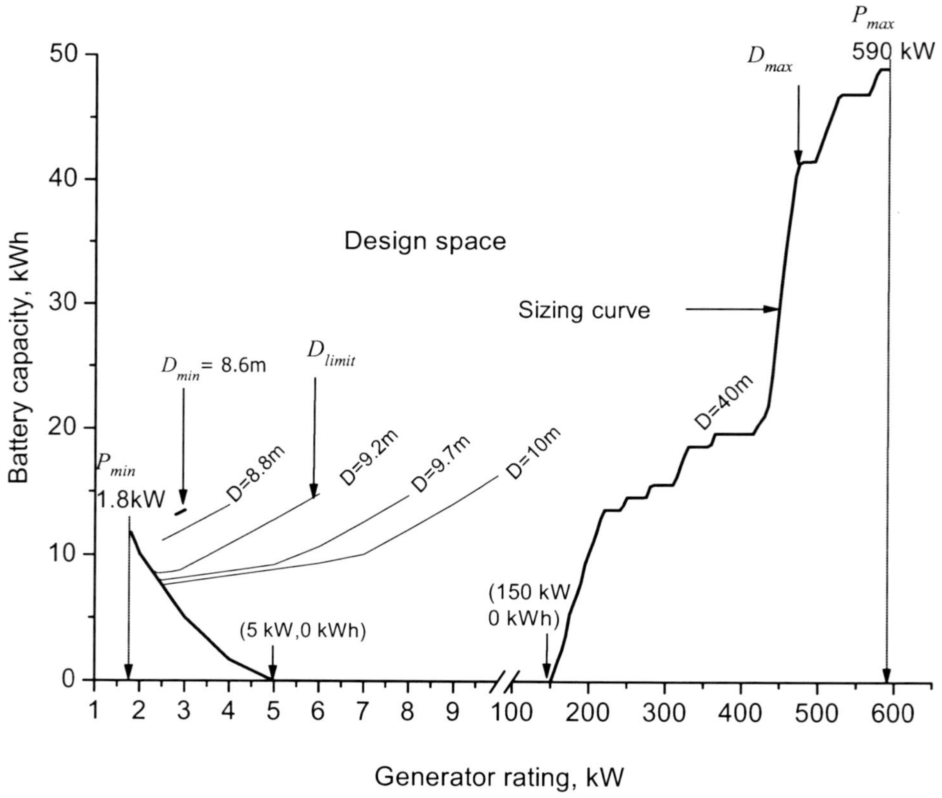

Figure 9: Sizing curve and constant rotor diameter lines in the design space.

Let us consider the curve corresponding to D_{max}. There exists a minimum generator size (P_{min}) which is necessary so that the demand can be met. This corresponds to a 1.8 kW generator along with an 11.8 kWh of storage capacity. It would be expected that an increase in the generator rating above Pmin would lead to an increase in the energy production over the time horizon. Since there would be surplus energy available at certain time steps the battery requirement would decrease. This is the reason for reduction in storage size from Pmin. However, the decrease in storage size continues up to a point at which the generator size is large enough that a wind turbine-only configuration is sufficient to fulfill the demand. This corresponds to a generator rating of 5 kW (Fig. **9**). A further increase in the generator rating beyond this value does not require storage up to a generator rating of 150 kW. Any further increase in the generator size leads to an increase in the battery requirement. It is unusual to note the increase in storage requirement with increase in generator size above 150kW. Increasing P_r (with a constant D), entails an operation with higher rated wind speed. As a result, the duration of time when the system operates on part-load also increases. To demonstrate this point, the power curve of two wind machines rated at 5 kW and 10 kW with the same rotor diameter are plotted in Fig. **10**.

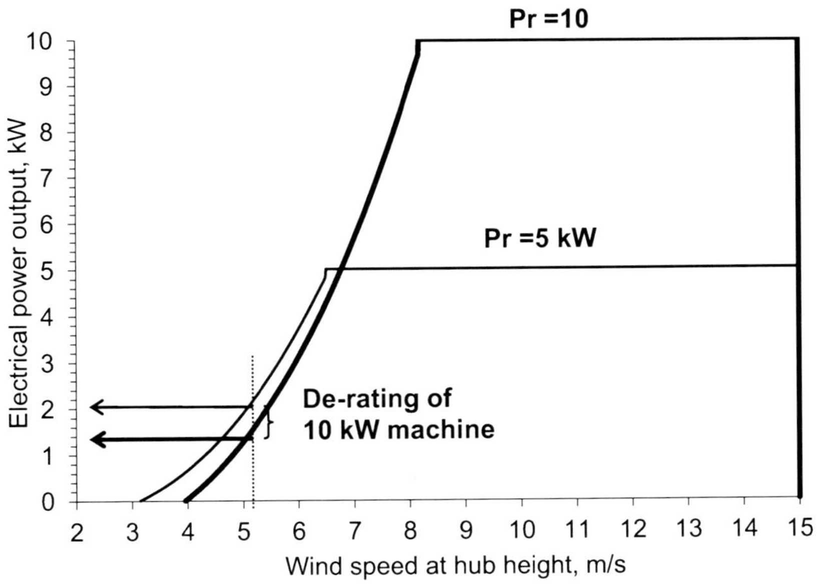

Figure 10: Power curves for different rated powers showing part load at lower wind speed.

If the operation of both wind machines at a wind speed of 5 m/s is considered, then employing the 10 kW machine leads to a loss of 0.6 kW in comparison to the

5 kW machine. This loss in energy is due to the lower generator efficiency associated with 10 kW generator which results in a higher degree of mismatch and an increased storage requirement. Therefore, in choosing a particular generator rating for a specific value of rotor diameter, following points become evident:

i. Increase in the generator rating leads to lower storage requirement provided the rotor diameter is greater than a limiting value (D_{limit}).

ii. For $D < D_{limit}$ increase in generator rating always leads to an increased storage size.

For the present example, D_{limit} is 9.2 m (Fig. **9**). In essence, the above discussion outlines a trade-off between the incremental energy captured by increasing the generator rating and the energy lost due to increased part load operation. This is also the reason for the existence of a minimum battery capacity on every constant diameter line in Fig. **9**. With further increase in rating; the rated wind speed would become so high that after a certain value of the rating, it may not be possible to satisfy the demand owing to a large part-load energy loss. Consequently, the rating of the generator is limited to a maximum possible value. The maximum generator rating for which the load may be met is 590 kW. The storage requirement curves corresponding to all different rotor diameter values are completely enveloped by the curve corresponding D_{max}. In other words, this curve represents the minimum storage size for all possible generator rating values. It is therefore referred to as the sizing curve. Any configuration lying above the sizing curve represents a feasible design option, and the entire region enclosed by the sizing curve is the design space for the given problem. A design point lying outside the design space will not be able to meet the load and is hence an infeasible design alternative. The sizing curve is similar to a design envelope within which a feasible design is assured.

The tradeoff between the system design variables D, P_r and B_r are represented on a battery capacity *vs.* generator rating diagram in Fig. **9**. A system designer/consumer would be interested in knowing the proper combination of D and P_r. If a system design corresponding to a given generator rating is sought such that the rotor diameter requirement is minimum, then the battery capacity may be treated an optimization variable. It may be observed from Fig. **9** that for a particular value of generator rating, it is possible to locate a design point for which the rotor diameter requirement is minimum. When plotted on a rotor diameter *vs.* generator rating diagram, a locus traced by all feasible combinations of generator rating and minimum diameter points forms the minimum diameter limit line. This is shown in Fig. **11**. Further, it is evident that for all feasible options of generator rating, the maximum possible diameter available for design is Dmax which is 40 m in the present example. The minimum rating of the generator that can be used for designing a feasible wind – battery system at the particular site is 1.8 kW.

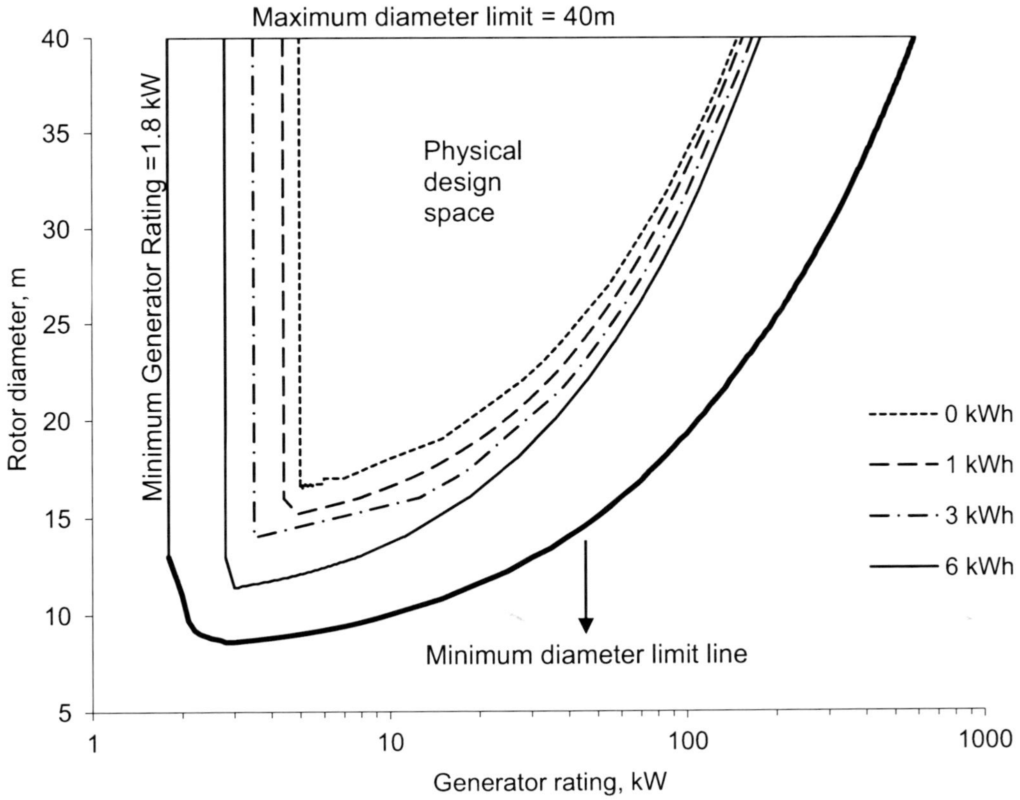

Figure 11: Physical design space showing limits of rotor diameter and generator rating.

These limits are graphically depicted in Fig. **11**. The closed space bounded by all these three limiting lines reflects the physical limits of the rotor diameter and generator rating in order to obtain a feasible design option. It is therefore designated as the physical design space of the system.

Fig. **11** also shows different constant battery capacity lines from 0 to 6 kWh. It is seen that within the physical design space, the battery capacity decreases as we move inward from the limiting lines.

Inside the zero battery energy line the storage requirement is zero, *i.e.* a standalone wind generator would be able to supply the demand load. A design which lies within the limits of the physical design space is guaranteed to meet the load. The physical design space depiction identifies the minimum and maximum possible rotor diameter, rated power and battery size for a given hub height of the turbine. Any combination of rotor diameter, rated power and battery size lying outside the physical design space leads to an infeasible design. It may be emphasized that besides the load and resource profile, the system design boundaries represented by the physical design space are specific to the battery and the wind turbine characteristics, operating strategy/dispatch as well as the arrangement of different system components (series/parallel).

CONCLUSIONS

Due to high degree of variability in the wind resource availability, it is important to design wind energy systems with appropriate size of energy storage. The power generation characteristics of a wind generator are inherently non-linear and depend on the appropriate choice of rotor diameter and generator rating. Further, the storage capacity is a function of the wind turbine power generation characteristic. A design activity can be viewed as a process of making decisions. An improper decision taken at particular step of the process can lead to failure of the system to supply the load which might bring an overall unacceptability to the technology. Given a load and resource data, the feasible options available for system design can be chosen by generating the sizing curve. The sizing curve distinguishes the feasible and unfeasible design regions in the design space while the physical design space determines the upper and lower limits for the rotor diameter, generator rating and battery bank capacity for serving the particular load. Such visualization is important from the system planner's perspective so that a design decision adheres to the physical system design limits. In the present study a deterministic approach and a time horizon of 24 hours was chosen for simplicity. In actual circumstances, the wind resource being highly stochastic by nature, a consideration of wind speed uncertainty in the modeling framework would obtain a more realistic picture. The design space serves as a tool for obtaining an optimum system based on appropriately chosen objective function. It may be noted that the design space is independent of the objective function and the cost coefficients. Therefore, the proposed methodology is suitable for multi-objective optimization of the entire system as well.

REFERENCES

[1] European Wind Energy Association (EWEA). Wind Energy - The Facts, Part-III - The economics of wind power, 2006. [Online] Available http://www.wind-energy-the-facts.org/documents/download/Chapter3.pdf. [Accessed Feb. 2010]

[2] Global Wind Energy Council (GWEC). Global wind 2008 report. [Online] Available:http://www.gwec.net/fileadmin/documents/Global%20Wind%202008%20Report.pdf. [Accessed Feb. 2010]

[3] E. Golding, The generation of electricity by wind power. New York: E. & F.N. Spon, 1976.

[4] G. L. Johnson, Wind Energy Systems, New York: Prentice Hall, 1985.

[5] T. Burton, D. Sharpe, N. Jenkins and E. Bossanyi, Wind energy handbook. Chichester: John Wiley and Sons Ltd., 2001.

[6] P.C. Putnam, Power from the wind, New York: Van Nostrand Reinhold, 1948.

[7] K H. Bergey, “The Lanchester-Betz limit”, Journal of Energy, Vol. 3, pp: 382-384, Nov.-Dec. 1979.

[8] I.H. Abbott and A.E. Von Doenhoeff, Theory of Wing sections: including a summary of airfoil data. New York: Dover Publications Inc.1959.

[9] D.A. Spera, Ed., Wind turbine Technology: Fundamental Concepts of Wind turbine Engineering. New York: ASME Press, 1994.

[10] M.L. Hansen, Aerodynamics of Wind Turbines: Rotors, Loads and Structure. London: James and James Publications. UK, 2000.

[11] A. Roy, S.B. Kedare and S. Bandyopadhyay, “Application of design space methodology for optimum sizing of wind-battery systems”, Applied Energy, Vol. 86, pp. 2690-2703, December 2009.

[12] T.K. Poddar and G.T. Polley, “Heat exchanger design through parameter plotting”, Chemical Engineering Research and Design, Vol. 74 A, pp.849-852, November 1996.

[13] G.N. Kulkarni, S.B. Kedare and S. Bandyopadhyay, “Determination of design space and optimization of solar water heating systems”, Solar Energy, Vol. 81, pp. 958-968, August 2007.

[14] G.N. Kulkarni, S.B. Kedare and S. Bandyopadhyay, “Design of solar thermal systems utilizing pressurized hot water storage for industrial applications”, Solar Energy, Vol. 82, pp. 686-699, August 2008.

[15] G.N. Kulkarni, S.B. Kedare and S. Bandyopadhyay, "Optimization of solar water heating systems through water replenishment", Energy Conversion and Management Vol. 50, pp. 837-846, March 2009.

[16] P. Arun, R. Banerjee, and S. Bandyopadhyay, "Optimum sizing of battery integrated diesel generator for remote electrification through design space approach", Energy, Vol. 33, pp. 1155-1168, July 2008.

[17] P. Arun, R. Banerjee, and S. Bandyopadhyay, "Optimum sizing of photovoltaic battery systems incorporating uncertainty through design space approach", Solar Energy, Vol. 83, pp. 1013 -1025, July 2009.

[18] P. Arun, R. Banerjee, and S. Bandyopadhyay, "Optimum design of battery-integrated diesel generator systems incorporating demand uncertainty", Ind. Eng. Chem. Res., Vol. 48, pp. 4908-4916, April 2009.

[19] A. Roy, P. Arun and S. Bandyopadhyay, "Design space for isolated power systems-a deterministic approach", SESI Journal, Vol. 17, pp. 54- 69, June and December 2007.

[20] P.K. Katti and M.K. Khedkar, "Alternative energy facilities based on site matching and generation unit sizing for remote area power supply", Renewable Energy, Vol. 32, pp. 1346-1362, July 2007.

[21] C.J. Rydh, "Environmental assessment of vanadium redox and lead-acid batteries for stationary energy storage", Journal of Power Sources, Vol. 80, pp. 21-29, July 1999.

APPENDIX: TERMS

a = Axial induction factor

a' = Rotational induction factor

a_c = Glauert's correction factor

A = Rotor swept area, m^2

B = Number of rotor blades

B_r = Required battery bank capacity, kWh

C_d = Coefficient of drag

C_l = Coefficient of lift

C_n = Coefficient of normal force

C_p = Coefficient of performance

$C_{p\,max}$ = Maximum coefficient of performance

C_t = Coefficient of tangential force

D = Rotor diameter, m

DOD	=	Depth of discharge
D_{max}	=	Maximum rotor diameter, m
D_{limit}	=	Limiting value of rotor diameter, m
D_{min}	=	Minimum rotor diameter, m
f	=	Loss factor in charging/discharging process
F	=	Prandtl's tip loss correction factor
GR	=	Gear ratio
h_o	=	Distance between the blade tip and the ground surface, m
H	=	Hub height, m
k	=	Constant of proportion for power loss per stage of gear transmission
L	=	Demand load, kW
P_r	=	Rated electrical power, kW
P_m	=	Mechanical power output of wind rotor shaft, kW
P_{mr}	=	Rated mechanical power of wind rotor shaft, kW
P_t	=	Power transmitted to the generator shaft, kW
P_w	=	Power generated by the turbine, kW
P_{du}	=	Power dumped, kW
q	=	number of gear stages
Q_B	=	Battery energy, kWh
R	=	Blade radius, m
r	=	Blade local radius, m
r_h	=	Radius at blade root, m
r_t	=	Blade tip radius, m
V	=	Instantaneous wind speed far upstream of the rotor, m/s
V_c	=	Cut-in wind speed, m/s

V_f	=	Cut-off wind speed, m/s
V_r	=	Rated wind speed, m/s
V_{rel}	=	Relative wind speed, m/s
t	=	Time, h
T	=	Total horizon, h
T_t	=	Torque transmitted to the generator shaft, Nm

Greek

α	=	Angle of attack, deg
β	=	Blade pitch angle, deg
β_t	=	Blade twist angle, deg
λ	=	Tip speed ratio
λ_{opt}	=	Optimum tip speed ratio
Δt	=	Time step for simulation, h
η_c	=	Charging efficiency
η_d	=	Discharging efficiency
η_o	=	Mechanical efficiency of gear box
η_o	=	Overall efficiency of generator and gear box
ρ	=	Air density, kg/m^3
Ω	=	Angular speed of rotor blades, rad/s
φ	=	Relative angle of flow, deg

Abbreviation

BEM	=	Blade element momentum

CHAPTER 6

Sustainable Electric Power System Based on Solar Energy

Z. Glasnovic[a,*] and J. Margeta[b]

[a]Faculty of Chemical Engineering and Technology, Dept. of Measurement and Process Control, Univ. of Zagreb, Croatia and [b]Faculty of Civil Engineering and Architecture, Univ. of Split, Croatia

Abstract: This chapter analyses the possibility of implementation of Sustainable Electric Power System (SEPS) as a totally green strategy of electric energy production for the world by the year 2040. The analysis presented in the paper is based on the EREC strategy which foresees the share of 82% of Renewable Energy Sources (RES). The problem of implementation of this strategy is that the more significant RES (Sun and wind) are characterized by intermittence of input energy, for which reason they cannot provide continuous and reliable supply of energy to consumers without electric storage. The solution to this problem and to creating conditions for achieving SEPS is an innovative concept of Solar Hydro Electric (SHE) power plant which is a basically combined photovoltaic power plant and pump storage which can produce and store relatively large quantities of energy and provide continuous supply of electric power and energy to consumers. In this way SHE is put into equal position with power plants using conventional power fuels, and because of that, SHE is presented in this paper as the main building element of the future SEPS. The conducted analysis and results clearly point not only to the reality, but the necessity for SEPS and to the exceptionally big achievements which the PV generator use will reach in the future. The proposed strategy of SEPS development could significantly contribute to realization of sustainability objectives, particularly to reduction of the problem of global warming.

Keywords: Sustainable development, sustainable energy supply, intermittent energy management, green energy, energy policy, renewable energy sources, solar energy, photovoltaics, solar power, hydroelectric power, PV generator, hydro energy, water storage, energy storage, hybrid energy systems, solar hydroelectric power plant.

SUSTAINABLE ELECTRIC POWER SYSTEM

In view of the obvious evidence of climate changes due to human activities (IPCC, 2007 [1]), it is logical to try and find solutions that would enable

***Address correspondence to Z. Glasnovic:** Faculty of Chemical Engineering and Technology, Dept. of Measurement and Process Control, Univ. of Zagreb, Croatia; Tel: +385 1 4597108; Fax: +385 1 4597260; E-mail: zvonglas@fkit.hr

Hatim Machrafi (Ed)

sustainable development. In energetic sense this means finding a solution of fully sustainable energy supply. There are numerous analyses and projections, but less technological solutions that could relatively quickly yield results. Different interests in the energy sector, but also in the same affiliation dealing with Renewable Energy Sources (RES), even in the affiliation for energy storage, create a special problem, additionally blurring the path towards the solution of the problem and complicating the matters to the extent that practically prevents "independent" decision makers to reach consistent and well founded decisions in the direction of sustainable development. When time limit in preventing the global warming problems and consequential melting of Arctic ice (Waldhams and Doble, 2009 [2]) are added, the priority objective is to find solutions that will attempt to forestall the problems mentioned. Bearing that in mind, in this chapter we wanted to establish the reality of achieving **Sustainable Electric Power System (SEPS)** based on the use of solar energy.

The energy policy of the European Union (EU) is interesting, foreseeing the share of 20% of RES in total energy production by the year 2020, of which 33.8% is for electric energy, EREC, 2004 [3]. Much more ambitious is the EU policy, which, based on the Advanced International Policies (AIP) scenario, foresees the share of 50% of RES by the year 2040, of which 82% is for electric energy only, EREC, 2006 [4]. Therefore, the AIP scenario of the EU policy with 82% of RES share is very close to implementation of SEPS. However, there are three significant problems in this ambitious scenario, due to which it is not realistically feasible:

1. Electric Power System (EPS), mostly based on RES, where solar, wind, tidal and wave energy would prevail, can't provide continuous and therefore safe energy supply to consumers, due to intermittence of input energy;

2. With the increase of RES (sun, wind) share, the need for conventional capacities/power plants does not decrease. On the contrary, with the increase of the RES share the insecurity of supply grows proportionately, along with the need for conventional power plants, Strbac *et al.* 2007 [5]. Therefore, it is very important to discard the misconception that the increase of the more significant RES will

automatically lead to reduction of the number of power plants using conventional fuel, because they are necessary for covering the needs when there are no inputs of RES into EPS;

3. Systems without energy storage, or very small storage, have greater need for peak power. Namely, considering that conventional EPS practically don't have energy storage, except in the case of hydroelectric (HE) storage (or it is very little), electric energy must always be consumed when produced. The consequence of this fact is that in conventional EPS generator capacities (power plants) should always cover energy consumption. This means that significant daily and seasonal power variations (oscillations) in EPS cause excessiveness of such systems, in the sense of installed capacities (electric power) of power plants. However, if sufficient electric energy storage capacities were planned, relatively smaller generator capacities (power plants) could be built (installed) in such EPS than would be the case with systems without storage, because energy storage balances out daily and seasonal surpluses and shortages of energy, Chen *et al.* 2009 [6].

All mentioned problems of implementation of RES in EPS can successfully be solved by adequate storage of electric energy. Numerous technologies of electric energy storage are known today (batteries, flywheel, pressure vessels, *etc.*), which differ in: size, energy storage costs, efficiency, lifetime, costs per cycle, *etc.* Chen *et al.* [6], ESA, 2009 [7]. Fig. **1** contains the comparison among various storage technologies by size, in linear and logarithm scale.

As can be seen, maximum values are obtained with **Pump Storage Hydro (PSH)** which is a mature technology with large volume, long storage period, high efficiency and reliability, while capital cost per unit of energy is low [6]. Therefore, when relatively big quantity of energy is required, the Pump Hydro Storage has no competition, because its rating is from about 600 MW and 10 h of operation (6 GWh of energy) to about 3.000 MW and 100 h of operation (about 30 GWh of energy) [7]. On the other hand, all new storage technologies have power of 10 kW and only 0.001 h (only 3.6 seconds), giving energy of only 10

Wh, maximum power of 100 MW and can give energy in the duration of 30 h (up to about 3 GWh of energy, mainly with CAES technology, while with other technologies the energy is below 300 MWh) [8]. This advantage specially applies when total, all year round, relatively highly reliable autonomy of local EPS is required (today it is unthinkable to leave a consumer without electric power), which is a necessity if these systems are supplied from RES, because PSH can easily balance not only daily and nightly, but partially seasonal surpluses and shortages of energy.

When looking into the issue of energy storage from ecological aspect, *i.e.* production of complex technologies (batteries, flywheel, pressure vessels, *etc.*) that require relatively high quantities of embodied energy, and decommissioning of these systems after the expiry of their lifetime, storage by means of PSH is in ecological sense a significantly better solution.

Therefore, not one of the new storage technologies can be compared to storage by PSH, in economical-technical sense, and especially in ecological sense. Precisely this is the reason why it is logical to choose this type of energy storage if SEPS is to be obtained as soon as possible.

Due to the mentioned advantages of PSH technologies (and development of foil technology and use of seawater, *etc.*), in the work of Glasnovic and Margeta, 2009 [9] they are in an original way combined with **Photovoltaic (PV)** power plants (with 2 pipelines and two electric lines), thus a new energy source is obtained, called **Solar Hydro Electric (SHE)** power plant that can provide continuous and very reliable electric energy and also power supply to its users (local consumer). Taking into account these characteristics of continuous and reliable energy supply to its consumers (with fast response), SHE (PV+PSH) is put in equal position with conventional/classic energy sources, *i.e.* power plants which use gas, oil, *etc.*

Furthermore, while conventional fuel power plants can provide the required energy and power when they have fuel, SHE does not depend on any energy supply, because its main power fuel is solar energy which is free of charge. Therefore, SHE is a naturally sustainable system that can be built practically at any location based on fresh or saline water, as well as close or far from water resources. It is an electric plant without CO_2 or methane emissions (in PSH system there is continuous recirculation of water and there are no problems with methane).

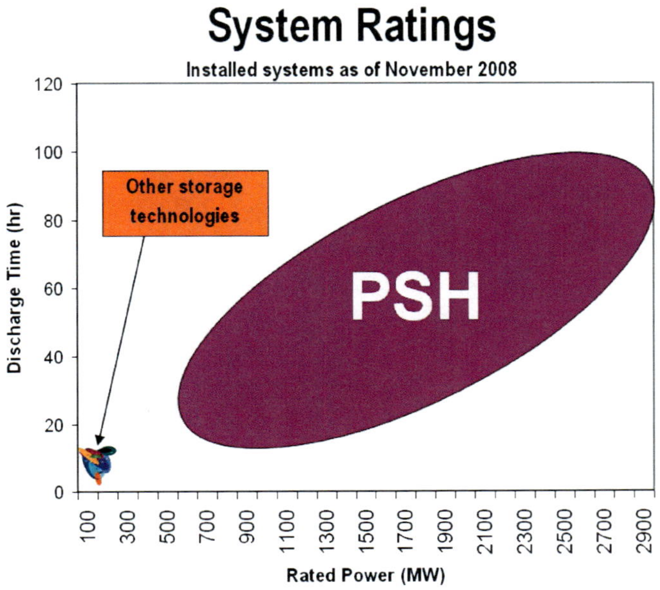

a) Linear scale

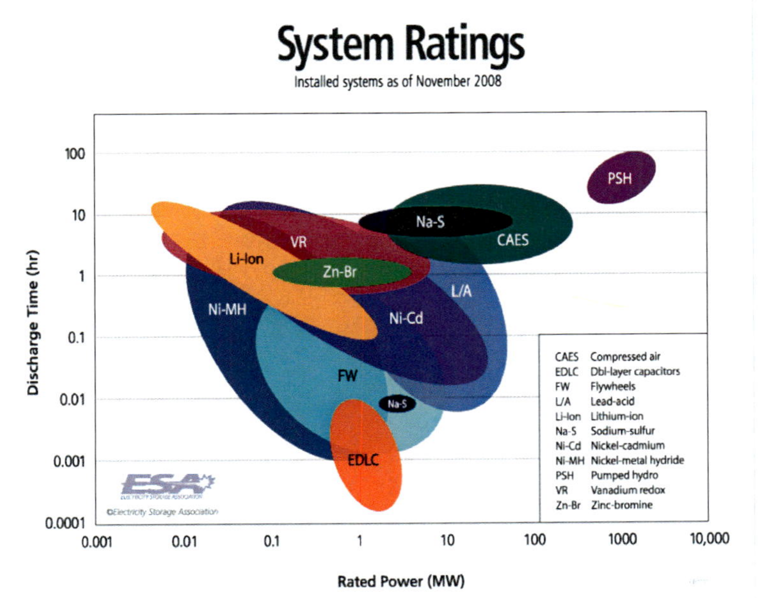

b) Logarithm scale

Figure 1: Storage system ratings: (a) linear scale, (b) logarithm scale (ESA, 2009 [7]).

Due to its exceptionally favourable characteristics SHE could have the key role in achieving such ambitious scenarios as is the AIP scenario of EREC forecast [4]. Even more, with such energy source, which can basically combine other RES and the existing hydro electric power plant reservoirs, it is logical to set the end theoretical objective, which is 100% of RES in EPS, because with energy storage and by proper dimensioning it could provide the required energy and power to its consumers at all times. In this chapter, which is based on the aforesaid, we wanted to show that with the already available technical-technological solutions of PV+PSH, *i.e.* RES+PSH systems, it is possible to activate the process which will enable SEPS or a significant part of the energy system in the future.

SHE – KEY BUILDING UNIT OF FUTURE SEPS

Background

Although RES power plants (mostly wind power plants) have been combined with Hydroelectric power plants (HE), the two power plants complemented each other in their operation (RES power plant was in operation when it had sufficient input energy and HE mostly covered peak loads), Bueno and Carta, 2006 [10]. The new way of using RES and HE power plants is when a local (isolated) consumer is being supplied, RES power plant gives its energy to PSH power plant which serves for daily and seasonal energy storage, while the local consumer is then supplied only from PSH power plant. In that way intermittent energy supply is converted into continuous energy supply and a new energy source is obtained, a power plant which is fully competitive to conventional sources, Fig. **2**, Glasnovic and Margeta, 2009 [9]. On the other hand, if the local consumers are connected to the regional EPS, it is logical that RES power plant will at the same time give its surpluses directly into the system.

PSH can be open or closed type, of bigger or smaller capacity, depending on local and regional building conditions. It can have one or two reservoirs, and use fresh water or sea water. Hybrid system RES+PSH is a typical distributed energy resource which could be installed close to the place of utilisation, which is a significant advantage, related to reliability of supply and reduction of losses due to transport.

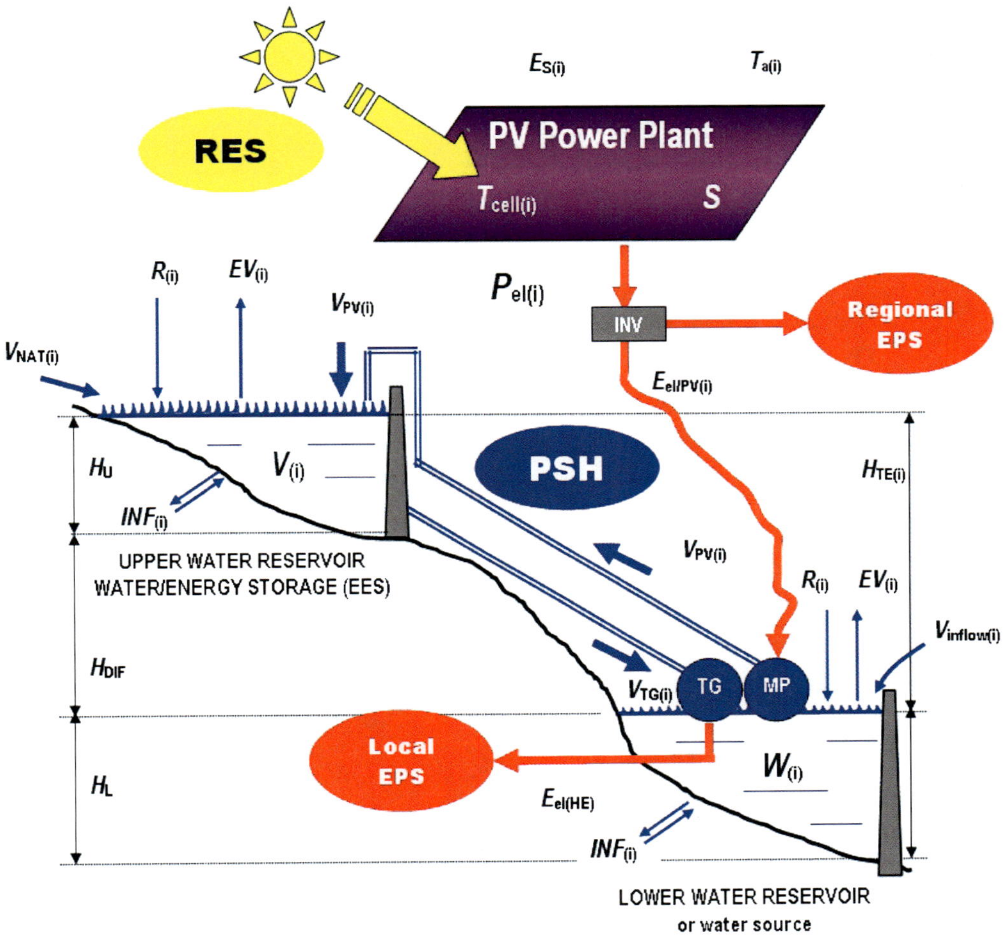

Figure 2: Concept of SHE as a new power unit.

On the other hand, the construction of SEPS is not conditioned by the construction of new RES+PSH units, but can be performed by upgrading RES units in the existing HE power plants, especially if there are favourable conditions for their upgrading into PSH power plants. An example of such unit is shown in Fig. **3**.

In both cases, whether it is the construction of new SHE, or reconstruction of the existing HE or PSH in order to obtain the SHE system, reservoir volume, available head and production of hydro energy, the power of the PV generator and HE power plant are determined by natural characteristics of the surroundings on one hand, and on the other hand by the consumers' needs.

General Concept of SHE Sizing

The main elements of sizing of SHE relate to defining the optimal size of PV generator as the most expensive part and defining the required volume of Electric Energy Storage (EES) and energy production.

Optimal sizing of PV power plant is a typical Dynamic programming problem, where operational decisions are carried out in a sequential manner in time. The optimization objective is to select the minimum required size of PV power plant in accordance with the needs and constraints during calculation period, *i.e.* it is min-max operation. The system state equation is mass balance equation.

Dynamic programming model for optimal sizing of the PV power plant in case the lower reservoir has sufficient capacity, or is positively unlimited (sea) is presented in the paper Glasnovic and Margeta, 2009 [9] and can be shown by the set of equations:

$$\left.\begin{aligned}
&f_{(i)}(V_{(i)}) = \underset{V_{PV(i)}}{\text{MIN}} \{ \text{MAX} [P_{el(i)}(V_{PV(i)}), f_{(i-1)}(V_{(i-1)})] \}, \\
&V_{(i)} = V_{(i-1)} + V_{PV(i)} + V_{NAT(i)} + R_{(i)} - EV_{(i)} - V_{TG(i)} \pm INF_{(i)} \\
&P_{el(i)} = \frac{1.36\left[H_U(V_{(i-1)}) + H_U(V_{(i)}) - H_L(W_{(i-1)}) - H_L(W_{(i)}) + 2H_F\right]}{\left[1-\alpha_c\left(T_{cell(i)} - T_0\right)\right]\eta_{MPI}\, E_{S(i)}} V_{PV(i)} \\
&V_{MIN} \le V_{(i)} \le V_{MAX} \\
&V_1 = V_{MAX} \\
&0 \le V_{PV(i)} \le Q_{MAX} \\
&i = 1, 2, ., N.
\end{aligned}\right\} \quad (1)$$

where state variables are expressed by water quantity in reservoir $V_{(i)}$ in stage i, *i.e.* $V_{(i-1)}$ in stage i-1, and decision variables $V_{PV(i)}$ are expressed by mean daily water quantity from the PV power plant in stage i; $f_i(V_{(i)})$ is optimal return function per state variable $V_{(i)}$ in time stage i; $f_{(i-1)}(V_{(i-1)})$ is optimal return function per state variable $V_{(i-1)}$ in time stage *i*-1; $P_{el(i)}$ is nominal electric power of the PV generator, *i.e.* PV power plant in time stage *i*; $R_{(i)}$ is total precipitation coming into the reservoir in *i* period; $V_{NAT(i)}$ is natural flow from the adjacent watershed; $EV_{(i)}$ is water from reservoir consumed by evaporation in *i* period; $V_{TG(i)}$ is water discharged from the upper reservoir into the turbine/generator unit for producing electric energy and $INF_{(i)}$ is infiltration in *i* period; $H_U(V_{(i-1)})$ and $H_U(V_{(i)})$ are upper water levels in function of upper reservoir volumes respectively; $H_L(W_{(i-1)})$ and $H_L(W_{(i)})$ are upper water levels in function of lower reservoir volumes ($W_{(i-1)}$ and $W_{(i)}$) respectively, and H_F represents linear and local hydrodynamical losses in the system; α_c is PV cell (generator) temperature coefficient; $T_{cell(i)}$ is temperature of PV cell (generator), T_0 is referential temperature of PV cell (generator) (25 ^{0}C), η_{MPI} is efficiency of motor-pump unit and inverter (%); E_S is mean daily radiation on horizontal plane (terrestrial radiation), V_{MAX} is maximum reservoir volume (m^3), V_{MIN} is minimum reservoir volume (m^3), Q_{MAX} is maximum water pumped by the motor/pump unit (m^3) in given period; increment *i* assumes the values *i*=1 to *N* (*N* is the total number of time stages, *e.g.,* months, decades or days). Fig. **3** also contains: H_{TE} is total head (m); H_{DIF} is difference between the lower level of upper reservoir and the upper level of lower reservoir (m); and V_{inflow} representing water inflow into the lower reservoir.

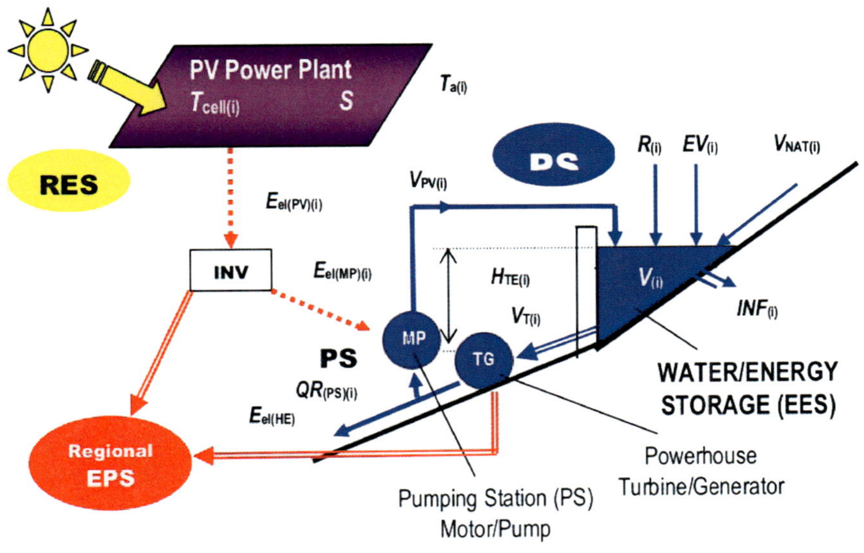

Figure 3: Concept of upgrading hydroelectric power (to PSH) with PV power plant towards SHE.

State variable values in observed time stages $V_{(i)}$ and in previous time stages $V_{(i-1)}$, and values of decision variable $V_{PV(i)}$, as well as returns per various stages $P_{el(i)}$, are calculated in the course of the process.

Optimal nominal value of electric power of the PV power plant $\boldsymbol{P_{el}} = \boldsymbol{f_i^*(V_{(i)})}$, working with PSH, is obtained as the output result, providing independence of energy supply to the consumer/settlement in the observed period, *i.e.* throughout the whole year.

The reality of this calculation was tested based on feasibility study for the island of Vis in Croatia, Glasnovic and Margeta, 2009 [9]. Table **1** contains monthly consumption of electric energy for the year 2007, Mucic, 2008 [11], which are key input data. As can be seen, total annual energy consumption is $E_{el(HE)} = 18.047$ MVAh/a. With this energy consumption and storage $V = 20$ hm^3, hydro electric power plant drop/height of $H_{TE} = 215$ m, and radiated solar energy (also as key input data), temperature, as well as other relevant influential values in the system (precipitation, evaporation, *etc.*), Table **1**, Hrabak-Tumpa, 2007 [12], the calculated electric power of PV power plant is $\boldsymbol{P_{el} = f_i^*(V_{(i)}) = 41}$ **MW.**

Table 1: Energy consumption, solar energy, air temperature, precipitation and evaporation for the island of Vis, Croatia[a].

Month	I	II	III	IV	V	VI	VII	VIII	IX	X	XI	XII
$E_{el(HE(}$ (MVAh/month)	1586, 29	1418, 34	1451, 26	1173, 53	1193, 61	1473, 25	2078, 84	2078, 77	1450, 39	1212, 78	1370.72	1559, 72
E_S (kW/m^{2}day)	1, 7	2, 8	4, 4	4, 8	6, 1	6, 8	7, 3	5, 7	4, 4	3, 2	1, 8	2, 5
T_a (^{0}C/day)	9, 8	9, 5	11, 4	14, 3	19, 4	23, 4	25, 9	25, 6	21, 4	18, 4	14, 0	11, 1
R (mm/day)	2, 1	0, 8	4, 0	1, 7	0, 1	0, 8	1, 7	3, 7	4, 2	3, 0	4, 4	2, 3
EV (mm/month.)	51, 3	48, 9	65, 9	82, 9	137, 7	191, 1	247, 3	233, 8	151, 8	115, 9	75, 5	58, 9

[a]Total annual electric energy consumption is: $E_{el(HE)} = 18.047$ MVA/a ≈ 18 GVA/a.

From the known electric power of PV generators and their operation efficiency η_{oc}, it is possible to calculate the required areas of the PV generator S (m^2) for reference conditions, according to Kenna and Gillett, 1985 [13]:

$$S = \frac{P_{el}}{1000 \cdot \eta_{oc}} (\text{m}^2). \quad \textbf{(2)}$$

For the PV plant power of 41 MW_p (410×100 kW_p), with PV generator efficiency η_{oc}=16%, a collector field of S =250.000 m^2 ($S \approx 25$ ha) should be foreseen, *i.e.* approximately 1.250x200 m^2.

Based on data on the PV power plant collector area and on radiated solar energy and temperature, total electric energy which the power plant could produce can be calculated, according to Kenna and Gillett, 1985 [13], by the equation:

$$E_{el(PV)} = \eta_{oc}\left[1-\alpha_c(T_{cell}-T_0)\right]\cdot\eta_I\cdot S\cdot E_S \tag{3}$$

where η_I is inverter efficiency, as well as the complete power conditioning unit, while other values have already been described.

Total electric energy that the sustainable SHE can produce in the observed climate region of the Island of Vis, Croatia, is $E_{el(PV)}$ = 55.447.234 kWh, and data for every day of the year, calculated by Equation (3), are presented in Table **2**.

Table 2: Electric energy production $E_{el(PV)}$ (kWh/day) from SHE power plant for the example of Vis, Croatia[a].

Day/month	I	II	III	IV	V	VI	VII	VIII	IX	X	XI	XII
1	60566	93733	157638	184040	190518	212160	231896	205180	162282	137225	68196	95088
2	59317	104876	126702	177037	207080	241085	227554	216235	154053	131804	73709	96590
3	69380	101922	146038	178817	216371	237584	220070	211477	158383	117672	82649	94123
4	73608	88695	159259	194243	207126	240000	220143	212573	152009	114516	80610	94770
5	84687	97182	150291	191531	197028	229090	230434	213300	137844	109745	78333	93380
6	65433	105748	167289	184551	203703	211067	223755	201700	158751	111091	60109	94377
7	57280	115816	149952	207463	214762	238150	210757	205108	141663	121620	65866	96143
8	51407	111710	150818	184777	225703	230498	218642	203727	144689	105946	79587	95814
9	56518	96736	134224	181161	203754	235484	242802	210336	146211	119394	81384	91567
10	58609	109431	160515	159685	230946	229674	232437	207703	144725	112286	67602	91115
11	54002	98532	153376	160141	201381	212892	225737	201579	158622	123414	74761	93219
12	61499	103754	154227	186855	222942	229693	241209	196357	148110	121813	68603	93834
13	66264	108217	141309	190453	233807	215083	245823	208084	138139	131801	71118	94034
14	66969	107248	165869	175164	221054	231604	243519	208121	142415	115370	74271	92137
15	73641	96217	138038	186471	218493	237241	239325	211548	149874	116326	68268	91141
16	80642	96977	141058	159488	214671	224853	245376	207144	136587	129133	57250	94502
17	67642	108239	170308	183194	214967	223950	238395	213322	128294	132387	75037	90629
18	62529	129089	158391	189738	205034	229520	244443	204780	135492	115604	78422	95429
19	67107	108375	159965	216337	203712	214520	239604	190318	144852	115156	65430	95646
20	72103	114269	151839	188701	203341	224112	239967	195210	149527	102371	78828	92677
21	57259	111275	163653	196420	214991	239662	248985	185768	144621	107945	68294	92756

Table 2: cont….

22	65510	102335	166038	191812	215741	212226	250052	200829	151293	112947	65921	97328
23	63724	101251	145841	189915	213555	237486	233437	220625	140536	123776	59810	95071
24	60687	105208	150819	185673	225403	243677	216483	216893	143392	99307	67841	96277
25	75894	100334	152396	193909	219954	231139	223597	195250	137953	125994	67759	92570
26	79069	118090	158549	186970	225338	241318	229205	200341	136529	119379	68618	92610
27	74334	105076	168055	165288	228771	245708	243526	197828	144084	118827	60850	91476
28	67989	112421	150868	141347	214228	233660	230460	189657	142970	118283	66319	92856
29	68749	118240	166305	182707	214955	221020	229518	190550	148600	104024	72130	91626
30	69122		159604	156642	236225	234151	230467	190468	144503	97119	61796	93140
31	66587		162356		223694		239777	174557		92057		91841
Total (kWh)	**2058127**	**3070996**	**4781590**	**5470530**	**6669248**	**6888307**	**7237395**	**6286568**	**4367003**	**3604332**	**2109371**	**2903766**

[a]Total annual electric energy production is: 55.447.234 kWh/a = 55.447 MWh/a ≈ 55 GWh/a.

In optimal sizing of the PV power plant the concept of SHE is described, in which energy is supplied to consumers from PSH, and the PV power plant is then used for pumping water into the upper reservoir. However, as can be seen in Fig. **2** and 3, the PV power plant can at the same time supply consumers of a regional EPS, according to demand and production in that system. This means that the role of PSH is not only constant energy production, but also backup power source and electricity storage. In this case study it is approximately 120 days.

Hydro energy, generated by the reservoir, stored in the upper reservoir, of volume V, at total head H, *i.e.* which it can produce, can be calculated according to:

$$E_{H(gross)} = \rho \cdot g \cdot H \cdot V \text{ (J, Ws)} \quad \textbf{(4)}$$

For the upper reservoir of PSH system on the Island of Vis, Croatia, this water volume would mean total capacity of stored electric energy of $E_{\mathrm{HE}} = 2,72 \cdot 20\mathrm{hm}^3 \cdot 215/1000 = 11.696.000$ kWh ≈ 12 GWh.

Energy production is bigger when accumulation and drop are bigger. Local conditions regarding reservoir construction (volume and elevation) will determine which combination of height difference H and water volume V is the better solution for the planned fulfilling of the consumers' demands – energy production. In addition, the selected drop H and flow Q will define the most efficient type of turbine (Ginocchio, 1978 [14]; Stojic, 1994 [15]). Net electric energy $E_{\mathrm{el(HE)}}$, produced by the HE is:

$$E_{el(HE)} = \rho \cdot g \cdot H_n \cdot V \cdot \eta_{TG} \text{ (J, Ws)} \quad \textbf{(5)}$$

where H_n is net available drop, and η_{TG} is total available efficiency of turbine and generator (0, 75 – 0, 92). However, in the work of Glasnovic and Margeta, 2009 [9], the starting point was the known electric energy consumption $E_{el(HE)}$, according to which other elements of the PSH were dimensioned. This means that Equation (5) was used in another way, not to define electric energy that can be produced by PSH, but to define the required daily water volumes V to be brought on turbines in order to satisfy the known energy consumption $E_{el(HE)}$.

The hydro accumulation power of the PSH system can then be calculated:

$$P_{H(gross)} = \rho \cdot g \cdot H \cdot Q \text{ (J/s, W)} \quad \textbf{(6)}$$

where V (m^3) is water volume in reservoir, H (m) is elevation difference between the lower and upper water level, g (m/s^2) is gravity constant, ρ (kg/m^3) is water density, and Q (m^3/s) is flow.

Therefore, the accumulated water, *i.e.* reservoir size V (m^3) and the available height difference H (m) determine the energy production and installed turbine capacity Q (m^3/s) determines the power. In this concrete case, the data on maximum daily consumption of electric energy is known (Mucic, 2008 [11]), *i.e.* maximum water volume V = 166.233m^3/day that flows to the turbines and duration of nominal values of solar energy of 7.3 h (with PV generators, energy of 7, 3 kWh/m^2day can be observed as product of 7, 3 h · 1000 W/m^2 = 7300 Wh/m^2), which defines the daily discharge flow of Q = 6, 3 m^3/s. Total hydraulic power of hydro accumulation, *i.e.* PSH is $P_{H(gross)} = \rho \cdot g \cdot Q \cdot H = 1000 \cdot 9.81 \cdot 6.3 \cdot 215$ = 13.287.645 W = 13, 3 MW. Therefore, this hydraulic power of PSH in SHE system is about three times smaller than the PV generator power. The pump station (pump) power is also determined based on Equation (6).

The power of PV generator is in direct dependence of volume storage for balancing energy production and consumption of any consumer. This connection has a logarithmic expression:

$$P = -a \cdot Ln(V) + b \quad \textbf{(7)}$$

For the case of island Vis, Croatia, it can be expressed:

$$P = -11.779 \cdot Ln(V) + 63.291$$
$$R^2 = 0.99 \qquad \textbf{(8)}$$

As the available volume is grater, the PV power is smaller. This fact allows optimization of the solution according to local conditions and storage construction, as well as conditions of solar energy production, *i.e.* local climate conditions.

In addition, possible local natural water flows (precipitation, surface runoff) in the upper storage, significantly improve production efficiency of SHE, which gives greater importance to the solution of green energy production, Margeta and Glasnovic, 2010 [16].

All of the above confirms the great importance of the proposed solution to achieve SEPS and reliable production of green energy of the power system.

Generalizing the SHE Characteristics

In order to implement the SHE power plant in various climate conditions (different countries), its characteristics have been generalized by variation of certain parameters for three typical climate areas in Croatia (north and south Adriatic and the continental part of Croatia which can, in view of radiated solar energy, be compared to North, Middle and South Europe), upon which the following was established:

- As the radiated solar energy increases, nominal electric power of the PV power plant decreases, because for equal consumption of electric energy and for more input (solar) energy, less electric power is obtained;

- As the air temperature rises, bigger power of the PV power plant is required, because temperature directly influences the decrease of the PV power plant efficiency;

- With increased precipitation and natural water input in the upper reservoir, water quantity in the reservoir increases, resulting in smaller required power of the PV power plant, as that precipitation is also used for production of electric energy;

- With increased evaporation in the reservoir (as well as other water losses), the required power of the PV power plant increases, because more evaporation means less water quantity in the reservoir, *i.e.* the PV power plant has to pump more water;

- With reservoir volume increase, the required power of the PV power plant decreases, because larger reservoir can more easily balance out the summer surplus and winter shortage of solar energy;

- Increased total head for the same energy consumption (electric power/water) clearly causes the increase of the required power of the PV power plant.

The stated parameter variations show that a significant advantage of SHE system in relation to other RES systems lies in the fact that the combination of two RES creates conditions where frequently smaller solar radiation is compensated by precipitation and water inflow into the upper reservoir. To put it more simply, the **SHE system can operate in different climate areas using simultaneously solar energy and energy from precipitation water and adjacent watershed, *i.e.* their advantages at various locations including existing accumulations.**

Precisely this fact enables generalization of the obtained results and observing the SHE system as the key building unit of the future SEPS. Therefore, the selected features of storage and working volume and certain level of supply security of the building unit are:

- Upper storage volume $V = 20$ hm^3;
- Working storage $V_o = 2$ hm^3;
- Nominal electric power of PV power plant: $P_{el} = f_i^*(V_{(i)}) = 41$ MW;
- Reserve of storage $V_r = 18$ hm^3;
- Stored energy reserve $E_r = 10, 5$ GWh.

Based on optimization results, the following generalization of the SHE characteristics have been determined in order to simulate a possible scenario for the SEPS development:

(a) Required electric power of the PV generator for unit energy consumption (from Equation (1) and (3)):

$$1\,GWh/a \Rightarrow 0.73\,MW \quad PV\ generator \tag{9}$$

(b) Required PV generator area for unit energy consumption (from Equation (2) and (3)):

$$1\,GWh/a \Rightarrow 4545\,m^2 \quad PV\ generator \tag{10}$$

(c) Required upper reservoir volume for unit energy consumption (from Equation (1) and (3)):

$$1\,GWh/a \Rightarrow 0.36\,hm^3 \quad \mathrm{Re}servoir\ volume \tag{11}$$

(d) Required storage energy for unit energy consumption (from Equation (3) and (4)):

$$1\,GWh/a \Rightarrow 0.21\,GWh \quad Energy \quad storage \quad (\rho gVH) \tag{12}$$

(e) Required hydraulic power of the PSH for unit energy consumption (from Equation (3) and (6)):

$$1\,GWh/a \Rightarrow 0.24\,MW \quad PSH \tag{13}$$

The stated Equations (9 – 13) will be used in balancing EPS with RES+PSH power plants, *i.e.* in covering that part of the future energy consumption which would be produced by the new type of SHE power plant.

Reduction of CO_2 Emission

Presuming that all energy of 55.447.234 kWh, produced by the SHE system, is produced in coal-fired power plant, which, according to US Energy Information Administration [17] typically emits:

$$1\,GWh/a \Rightarrow 950\,t\,CO_2 \tag{14}$$

SHE would then reduce CO_2 emission by:

$$55.447\ \text{GWh/a} \cdot 950\ \text{t of } CO_2/\text{GWh} = 52,6\ \text{kt } CO_2/\text{a}.$$

The current analysis neglects the fact that the construction process for the first SHE will be driven by fossil fuels. However, considering even that fact, it can be seen that the new power plant can annually save relatively high quantity of CO_2 which it would emit, in comparison to *coal-fired power plant.*

DEVELOPMENT OF THE FUTURE SEPS

EREC Forecast and its Upgrading towards SEPS

The electric energy consumption according to AIP Scenario of EREC forecast is shown in Table **3** (EREC, 2006 [4]). As can be seen, the scenario foresees the use of: *Biomass (BIO) 4290 TWh, Large Hydro (LHE) 4165 TWh, Small Hydro (smHE) 2200 TWh, Wind (W) 8000 TWh, PV 9113 TWh, Solar Thermal (ST) 790 TWh, Geothermal (GT) 1020 TWh and Marine (M) 230 TWh.* 18% of the required energy is intended for covering from Non-Renewable Energy Sources, *i.e.* NRES. Therefore, for the year 2040 such EREC scenario can be expressed by the following relation:

$$TOTAL = 12\%BIO + 11\%LHE + 6\%smHE + 22\%W + 25\% \\ PV + 2\%ST + 3\%GT + 1\%M + 18\%NRES \quad \textbf{(15)}$$

or, more simply:

$$TOTAL = 82\%RES + 18\%NRES\,. \quad \textbf{(16)}$$

Table 3: AIP scenario, share of renewables in world total electricity consumption (EREC, 2006 [4]).

	Years	2001	2010	2020	2030	2040
	Total Consumption in TWh (IEA)	15570	19973	25818	30855	36346
1	Biomass (BIO)	180	390	1010	2180	4290
2	Large Hydro (LHE)	2590	3095	3590	3965	4165
3	Small Hydro (smHE)	110	220	570	1230	2200
4	Wind (W)	54, 5	512	3093	6307	8000
5	PV	2, 2	20	276	2570	9113

Table 3: cont....

6	Solar Thermal (ST)	1	5	40	195	790
7	Geothermal (GT)	50	134	318	625	1020
8	Marine (M)	0, 5	1	4	37	230
9	**Total RES**	**2988, 2**	**4377**	**8901**	**17109**	**29808**
10	***RES Contribution (%)***	***19***	***22***	***34***	***55***	***82***
11	**NRES**	**12581, 8**	**15596**	**16917**	**13746**	**6538**
12	***NRES Contribution (%)***	***81***	***78***	***66***	***45***	***18***

However, as has been said at the beginning of this chapter, regardless how ambitious the EREC forecast is, it is not easily feasible, because such share of RES without adequate energy storage means that they can't provide the necessary energy supply to consumers (for example, in case of a cloudy day without wind, which is not rare, half of the remaining RES that would be available after total energy consumption is not sufficient to cover the consumers' needs for electric energy). In order to solve the resulting problem, the authors of this paper **propose that the share foreseen only for the PV be replaced with SHE systems, *i.e.* PV+PSH power plants,** because one part of energy storage would be provided. However, as the 25% share of PV together with storage in the SHE system wouldn't be sufficient for reliable electric energy supply to consumers, step 2 proposes that **instead of 18% of NRES, PV+PSH systems are also used** as a long-term strategic goal towards which to strive.

The reason for this proposal is that of all available RES, solar energy is the most sustainable and the least harmful source for the Earth. Solar energy is the only sustainable RES, external and independent, permanent, regular on daily basis and global. All other RES are local energy sources on Earth and more or less depend on local conditions. Local use of these sources more or less affects local surroundings. Unlike solar energy, other RES can't be distributed everywhere or be evenly available on most part of the Earth surface, especially where people (consumers) live. Therefore, the reliability of energy production depends on local conditions and changes that occur continuously due to human activity, so these other RES are less reliable than solar energy. Due to all aforesaid, solar energy is given much greater importance in many energy forecasts (World Energy Council 2007, 2007 [18], The Club of Rome, 2009 [19, 20]).

Therefore, if NRES are replaced with SHE systems, the share of PV+PSH system, the total share of SHE systems would be:

$$25\%PV + 18\%NRES = 43\%SHE \quad \textbf{(17)}$$

Therefore, based on the stated proposals of upgrading the EREC forecast from 82% to 100% of RES share and by using SHE instead of PV and NRES, **Total Renewable Electric Power System (TREPS)** would be obtained by the year 2040 with the following share:

$$\begin{aligned} TREPS &= 12\%BIO + 11\%LHE + 6\%smHE + 22\%W + 2\%ST \\ &+3\%GT + 1\%M + 43\%SHE \end{aligned} \quad \textbf{(18)}$$

or more simply:

$$TREPS = 43\%SHE + 57\%ORES \quad \textbf{(19)}$$

where ORES are all RES outside the PV, *i.e.* SHE system.

This scenario and other corresponding values of the SHE system are presented in Table **4**.

Table 4: Characteristics of SHE system implementation in construction of SEPS until 2040.

		1	2	3	4	5	6	7
	SHE	**Electric energy consumption (TWh/a)**	**PV generator power (TW)**	**PV generator area (km^2)**	**Reservoir volume (km^3)**	**Reservoir energy (TWh/a)**	**PSH power (GW)**	**CO_2 saving (Gt/a)**
1	**PV → SHE**	9113	7	41419	3281	1914	2.2	9
2	**PV+NRES → SHE**	15651	11	71134	5634	3287	3.7	15

Column 1 contains energy consumption according to AIP scenario of EREC forecast [4]; Column 2 shows the required power of SHE system, calculated based on Equation (9); Column 3 shows total required area of PV generator, calculated by Equation (10); Column 4 shows reservoir volumes, calculated by Equation (11); Column 5 contains energy stored in reservoirs, calculated by Equation (12);

Column 6 contains hydraulic power of PSH, calculated by Equation (13); the last, seventh, column contains saving of CO_2 calculated by Equation (14).

The obtained values of the total required PV generator areas, calculated based on current efficiency values, presented in Table **4**, of 41.419 and 71.134 km^2, can be compared to areas of, for example, Switzerland (41.284 km^2) and Ireland (70.273 km^2). Naturally, the technology development of PV generators (efficiency, concentrating systems, *etc.*) will lead to decrease of their area. The total required reservoir volume of 3.281 and 5.634 km^3 can be compared to *e.g.,* the size of Lakes Huron, North America (3.540 km^3) and Michigan, North America (4.900 km^3).

As has been said, the key element of energy production reliability is hydro accumulation that can in this concrete case store energy of 1.914 and 3.287 TWh. These values would be about 5, 2% and 9% of total energy consumption. Therefore, SEPS as an integral system could have energy reserves for about 19 or 32 operating days without input of any energy from any RES in the system. However, as RES such as: BIO, LHE, smHE, GT and M, are basically relatively reliable systems (that can supply energy and power to consumers – their total is 11.905 TWh, with remaining fluctuating only 24, 441 TWh), this reserve is even bigger and gives the possibility of SEPS operating without sun and wind for 7.8% and 13.4% of interminated energy production, *i.e.* 28, 5 and 49 days (roughly from one to two months). The power of 2, 2 and 3, 7 TW PSH correspond to PV generator power of 7 and 11 TW, whereat can be seen that the PV generator power is about three times bigger than PSH power.

These values are expected to be different for different working volume storages. Larger volumes result in less power, but larger volumes also result in higher reserves of power supply. This was expected because the operating period of PV power plant is significantly shorter, as it depends on daily insolation.

If all energy produced by SHE units were to be replaced with energy produced from coal power plants, 9 to 15 Gt of CO_2 would be saved per year.

System Development

Obviously, the objective values of energy production from RES cannot be achieved abruptly until the year 2040. Therefore, it is logical that this development and increase of all RES will follow the *learning curve*. To illustrate, the authors of this paper have retained all growth trends of RES share as in the EREC forecast, applying the *learning curve* only to **energy production** X**(t) for** SHE systems, *i.e.* PV+NRES energy, expressed by the equation:

$$X(t) = a(1 - e^{-b(t-c)^d}) \quad \textbf{(20)}$$

where a, b, c and d are constants derived from the curve. Development by learning curve is usual for new technologies, therefore also for the SHE.

The new forecast of achieving SEPS with drawn learning curve for SHE systems can be seen in Fig. **4**. According to this scenario, the development of SHE system would start in 2010 and the constructed system would be very reliable, because 43% of total energy production would be from PV+PSH or from SHE units with energy storage.

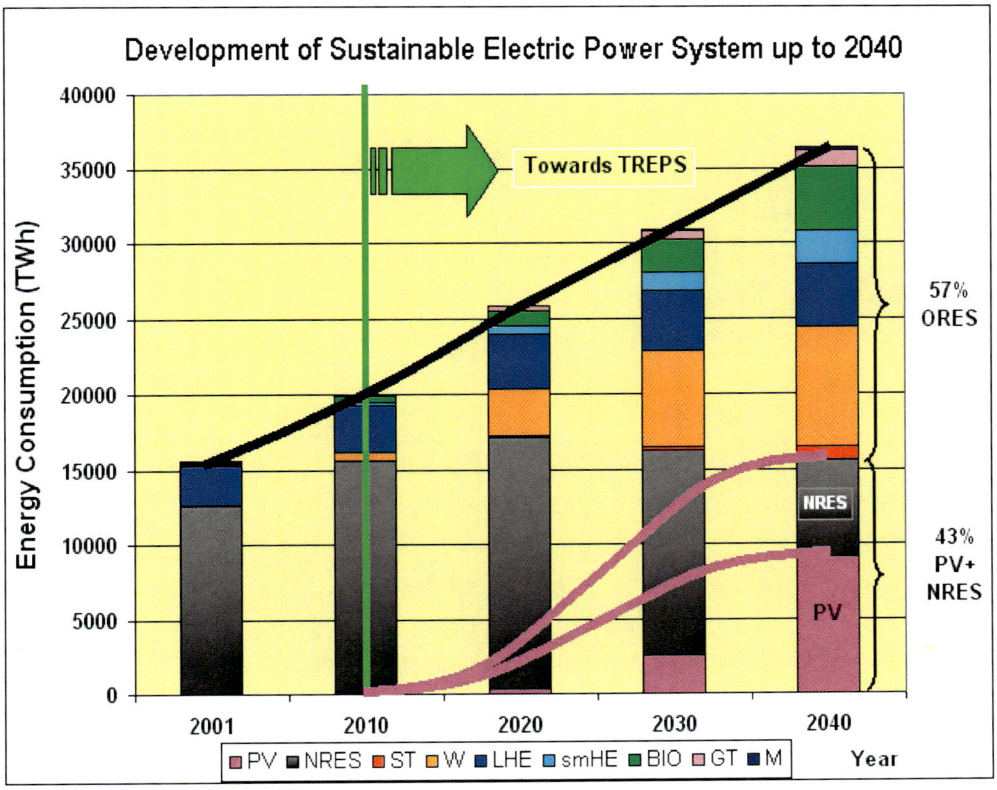

Figure 4: Development of sustainable electric power system up to 2040.

Necessary Policy Measures to achieve SEPS and Economical Aspect

The long-term goal is a sustainable energy system (SEPS), developed by systematic introduction of SHE, by improving the use of hydro energy, minimizing energy losses within the energy system, all for the purpose of maximizing the reduction of CO_2 emission. The SEPS system should be developed systematically, in accordance with ecological and economic criteria. The same can be achieved by integrated planning, using tools of system engineering.

This approach includes a thorough approach to the problem, taking into account all characteristics and constraints of the problem. An important step of this procedure is the analysis of all available locations suitable for construction of SHE and similar plants, and their ranking based on sustainability criteria (economic, ecological, social). Project values and realization should be determined upon detailed analysis of the best ranked.

To make this scenario become a reality, advanced policy measures have to be adopted. The starting steps have already been made by EU requirements (EREC 2006 [4]), which guarantee the deployment of renewable energy technologies.

Naturally, the policy will be realized only after the appropriate legal framework has been established. The targets should be complemented by financial incentives. This too would be an effective policy to address security of supply, technology development, employment and climate objectives.

Of course, there are still skeptics among the experts, especially among decision-makers who are not aware of the many results that can be derived from renewables and especially related to supply security. In this paper we wanted to point out that the RES+PSH combination can achieve a sustainable energy system and thus solve significant problems of atmosphere pollution and climate changes.

In the current development policy, the developing countries' governments put little emphasis on RES, although application of RES technology is especially suitable in areas with undeveloped reliable energetic network. The proposed SHE can be realized close to users and in general it does not require the existence of national energy network to be a reliable source of energy without putting pressure on the environmental equilibrium.

Those opposed to RES will always say that introduction of RES is more expensive than using conventional energy sources, which is basically true. The current prices show that, for example, energy from PV power plants is 5-6 times more expensive than that from conventional power plants. However, this price is in the first place conditioned by the fact that development of RES power plants is still in infant stages and that strong technological improvement will take place over the time, and the price will continue to decrease. On the other hand, the price can already be compensated to some extent by avoiding expenses for fuel required for operation of conventional power plants. Precisely this is the reason for building relatively big PV and wind power plants, Lenardic, 2009 [21].

However, in an extremely rough approximation it is possible to estimate the necessary investment in SEPS, based on comparison with investment in EPS which, for example Croatia has foreseen in the following years and which would practically upgrade the existing EPS based on the so-called *Update/upgrade of the energy strategy and of the implementation programme of the Republic of Croatia (Update/Upgrade), 2009* [22]. Namely, this investment has been estimated to about 1.5% of Croatian GDP until 2030, whereat necessary capacities would be built, providing sufficient energy for electric power independence of Croatia. Presuming that current investment in RES, *i.e.* PV power plants, exceeds investment in conventional power plants by 5-6 times, based on simple analogy, it could be deducted that by allocation of about 2 – 3% of GDP Croatia could achieve complete supply from RES by 2040. This is a very rough (orientation estimate) because it disregards the very big cost reduction expected in that long period (especially in case of such big quantities), as well as significant technological improvements. Also, it doesn't take into account fuel costs, which is a significant item, external costs (such as reduction of CO_2 emission, *etc.*), reduction of electric energy transmission costs, because extensive construction of RES is planned (in direct vicinity of energy consumption), security of supply which is important for the economy and all other elements which exceed the framework of this paper. By this comparison of investment in RES with conventional systems the authors wanted to show that investment in RES is within the limits which are realistically feasible (tolerable).

However, there is the question of benefits of use of RES, *i.e.* ecological values – damages, environmental costs. This has never been analysed in detail, so the real price of further intensive exploitation of conventional energy sources is basically unknown. Furthermore, there is the question of the extent of threat to mankind if the current trend continues. These are the key issues that determine the acceptability of the proposed strategy which undoubtedly reduces negative impacts on the environment.

Regarding the higher price of energy from RES, a thesis can be set that the issue of sustainability is the key issue for the survival of mankind. The main contribution of implementation of the proposed strategy lies in the increase of life sustainability, and most concretely in reduction of greenhouse gases. The benefit of CO_2 reduction should by all means be investigated and financially quantified in order to recognize the real value of change in the current trend of fulfilling energy requirements. With reduction of the existing direct or indirect subsidies to conventional energy, renewable energy sources will be even more competitive.

A special focus needs to be set on financial institutions so that financial resources could be mobilized to help developing countries carry out their obligations in the field of sustainable development. A significant part of financial institutions' resources should go to funding of RES projects for climate change purposes as is SHE.

Naturally, insisting on research and development in the renewables and energy efficiency is the most important issue, and the authors of this chapter hope that the proposed vision is at least a small step forward.

CONCLUSIONS

This chapter proposes the vision of the development of a **new energy policy** of full or prevailing supply of electric power systems from RES which should, in view of current environment and energy production trends, be considered as one of the possible ***green alternatives***, *i.e.* a solution for sustainable energy supply based solely on renewable SEPS. This paper and the presented opinions are based on the AIP scenario of EREC forecast which foresees covering of 82% of total consumption with energy from RES until 2040 [4], which has theoretically been extended to 100% of RES share.

As can be seen, the realization of the proposed EREC vision is real even for the current state of technology development, especially if we take into consideration that most progress is expected in the use of solar energy that has been intensively developing in recent years. A more efficient use of solar energy is expected, because the development of this technology has not reached the end, which is not the case for some others technologies (wind). Increase of production efficiency leads to reduction of energy storage capacity, *i.e.* water storage volume according to the proposed solution of SHE. This means that as the implementation of the vision becomes more realistic it will take less area of land and water (volume in storage).

While it is understandable that in the proposed strategy of SEPS development, 25% of energy consumption foreseen from the PV generator will be covered by SHE power plants (which are already combined PV and PSH power plants), the question remains whether 18% of energy, foreseen according to EREC for non-renewable energy sources (NRES), could somehow be covered from RES until 2040. The authors of this paper deem that in order to achieve SEPS, NRES, which will sooner or later be replaced with SHE systems, will most likely be based on PV generators of newer generations, as the biggest progress is to be expected in that technology. Due to this, total share of PV generators, combined with PSH systems into SHE systems, is foreseen in this paper with application potential of 43%. Therefore, there are exceptionally good conditions for implementation of PV systems in big quantities, but it is very important to combine these systems with PSH systems in order to secure electric energy storage in the future SEPS.

Another advantage is that SEPS could be planned systematically and extensively, *i.e.* new SHE units could be constructed extensively, as close as possible to the consumers, which would avoid high costs of transportation and distribution of energy. It would significantly reduce energy losses which, according to IEC (2010), could be between 8 % and 15%, [23]. Naturally, such sustainable system would have to be optimized, which would lead to full energy independence of certain locations and settlements.

The disadvantages of SEPS with big participation of PV generator lie in the first place in their price which is today 5 to 6 times higher than the price of conventional power plants of equal capacity. Another problem is that a relatively small number of PV power plants have been installed in the world to the day, because production of PV array and putting the power plants into operation requires a relatively long time, Lenardic, 2009 [21]. Another problem of SHE is embodied energy necessary for their production (particularly PV generator), but the problem could also be relatively big areas required for both PV power plant and hydroaccumulation.

Nevertheless, the following arguments can be opposed to all these disadvantages: timely planning by implementation of RES can provide full sustainability of most Electric Power Systems in the world, until SEPS is achieved; investment costs could be within the range of 2 to 3% of GDP in the first years; intensifying the production of PV generators will result in decrease of their price, while efficiency increase of PV power plants and concentrating systems will lead to significant decrease of their area (naturally, this also applies to technological improvement of ST and W power plants); it is to be expected that new technological breakthrough will take place in that field; embodied energy is relatively small in comparison to the energy which PV power plants can produce during their lifetime; PV power plants should be constructed on locations which are otherwise not in use (avoiding arable areas and similar), which is frequently the case with "sloping sunny ground", while multi functionality of hydro accumulation should somehow compensate for taking relatively large areas; and the strongest argument related to cost effectiveness of the SHE system – that today sustainability is an absolute priority in the development and solar energy is free and inexhaustible. The expected rise in prices of conventional fuels should not be disregarded.

Therefore, over a long term it would be justified to contemplate on the possibility of development of SEPS, *i.e.* on mostly or fully covering the needs for electric energy from RES, where SHE units of various types would have an important place.

The authors deem that, apart from the measures of energy efficiency that should constantly be implemented, development of SEPS is an important step, because globally important and desirable results could be relatively quickly obtained (slowing of environment pollution).

REFERENCES

[1] IPCC, 2007. *Intergovernmental Panel on Climate Change: Fourth Assessment Report*, Synthesis report, 52p. Available from: http://www.ipcc.ch.

[2] P. Waldhams, M.J. Doble, *Sea ice thickness measurement using episodic infragravity waves from distant storms,* Cold Regions Science and Technology, Elsevier Science Inc., 56 (2009) 98-101.

[3] EREC, European Renewable Energy Council. *Renewable Energy Target for Europe 20% by 2020*; 2004. Available from: http://www.erec-renewables.org/fileadmin/erec_docs/Documents/Publications/EREC_Targets_2020_def.pdf.

[4] EREC, European Renewable Energy Council. *Renewable Energy Scenario to 2040, Half of the global energy supply from renewables in 2040,* 2006. Available from: http://www.censolar.es/erec2040.pdf.

[5] G. Strbac, A. Shakoor, M. Black, D. Pudjianto, T. Boop, Impact of wind generation on the operation and development of the UK electricity systems, *Electric Power System research* 2007; 77(9): 1214-1227.

[6] H. Chen, T.N. Cong, W. Yang, C. Tan, Y. Li, Y. Ding, Progress in electrical energy storage system: A critical review. *Progress in Natural Science* 2009; 19(3): 291-312.

[7] ESA - Electricity Storage Association, Energy storage technologies. *Technologies & Applications – Technology Comparisons,* 2009. Available from: http://www.electricitystorage.org/tech/technologies_comparisons.htm.

[8] J.K. Kaldellis, D. Zafirakis, K. Kavadias, Techno-economic comparison of energy storage systems for island autonomous electrical networks, *Renewable & Sustainable Energy Reviews* 2009; 13(2): 378-392.

[9] Z. Glasnovic, J. Margeta, Optimal Sizing of Photovoltaic-hydro Power Plant, *Progress in Photovoltaics: Research and Applications*, 2009; 17:542–553.

[10] C. Bueno, J.A. Carta, Wind powered hydro storage systems, a means of increasing the penetration of renewable energy in the Canary Islands. *Renewable & Sustainable Energy Reviews,* 2006; 10(4): 312-340.

[11] D. Mucic, *Consumption of electric energy on the Island of Vis in the year 2007,* Croatian Electricity Company, Elektrodalmacija Split, 2008.

[12] G. Hrabak-Tumpa, *Climate Data for Vis, Croatia,* Meteorological and Hydrological Service of Croatia, 2007.

[13] J. Kenna, B. Gillett, *Solar Water Pumping - Handbook*, Intermediate Technology Publications: London, UK; 1985.

[14] R. Ginocchio, *L'Energie hydraulique,* Paris, 1978.

[15] P. Stojic, *Hydroenergy,* Faculty of Civil Engineering, University of Split, Croatia; 1995, (in Croatian).

[16] J. Margeta, Z. Glasnovic, *Feasibility of the green energy production by hybrid solar + hydro power system in Europe and similar climate areas*. Renewable and Sustainable Energy Reviews, doi:10.1016/j.rser.2010.01.019, 2010.

[17] US Energy Information Administration; 2006. Available from: http://www.eia.doe.gov/cneaf/electricity/page/co2_report/co2report.html#electric.

[18] World Energy Council, 2007. *2007 Survey of Energy Resources,* Available from: http://www.worldenergy.org/documents/ser2007_final_online_version_1.pdf.

[19] The Club of Roma. *Clean power from desert*, The DESERTEC Concept for Energy, water and Climate Security, WhiteBook, 4th Edition, 2009. Available from: http://www.desertec.org/fileadmin/downloads/DESERTEC-WhiteBook_en_small.pdf.

[20] The Club of Roma. *Red Paper*, An Overview of the DESERTEC Concept, 2009, Available from: http://www.desertec.org/fileadmin/downloads/DESERTEC_RedPaper_2nd_en.pdf.

[21] D. Lenardic, *Large-scale photovoltaic power plants,* Annual Review 2008, Available from: http://www.pvresources.com/download/AnnualReview_FreeEdition.pdf.

[22] *Update/upgrade of the energy strategy and of the implementation programme of the Republic of Croatia*. Ministry of Economy, Labor and Entrepreneurship and United Nations Development Programme, 2009, Available from: http://www.energetska-strategija.hr/doc/GreenPaperDraft.pdf.

[23] IEC, International Electrotechnical Commission, 2007. *Efficient Electrical Energy Transmission and distribution*, 3, rue de Varembé, P.O. Box 131, CH-1211 Geneva 20, Switzerland, Available from: http://www.iec.ch/news_centre/onlinepubs/pdf/transmission.pdf.

CHAPTER 7

Homogeneous Charge Compression Ignition (HCCI) Combustion

N.P. Komninos*

School of Mechanical Engineering, National Technical University of Athens (NTUA), Greece

Abstract: Homogeneous charge compression ignition (HCCI) combustion is a distinct combustion concept, which can be implemented in internal combustion engines. Its development began thirty years ago and is still the focus of many researchers worldwide. The main features which attract attention to HCCI engines are of both environmental and energy-saving character. Due to the premixed nature of HCCI combustion and the relatively lean mixtures used, NO_x and soot emissions are but a fraction of the ones emitted by conventional spark ignition (SI) or compression ignition (CI) engines. Moreover, the relatively rapid combustion process and the unthrottled operation provide the potential for high thermal efficiency. Apart from these favorable attributes of HCCI combustion, significant issues have to be resolved. These issues are related to the high unburned hydrocarbons and carbon monoxide emissions, which are emitted during HCCI operation. Moreover, technical issues have arisen regarding the implementation of the HCCI combustion concept to actual engines. The latter is related to difficulties in controlling the ignition timing and the combustion rate over a wide load-engine speed range. The ignition timing must be adequately controlled if the thermal efficiency is to be kept high; the combustion rate control is of importance, since the high combustion rates encountered in HCCI combustion increase the peak combustion pressures and the pressure rise rates, thereby limiting the maximum attainable load. The present chapter presents the main features of HCCI combustion, namely its characterization based on experimental data, the pollutant emissions formation processes, the effect of major operating parameters on HCCI combustion and the various strategies used for the realization of the HCCI combustion concept to gasoline or diesel HCCI engines.

Keywords: HCCI combustion, combustion characterization, hydrocarbons, CO, emissions, HCCI implementation, gasoline HCCI, external EGR, exhaust gas retention, exhaust gas rebreathing, diesel HCCI, port injection, early injection, multiple injection, auto-ignition.

INTRODUCTION

Internal combustion engines have been extensively developed since their first

***Address correspondence to N.P. Komninos:** Internal Combustion Engines Lab., Thermal Engineering Section, School of Mechanical Engineering, National Technical Univ. of Athens; Tel: +30 210 7721710; E-mail: nkom@central.ntua.gr

Hatim Machrafi (Ed)

appearance. Diesel and spark ignition engine systems have been improved to provide for better fuel economy and lower pollutant emissions [1]. The emission standards become more and more stringent, placing demands on the research and design related to the internal combustion engine processes. Moreover, the dependence on fossil fuels has rendered minimum fuel consumption as a main design criterion, which becomes even more significant than the power generated by the internal combustion engine: it is not how much power an engine can deliver that is the main concern, but how much fuel is needed to deliver it and what the environmental consequences are for this power production. In view of the stringent emissions standards and the low availability of fossil fuels, various alternatives have been and are being examined, *e.g.,* using different fuels which are more environmental friendly, such as biofuels [2-7] and hydrogen [8, 9]. Moreover, systematic advances have been achieved in the spark ignition and compression ignition engines which have improved the fuel conversion efficiency and have reduced the emitted pollutants.

Apart from these combustion modes, a relatively new combustion concept has been developed in the last 15 years, receiving intense and worldwide attention. This concept is the *Homogeneous Charge Compression Ignition (HCCI)* combustion concept. The main benefits, which justify the enormous attention that this concept has received, are simultaneous reduction in both smoke and NO_x emissions and a potential for high fuel conversion efficiency relative to the conventional combustion modes. It is well known that the simultaneous reduction in NO_x and Soot emissions emitted from Diesel engines has been a major challenge. On the other hand, spark ignition engine operation at part load, is inefficient due to high pumping losses induced by the throttled operation. However, HCCI operation is not without drawbacks: the most significant problems associated with HCCI combustion are an inability to directly control the ignition and combustion processes and also the high production of CO and HC emissions relative to the conventional combustion modes. Both the beneficial and the detrimental effects of HCCI combustion can be understood, considering the details of this combustion concept.

In HCCI combustion a homogeneous mixture of air and fuel is compressed and auto-ignites producing work during expansion. The preparation of the mixture, *i.e.* the fuel-air mixing is usually achieved in either of two ways: (a) introducing the fuel into the incoming air stream at the inlet manifold or (b) injecting the fuel directly into the combustion chamber early during the compression stroke. In the latter case a higher degree of inhomogeneity is achieved due to the limited time available for the air-fuel mixing and the possibility of impingement of the fuel jet on the combustion chamber wall. Regardless of the specific method used for the preparation of the charge, a sufficiently homogeneous mixture is compressed within the combustion chamber and auto-ignites when the in-cylinder temperature exceeds the auto-ignition temperature of the mixture. These stages are shown in Fig. **1**. The attainment of the desired output is achieved by varying the relative amount of fuel introduced into the combustion chamber, since HCCI engines operate mostly on wide open throttle mode. According to the aforementioned, HCCI engines share common features with both types of conventional engines: they are similar to spark ignition (SI) engines, since they require the preparation of a homogeneous mixture, but rely on the auto-ignition of the fuel as is the case in compression ignition (CI) engines.

Apart from these similarities, HCCI combustion demonstrates quite distinct characteristics. The absence of any external means for the adjustment of ignition timing or for the combustion rate control are key elements of HCCI combustion and determine, to a great extent, the essential features of the HCCI engine operation: since HCCI relies upon auto-ignition, the combustion rate is usually higher and the combustion duration shorter relative to conventional types of combustion. This may be thermodynamically favorable, since HCCI combustion resembles constant volume combustion; however, high combustion rates induce high pressure rise rates and increase the peak combustion pressure. The former leads to abnormal combustion and noise, while the latter is detrimental to the engine durability. Both of these factors impose a limit to the maximum amount of fuel that can be utilized in HCCI engines, thereby restraining the maximum attainable load under HCCI operation. Moreover, the ignition timing, which significantly affects the fuel conversion efficiency, depends upon the auto-ignition chemistry of the specific fuel used, and is sensitive to operating conditions such as the initial temperature of the mixture, the engine speed and load. Thus, it becomes obvious that HCCI combustion is limited in a relatively narrow load-engine speed region, the expansion of which is one of the driving forces that motivates the current research.

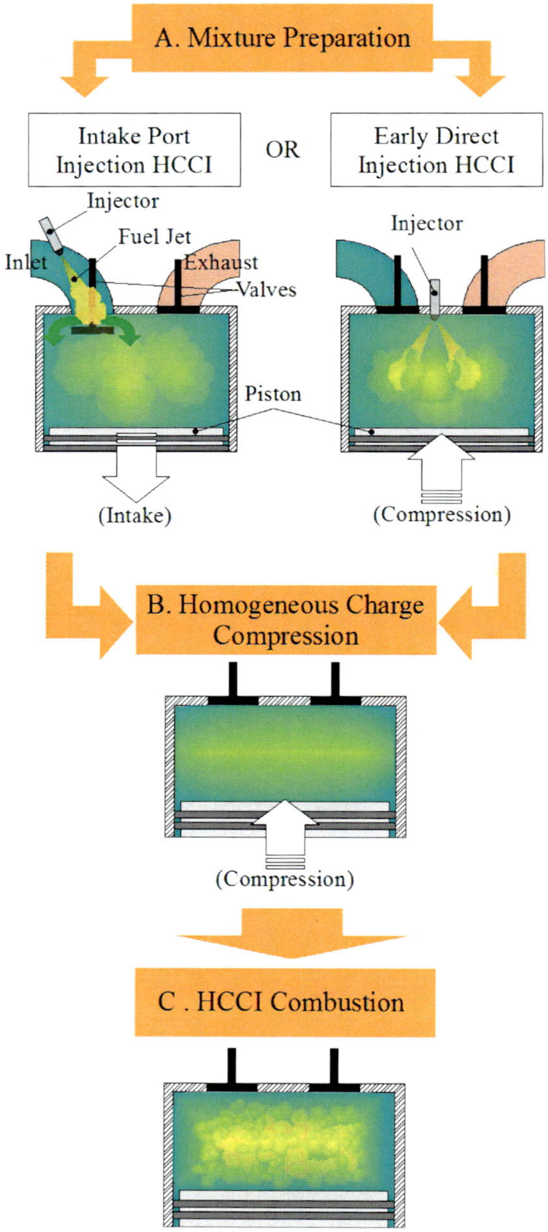

Figure 1: Successive stages of HCCI combustion.

Apart from the problems in the implementation of HCCI combustion, this combustion mode has become popular due to its potential to simultaneously reduce NO_x and Soot emissions, a feature that is quite attractive in view of the stringent future emissions standards. The reduction in NO_x emissions is due to the lean mixtures used and the premixed nature of combustion, which reduce the peak combustion temperatures developed within the combustion chamber. Furthermore, the uniformity of the fuel-air mixture and the absence of fuel rich regions within the combustion chamber results to a significant reduction of Soot and particulate matter (PM) emissions. The aforementioned potential for high thermal efficiency, due to the rapid heat release located near the top dead center (TDC), and the unthrottled operation, which reduces the pumping work of the inlet and exhaust strokes, are additional attractive attributes of HCCI combustion which motivate the ongoing HCCI research.

ORIGINS OF THE HCCI COMBUSTION CONCEPT

The origins of HCCI combustion date back in 1979 with the research work of Onishi *et al.* [10]. These researchers studied the behavior of small two-stroke gasoline engines during part load operation. The main problem was the high cyclic variability observed, which was due to combustion instability. While experimenting with methods to overcome this problem, the researches discovered a new combustion mode, which relied upon the auto-ignition of the fuel. The name used for this new combustion mode was Active-Thermo-Atmosphere Combustion (ATAC). The research group, studied photographs taken during the combustion event and proceeded to an analysis of the pressured traces. The results of this study showed that this new combustion mode did not rely upon flame propagation, but on the gradual, simultaneous consumption of fuel at a wide region throughout the combustion chamber. With this new combustion concept, the cyclic variability was reduced; the combustion event was more stable, resulting to a decrease in fuel consumption and unburned hydrocarbon emissions. The prerequisites to achieve ATAC were the high percentage of residual gases and high initial charge temperature.

Nogushi *et al.* [11] also presented the same combustion mode, *i.e.* relying on the auto-ignition of the mixture, in a two stroke engine. They called this combustion mode "TS (Toyota-Soken) Combustion". These researchers confirmed the results obtained from Onishi *et al.* and also identified some of the chemical species that take part in the combustion event. They concluded that during this combustion mode operation species such as OH, HO_2 and O are found during the combustion event.

In 1983 the homogeneous charge compression ignition concept was applied to a four stroke engine by Najt and Foster [12]. The researchers used mixtures of isooctane and n-heptane to study the ignition timing and the heat released during combustion. They concluded that ignition is governed by the low temperature (<950K) hydrocarbon oxidation reactions, while the main combustion event is governed by the high temperature (>1000K) hydrocarbon oxidation reactions. They also examined the effect of the initial temperature and compression ratio variation on this combustion mode, which they named Compression Ignited Homogeneous Charge (CIHC) combustion.

Thring [13] studied the conditions under which HCCI combustion can be achieved in four stroke engines. A schematic of the experimental set up is shown in Fig. **2**. This set up allowed for the variation of inlet temperature, air-fuel ratio and exhaust gas recirculation (EGR) amount, all of which affected the ignition timing and the combustion rate. Thring used port fuel injection for the mixture formation, combined with high levels of EGR and high intake temperature to achieve in-cylinder conditions similar to the ones encountered in two stroke engines.

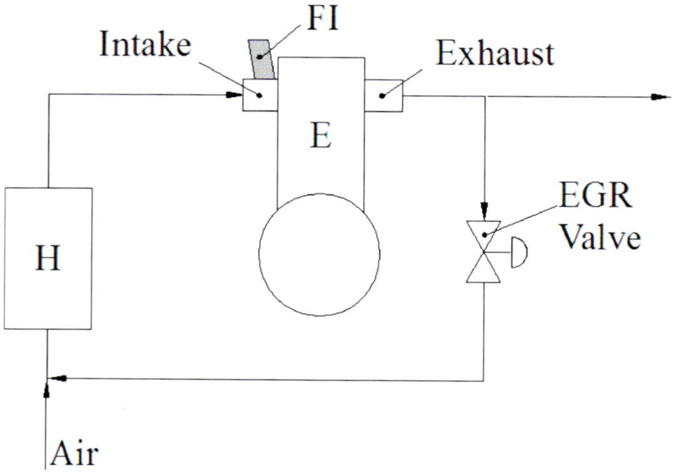

Figure 2: Representation of Thring's experimental set up for the implementation of HCCI combustion to four-stroke engines. FI: electronic fuel injector, H: heater, E: engine.

Three distinct regions of abnormal HCCI combustion were found: the power limited region, which corresponds to fuel lean mixtures that are unable to produce enough power to overcome the engine friction; the misfire region, which was encountered at high EGR rates and/or fuel rich conditions and the knocking region, which corresponds to low EGR and/or close to stoichiometric mixtures. In the knocking regime the pressure rise rate was very high. Thring also studied the effect of engine speed on the HCCI combustion and found that increasing engine speed retards ignition, by decreasing the time available for pre-ignition reactions to occur.

From the aforementioned and other studies [14], it had become obvious that HCCI combustion:

- improved the cyclic variability of two stroke engines at part load,
- can be operated unthrottled, thereby reducing the pumping losses,
- has the potential of high thermal efficiency due to the short combustion duration,
- reduces NO_x emissions dramatically (by 90-98%), due to the fuel lean mixtures used,
- produces negligible particulate emissions.

However, despite these favorable attributes, several challenges arose such as:

- the ignition control over a wide range of operating conditions,
- abnormal combustion at high load,
- satisfactory operation only at a limited load and engine speed range,
- high levels of unburned hydrocarbon and carbon monoxide emissions.

These characteristics are the result of the specific physicochemical processes that occur during HCCI combustion. In the following sections these physicochemical features of the HCCI combustion mode are described in detail.

HCCI COMBUSTION CHARACTERIZATION

As already mentioned in the Introduction, there are two main ways of preparing a homogeneous mixture within the combustion chamber: (a) injecting the fuel in the inlet system prior to or at the inlet valve port during the induction stroke or (b) injecting the fuel directly within the combustion chamber early during the compression stroke. In this latter case it is of paramount importance that the fuel jet wall impingement be avoided. Compared to the port fuel injection HCCI, direct fuel injection induces a certain degree of inhomogeneity to the mixture, the extent of which depends upon the time available for the fuel-air mixing to occur.

Regardless of the degree of homogeneity, the resulting mixture of fuel, air, residual gases and sometimes recycled exhaust gases is compressed and auto-ignites at regions were the thermodynamic conditions – primarily temperature and mixture composition – are most favorable. Therefore, ignition relies merely upon the auto-ignition of the fuel; it is not stimulated by any external ignition means such as a spark plug or an injector. It is the mixture preparation and the fact that ignition relies upon autoignition that determine the essential features, which differentiate HCCI combustion from combustion *via* flame propagation (SI combustion) or *via* diffusion flames (CI combustion). These features manifest themselves in the pressure traces and heat release rates obtained from HCCI engines and have been observed in optically accessible HCCI engines using various experimental techniques. The pressure traces and heat release rate analysis provide information on the temporal evolution of the chemical energy release within the combustion chamber; the observations obtained from engines with optical access provide information for both the temporal and local evolution of combustion at various regions within the combustion chamber. The following sections summarize some of the most essential findings on HCCI combustion, obtained using the aforementioned methods.

Pressure Trace and Heat Release Rate Analysis of HCCI Combustion

Pressure traces and heat release rates (HRR) extracted from HCCI engines provide significant information on the ignition timing, the chemical energy release and the combustion rate and duration. A typical HCCI pressure trace and heat release rate for diesel-like fuels is shown in Fig. **3**. Compared to the pressure trace and heat release rate from SI or CI engines, HCCI combustion presents a peculiar two stage ignition, *i.e.* heat is released in two consecutive stages. These stages correspond to the two peaks of the HRR diagram and divide it to three distinct regions. The first region corresponds to the first observable heat release and takes place at relatively low temperature; hence the term low temperature heat release (LTHR), which is quite often used to describe this stage. The heat released during this period equals the shaded area under the LTHR peak in Fig. **3** and constitutes a fraction of the total heat released during the entire combustion process. The chemical reactions, which transpire during this stage, are connected to the mechanism leading to autoignition and abnormal combustion (knock) in SI engines.

As compression evolves, the LTHR diminishes and the combustion rate decreases, despite the increase of gas pressure and temperature. This designates the transition towards the second stage of HCCI combustion, which is characterized by low energy release and extends up to the initiation of the second observable heat release. Due to this apparent negative effect of temperature on the heat release rate, this region is referred to as the negative temperature coefficient region (NTC). After the NTC stage, the heat release rate increases again and leads to the main combustion event, during which the greatest portion of the fuel chemical energy is released. Since this stage occurs at a higher temperature relative to the first heat release, it is denominated the high temperature heat release (HTHR) region.

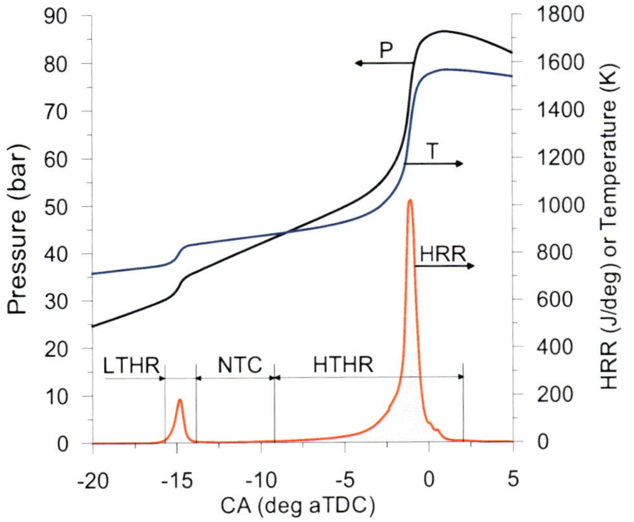

Figure 3: Typical heat release rate encountered in HCCI engines with diesel-like fuels.

The two-stage ignition pattern shown in Fig. **3** does not seem to be affected by the specific mixture preparation process, since it has been observed in HCCI engines with port fuel injection [15], as well as in direct injection HCCI engines [16]. However, the fuel used significantly affects the ignition timing and the shape of the heat release rate: for diesel-like fuels, such as n-heptane, mixtures of n-heptane-isooctane or other normal parafins, the low temperature heat release is quite pronounced [14, 16-21]. Fuels such as isooctane, ethanol and methane do not exhibit two-stage heat release [22, 23]. The ignition timing (initiation of the LTHR), and the amount of heat released during the LTHR reactions vary significantly with the mixture composition and depend, to a certain extent, upon the octane rating of the fuel [21]. The amount of LTHR increases the in-cylinder temperature and affects the initiation of main combustion. These characteristics play an important role in the behavior of HCCI combustion under varying load and/or engine speed conditions; fuels that exhibit strong LTHR are characterized as low-torque, high-speed, knock-prone fuels. These fuels usually have a low octane rating. On the contrary, fuels with high octane rating usually present low LTHR and they are referred to as knock-resistant, low-speed and high-load fuels [21, 24, 25]. Considering the aforementioned, it is of no surprise that the use of multi-fuel mixtures is one of the strategies for HCCI ignition and combustion control.

Regardless of the fuel used or the shape of the HRR, a general characteristic of HCCI combustion is its high combustion rate and short combustion duration relative to conventional SI or CI combustion [14]. Typical values for the combustion duration under normal HCCI operation are 10-20° crank angle (CA). Shibata *et al.* found that the combustion duration in terms of time is not affected significantly by variations in engine speed [21]. This implies that as engine speed increases, the combustion duration increases, in terms of crank angle degrees. Indeed, the upper limit to the maximum attainable engine speed under HCCI operation is imposed by the time available for the chemical reactions to occur.

The short combustion duration encountered in normal HCCI operation, is favorable at low load operation. However, at higher loads, *i.e.* at higher fueling rates, the peak pressure rises appreciably and the pressure rise rate becomes unacceptably high, leading to abnormal combustion and knock. It must be emphasized that the term *abnormal combustion* is used differently in HCCI combustion than in SI or CI combustion. In the latter two combustion modes, abnormal combustion (knock) is related to a shift in the nature of the combustion process: in CI combustion the normal combustion mode is primarily one of diffusion flame; premixed combustion also occurs early during normal CI combustion but does not dominate it, as is the case during knocking CI combustion. Normal SI combustion evolves *via* flame propagation; knock occurs when a part of the mixture auto-ignites prior to the flame arrival. On the contrary, in both normal and abnormal HCCI combustion the underlying combustion process is the same, *i.e.* the autoignition and the subsequent combustion of the mixture. The element that differentiates normal from abnormal HCCI combustion is not a shift in the combustion process, but rather a change in the rate and intensity of the auto-ignition and combustion processes. Knocking HCCI combustion restrains the highest attainable load, the increase of which is the goal of many researchers in the field. The lowest attainable load in HCCI operation is determined by the ability of the mixture to auto-ignite: too low fueling rates decrease the reaction rates, resulting in partial combustion and eventually misfire.

The general characteristics of the pressure traces and heat release rates obtained from HCCI engines provide some insight into the underlying mechanism of HCCI combustion. In particular, the dependence of the shape of the heat release on the fuel properties and the negative temperature coefficient observed for fuels exhibiting two-stage ignition, indicate a strong dependence of HCCI combustion on the details of the fuel oxidation processes. The two-stage combustion cannot be described by a simple one-step chemical reaction scheme, since this would always lead to an increasing combustion rate with increasing temperature. Another indirect indication of the dependence of HCCI combustion on the chemistry of the fuel oxidation process is the fact that the combustion duration in milliseconds is little affected by the engine speed. This is not observed in SI engines, in which the combustion rate is essentially determined by the flame propagation across the combustion chamber. In this case, an increase in engine speed enhances the in-cylinder turbulence and increases flame speed almost in proportion to the engine speed increase. Thus, in SI engines, the combustion duration, expressed in crank angle degrees, is little affected by the engine speed [26]. The aforementioned are merely indications of the fact that HCCI combustion is governed by chemical kinetics. Such conclusions cannot be drawn solely by examining the pressure traces and heat release rates, but by observing the prevailing physicochemical processes within the combustion chamber. This is achieved with the use of sophisticated optical diagnostic techniques, which reveal the phenomenological characteristics of the HCCI combustion process.

Phenomenological Description of HCCI Combustion and Emissions Formation

The information extracted from pressure traces and heat release rates although significant, do not provide insight into the spatial evolution of combustion within the combustion chamber or into the predominant chemical species encountered in HCCI combustion. Such information, which is necessary for the characterization of the HCCI combustion process and the determination of the prevailing physicochemical processes, can be provided by the utilization of optical diagnostics applied to HCCI engines. The relevant studies revealed the essential differences between HCCI combustion and the conventional combustion modes and contributed to the interpretation of the peculiar two stage ignition pattern of the heat release rate of diesel-like fuels.

Comparison of HCCI Combustion to Conventional Combustion modes

Photographic studies conducted in SI, CI and HCCI engines [27, 28] reveal the main characteristics, which distinguish HCCI combustion from the other combustion modes.

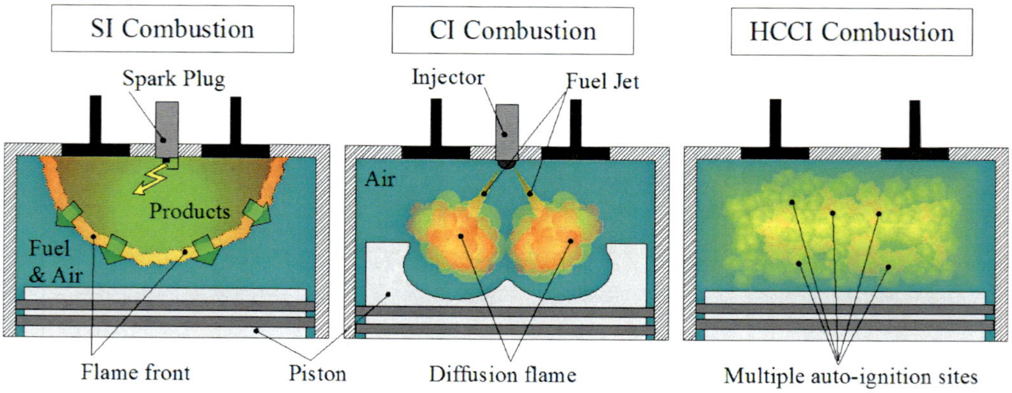

Figure 4: Comparison of spark ignition (SI), compression ignition (CI) and homogeneous charge compression ignition (HCCI) combustion.

The three combustion modes and the special phenomenological characteristics of each are shown schematically in Fig. **4**.

Spark ignition combustion relies upon flame propagation. This is observed in spark ignition engines with optical access, in which the evolution of the combustion event can be monitored [27]. In such studies, the flame front initiates at the vicinity of the spark plug, in which an electric discharge induces chemical reactions within the unburned mixture. A roughly spherical, thin, luminous layer, *i.e.* a flame, develops and progressively increases in size as it moves away from the spark plug to consume the remaining unburned mixture. Although this flame front is microscopically highly irregular due to the turbulent in-cylinder conditions, a distinct luminous flame front is usually observed, moving away from the spark plug toward the unburned regions of the combustion chamber. This flame travel is depicted schematically at the left frame of Fig. **4**.

In similar studies with diesel engines equipped with optical access [28], the essential features of diffusion flame combustion are evident: the fuel is injected within the combustion chamber in liquid form; the fuel jet breaks up in droplets as it penetrates the air within the combustion chamber; the droplets vaporize gradually; the vaporized fuel is mixed with air and auto-ignition occurs in the regions where the local conditions – primarily local temperature and fuel-air ratio – are favorable. Since this complicated procedure requires sufficient time and space to evolve, the auto-ignition reactions are observed in regions surrounding the fuel jet. The flame structure is highly irregular, due to the irregularity of the fuel distribution within the combustion chamber. The general structure of the flame depends upon various parameters such as fuel injector characteristics (injector type, injection pressure, number of holes, spray angle *etc.*), combustion chamber design and in-cylinder flow conditions. The luminosity of this diffusion flame is enhanced by the partial burning of soot particles, which manifests itself with an intense yellow-white light emission [26].

In HCCI combustion studies [27, 28], the luminosity observed during the combustion event is quite mild and sometimes not easily discernible. HCCI combustion lacks a specific structure and seems to take place simultaneously in various regions within the combustion chamber, in which the luminosity simultaneously increases and decreases as combustion progresses. The non-localized nature of HCCI combustion and the multiple autoignition sites observed in HCCI engines are shown in Fig. **4**.

Optical Diagnostics Applied to HCCI Engines

Apart from the photographic studies, other optical diagnostic techniques have been used for the characterization of HCCI combustion, *i.e.* chemiluminescence imaging, spectroscopic analysis and planar laser-induced fluorescence (PLIF). In chemiluminescence imaging the light emitted from excited chemical compounds is monitored. Moreover, using appropriate optical filters, this method yields information regarding specific chemical compounds, which take part in the combustion event. With chemiluminescence imaging the distinct stages of HCCI combustion are observed [29]. These stages are depicted in Fig. **5**. The first stage involves the emission of light at the latter stage of compression. This is a low intensity light emission, which is located at a wide central region of the combustion chamber and it is termed *cool* or *blue flames*. The structure of such light emission is macroscopically homogeneous but microscopically inhomogeneous, with a grained morphology. This light emission initially increases and progressively disappears within a few ^{o}CA [29]. The emission spectra of cool flames indicate the existence of exited formaldehyde [29], which is one of the main intermediates of the hydrocarbon oxidation process [30]. The reactions occurring during the appearance of cool flames are also responsible for the low temperature heat release observed in the pressure trace (see Fig. **3**) and depend upon the fuel chemical properties. Specifically, cool flames are more intense in n-heptane combustion [31] and less pronounced in isooctane, methanol or ethanol HCCI combustion [18].

As compression progresses, the luminosity of cool flames diminishes and essentially disappears for a certain time interval. However, as temperature rises due to compression, the mixture enters the high temperature heat release regime, which is accompanied by a light emission that is more intense than the one observed in cool flames. Once again, the luminosity appears to be widespread within the combustion chamber, implying multiple ignition and combustion sites during the main combustion event (Fig. **5**). The interesting observation made in Ref. [29] was that after the light emission from the core of the combustion chamber, a delayed light emission was observed in the regions near the combustion chamber wall. This implies delayed chemical reactions in the near-the-wall region, which can be attributed to the existence of a thermal boundary layer in the vicinity of the wall. This is shown on the right side of Fig. **5**. In Ref. [29] the thickness of this thermal boundary layer was estimated to be about 2-3 mm.

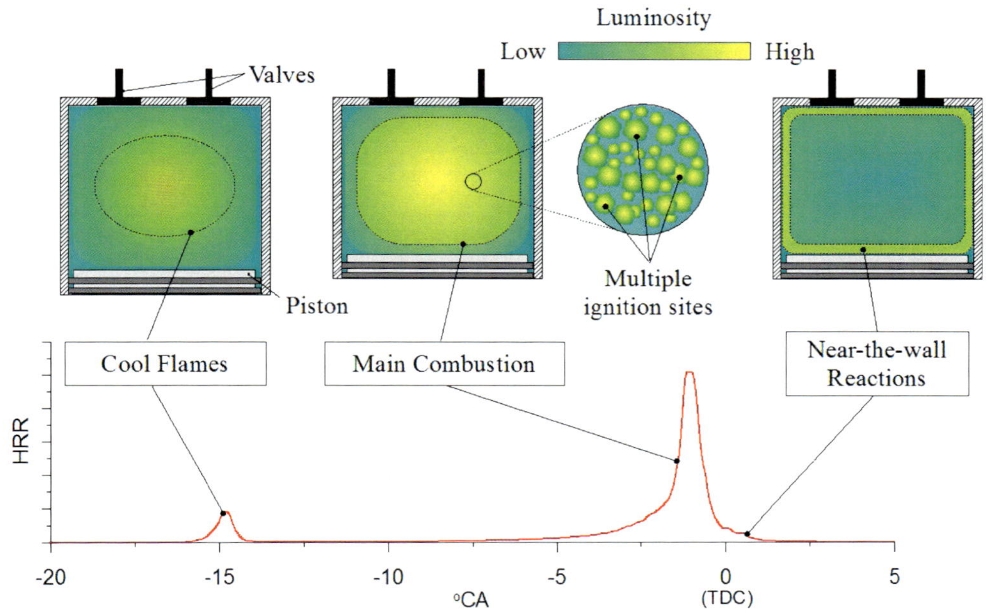

Figure 5: Schematic representation of in-cylinder gas luminosity during HCCI combustion.

The existence of the thermal boundary layer was also confirmed using another technique, *i.e.* the PLIF, in which a flat laser beam crosses the combustion chamber of an optically accessible HCCI engine. In this technique, a fluorescent fuel tracer is usually added to the mixture, in order to track it within the combustion chamber, thereby gaining insight into the fuel consumption during the combustion process. The results from such studies confirmed the delayed fuel consumption in the regions near the wall [19] and also led to the conclusion that HCCI combustion cannot rely upon flame propagation, since the required flame speed was estimated to be significantly greater (about 82 m/sec) than the one encountered in SI engines [32]. This conclusion is also supported by the fact than in HCCI engines the mixture is too lean to sustain a propagating flame. Therefore, the combustion process must rely upon simultaneous ignition at multiple sites, where the fuel is consumed gradually.

This progressive heat release from the bulk of the mixture observed in HCCI engines was described by Onishi *et al.* [10]. A schematic of their interpretation of HCCI heat release and the differences between HCCI and SI heat release is depicted in Fig. **6**. As is shown in the figure, in SI combustion the chemical energy released at any time interval corresponds to the essentially complete combustion of a relatively small portion of the mixture, through which the flame propagates. The heat released during any time interval is represented by the shaded area of the corresponding heat release rate diagram and equals the product of the heating value of the mixture per unit mixture mass q_{hv} times the portion dm of the mixture mass, through which the flame propagates. On the contrary, in HCCI combustion the chemical reactions are dispersed in a wide region within the combustion chamber with numerous reaction sites. In this case, heat is released simultaneously throughout this wide region and progresses to completion.

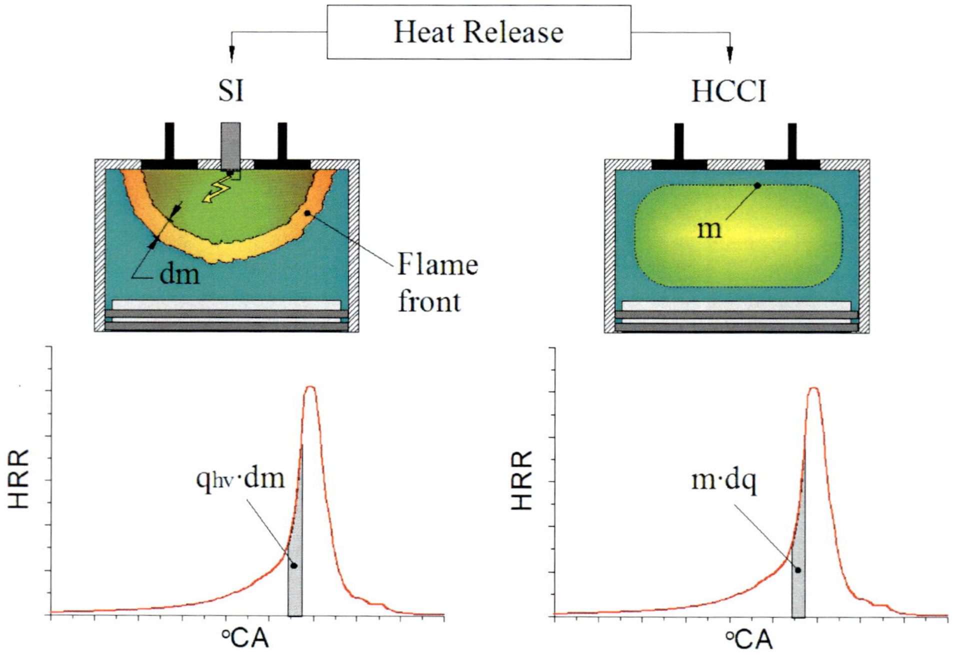

Figure 6: Conceptual representation of the heat release procedure in (a) SI combustion and (b) HCCI combustion as described in Ref. 10. q_{hv}: heating value per unit mixture mass, dq: heat release per unit mixture mass, m: mixture mass.

The amount of heat released at each time interval is approximately equal to the amount of heat released per unit mixture mass dq, multiplied by the mass m of a great portion of the mixture.

In summary, HCCI combustion does not rely upon flame propagation, but it is characterized by chemical reactions, which simultaneously initiate at multiple regions within the combustion chamber and progress towards completion. Cool flames may be observed prior to main combustion, depending on the fuel characteristics. Both cool flames and the main combustion event seem to be dominated by the fuel chemistry and chemical kinetics, rather than by some physical mechanism. After the main combustion event, delayed reactions are observed in the colder, near-the-wall regions, due to the existence of a thermal boundary layer.

Pollutant Emissions Formation in HCCI Combustion

As already mentioned, in HCCI combustion a premixed charge is used. The charge is lean in order to avoid excessive heat release rates, which lead to abnormal combustion and knock. It is this premixed nature of combustion and the lean mixtures used that are responsible for the significant reduction of NO_x and soot emissions. NO_x emission formation is suppressed due to the absence of near-to-stoichiometric regions within the combustion chamber. This results to lower local peak combustion temperatures, which remain below the temperature threshold of NO_x production (approximately 1800 K). Increasing the fuel introduced into the combustion chamber, may lead to an increase in NO_x emissions due to the higher combustion temperature [33]. However, this NO_x production threshold is not usually encountered, since the increase of fuel quantity also increases the combustion rate and results to knocking combustion. Moreover, in the case of direct injection HCCI combustion, NO_x emissions can be maintained at low levels provided the ignition delay is adequate for the fuel-air mixing to be completed: if combustion begins before a sufficient dispersion of the fuel is achieved, then NO_x emissions can be formed in near-to-stoichiometric regions, despite the over-all lean mixture used. Under sufficiently homogeneous conditions, NO_x emissions are formed in the hotter central regions of the combustion chamber during the main combustion event, as shown in Fig. 7.

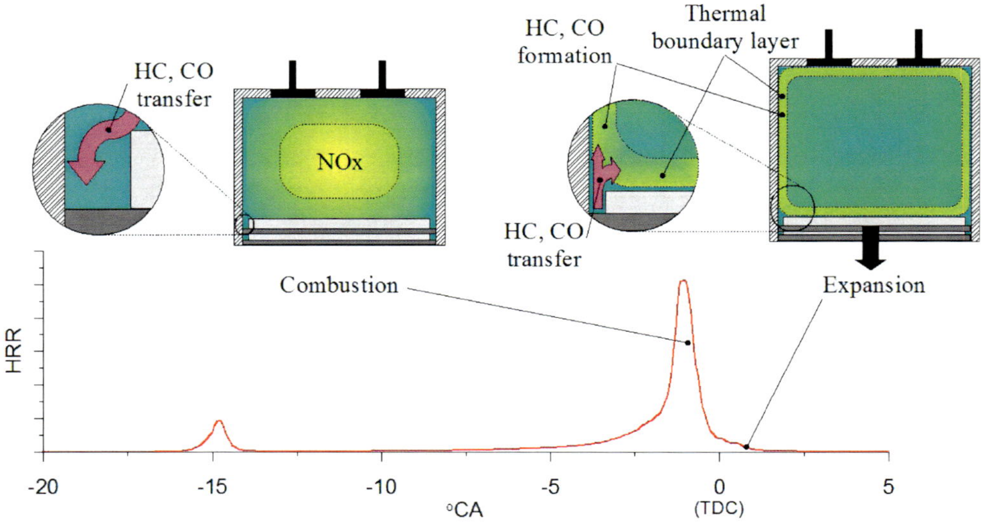

Figure 7: NO_x, HC and CO emissions formation in HCCI engines.

Soot emissions are also reduced greatly with HCCI combustion mode, mainly due to the absence of fuel-rich regions and, to a lesser extent, due to the low combustion temperatures encountered. However, in order to achieve smokeless operation, the introduction of the fuel within the combustion chamber must ensure that the spray-wall interaction is avoided. This is especially true when port fuel injection of low volatility fuels is used or when the fuel is injected directly into the combustion chamber early during or prior to the compression stroke.

The main pollutants produced in significant quantities in HCCI combustion mode are unburned hydrocarbons (HC) and monoxide emissions (CO). These pollutants are the result of incomplete combustion within the combustion chamber; they indicate either that the combustion temperatures are overall too low throughout the combustion chamber or that a portion of the fuel has entered low temperature regions, in which complete oxidation is prohibited. Low overall combustion temperatures are observed at very low fuel loading and this phenomenon is described as *bulk gas quenching* [34]. In this case HC and CO emissions are formed throughout the combustion chamber.

The regions of HC and CO formation at higher loads are shown in Fig. 7. Under high load conditions mass transfer between the hot core and the colder crevice and boundary layer regions becomes significant: multi-zone model simulations in which the mass transfer is neglected resulted to a severe underestimation of HC and CO emissions (up to 50% and 45% respectively), relative to the simulations in which mass transfer is considered for [35, 36]. At high loads, HC emissions are formed in the cold regions of the combustion chamber, which are located near the combustion chamber wall (thermal boundary layer) and in the crevices [22, 35, 36]. The relative effect of the thermal boundary layer and the crevices in not clear; it is presumed that the thermal boundary layer contribution becomes significant as load decreases [37, 38]. This can be inferred from the general trend of HC and CO reduction at increasing load [22, 38]: as load increases, the fuel introduced into the combustion chamber increases almost in proportion. This implies that more fuel is located into the crevice region, a phenomenon which tends to increase HC emissions. However, the fact that HC emissions are lower at the higher loads implies that overall the HC oxidation rates increase, probably due to the higher temperatures at the thermal boundary layer. At lower loads the local temperature in the combustion chamber is low, especially at the thermal boundary layer, which becomes a significant source of HC. Thus HC emissions increase at low load despite the lower fuel quantity introduced in the combustion chamber [37, 38]. CO emissions are the result of partial HC oxidation and are produced at the thermal boundary layer, which is located near the combustion chamber walls. Moreover, it has been suggested [35, 36] that CO formed at the beginning of – and during the – combustion event can enter the crevice regions during combustion and flow back into the combustion chamber during the expansion stroke.

Effect of Operating Parameters on HCCI Combustion and Emissions

The strong dependence of HCCI combustion on the chemical kinetics of the fuel oxidation process has rendered HCCI ignition and combustion rate control difficult. No external means, *e.g.,* a fuel injector or a spark plug, are used as instigators of the combustion event. Moreover, the combustion rate cannot be directly controlled by modifying physical mechanisms, such as a diffusion or a turbulent flame. This fact has led the research towards finding appropriate indirect means for the HCCI combustion control, by studying the effect of various operating parameters on the HCCI combustion process. These parameters affect the combustion phasing and/or the combustion rate. Although the ignition controlling parameters are diverse, the basic principle underlying their effect on HCCI combustion falls within two categories: they affect the reactivity of the inducted mixture and/or alter the temperature history of the mixture [14]. Some of these ignition controlling parameters also affect the combustion rate. The aforementioned parameters and the characteristics of HCCI combustion under varying load and engine speed are presented in the following sub-sections.

Mixture Reactivity

The mixture reactivity is affected by the *fuel properties*. The fuel cetane (CN) or octane numbers (ON) affect the HCCI ignition but a specific correlation has not been established. By appropriate selection of the compression ratio of the engine and the inlet conditions – mainly inlet temperature, various fuels can be used in HCCI engines, such as methanol, ethanol, dimethyl ether (DME), diethyl ether (DEE), isooctane, n-heptane, diesel, gasoline, liquefied petroleum gas (LPG), natural gas, methanol-reformed gas (MRG), biodiesel, hydrogen *etc.* and blends thereof [23, 24, 33, 39-48]. In particular, the use of biofuels in HCCI engines and in internal combustion engines in general has received great attention lately due to their environmental and socioeconomic benefits [43, 49-55]. The *fuel blends* used in HCCI combustion usually consist of fuels with different chemical reactivity (auto-ignition temperature, ON, CN *etc.*). By altering the relative amount of each fuel in the blend, the ignition timing and combustion duration can be significantly affected, especially when the auto-ignition characteristics of the fuels constituting the blend differ to a great extent. [25, 46-48].

The chemical effect of *exhaust gas recirculation (EGR)* on the mixture reactivity is twofold: it reduces the oxygen concentration, a feature which tends to delay ignition and suppress combustion rates, and it contains several chemically active species [56, 57]. The effect of the chemically active species is not universal and depends not only on the specific species present, but also on its concentration in the mixture: NO at low concentration can promote ignition, thereby decreasing HC and CO emissions; at higher concentration NO tends to inhibit the fuel oxidation, thereby increasing HC and CO emissions [58, 59]. Depending on the specific EGR strategy employed (internal or external, hot or cooled EGR) the recycled exhaust gases also induce a thermal effect on HCCI combustion, as described in the following sub-section.

In-Cylinder Temperature History

The temperature history of the mixture directly affects the HCCI ignition timing and the combustion rate. High in-cylinder gas temperature results to earlier ignition and faster combustion. Various parameters have been used to alter the mixture temperature history, including exhaust gas recirculation (EGR), variable compression ratio, inlet charge temperature, water injection, coolant temperature and combustion chamber geometry.

The chemical effect of EGR has already been mentioned. The *thermal effect of EGR* is due to the temperature of the EGR and also due to its higher heat capacity relative to the heat capacity of air [56, 58-60]. The EGR temperature depends upon the specific method of EGR implementation: external, cooled EGR tends to suppress the reaction rates – due to its diluting effect and higher specific heat – thereby retarding ignition. The dilution effect also results to lower combustion temperatures and lower combustion rates [13]. Internal EGR applied through the use of variable valve timing (residual gas trapping) demonstrates a different behavior: the hot internal EGR heats the incoming fresh charge, thereby enhancing the reaction rates, despite the reduction in O_2 concentration [57].

Inlet charge temperature and compression ratio directly affect the compression temperature history; increase in any of these parameters results to advanced ignition and higher combustion rates [25]. The inlet charge temperature regulation is usually achieved with an external heater [13, 58] or by recovering heat from the exhaust gases in a heat exchanger [61]. This preheating is particularly required when the compression ratio is low for auto-ignition to occur under ambient inlet air conditions. Moreover the air preheat is necessary when low volatility fuels are injected into the inlet air stream, in order to promote the fuel vaporization and the homogeneous mixture formation [46]. High inlet temperatures result to high peak combustion temperatures and reduced CO and HC emissions.

Compression ratio mainly affects the compression temperature history and especially the peak compression temperature. As already mentioned the compression ratio and inlet temperature selection are coupled, since both affect the ignition delay significantly. With the proper adjustment of the compression ratio and inlet temperature, a wide variety of fuels can be used in HCCI combustion mode [46]. Moreover, engines equipped with variable compression ratio (VCR) and heat exchangers for exhaust gas heat recovery utilize CR-load-engine speed maps to achieve HCCI combustion over a wide load-engine speed operating range [61]

A way of decreasing the in-cylinder gas temperature is direct in-cylinder *water injection* [62]. Direct water injection dilutes the charge and also reduces its temperature due to water vaporization. Increasing the water quantity decreases the local temperature, retards ignition and decreases the heat release rates. This is beneficial, since knocking can be avoided at higher loads, thereby enlarging the load operating range. However, HC and CO emission increase with water injection.

Coolant temperature affects the in-cylinder mixture temperature, due to its effect on heat transfer from the gas to the combustion chamber wall: high coolant temperatures induce high wall temperatures. This tends to decrease heat losses to the wall, since the wall heat flux is proportional to the difference between the mean gas temperature and the combustion chamber wall temperature. Accordingly, high coolant temperatures (~125 ^{o}C) have been used for extending the HCCI operation to low loads [63]. Conversely, low coolant temperature has been utilized as a means to suppress the heat release rate and avoid knocking combustion at high load operation, thereby extending the knocking combustion limit.

The *combustion chamber geometry* effect on HCCI combustion and emissions was studied by Christensen *et al.* [64]. In their study, the researchers compared HCCI combustion using two different combustion chambers: a low-turbulence disc-shaped combustion chamber and a high turbulence square bowl-in-piston combustion chamber. The modification of the combustion chamber was achieved by mounting each of the piston crowns, which are shown schematically in Fig. **8**, on the same piston body.

The resulting combustion chambers had the same clearance volume and consequently identical compression ratios. Relative to the disc shaped combustion chamber, the square bowl-in-piston configuration:

- enhances the turbulence,
- increases the combustion chamber wall area,
- increases the crevice volume, and
- decreases the minimum clearance height.

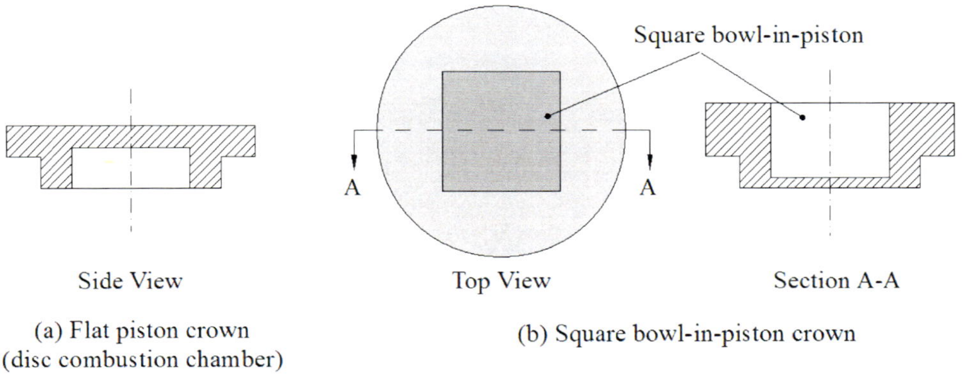

Figure 8: Schematic of the piston crown configurations used in Ref. [64] for the investigation of the combustion chamber geometry effect on HCCI combustion.

As noted by the researchers, the effect of *turbulence* on HCCI combustion has been much debated [64]. Some researchers believe that there is no direct effect of turbulence on the reaction rates [65], while others claim that such an effect exists [66]. Despite of this controversy, turbulence does affect the HCCI ignition and combustion process indirectly, *via* heat transfer [38, 67]. In the square bowl-in-piston configuration, the heat flux was greater due to the enhanced turbulence and the greater heat transfer area. These effects, combined with the deteriorated heat conduction through the piston rings, resulted in higher wall temperature during the engine operation, which advanced ignition. For the same combustion phasing, the disc-shape combustion chamber produced higher heat release rates and shorter combustion duration relative to the square bowl-in-piston. This was assumed to be the result of higher heat losses during combustion and the enhancement of turbulent mixing between hot-cold and lean-rich regions, which tends to suppress the heat release rates in the square bowl-in-piston design. However, another factor that affected the combustion rate was the higher intake temperature required for the disc combustion chamber to achieve the same combustion phasing with the square bowl-in-piston. The increased combustion chamber area and crevice volume and the lower top clearance height contributed to the higher HC and CO emissions of the square bowl-in-piston combustion chamber. The prolonged combustion duration, which resulted to lower combustion temperatures, also contributed to the increase of these emissions.

Engine Speed and Load

One of the main concerns of HCCI researchers is the realization of the HCCI combustion concept over a wide speed-load operating range. *Engine speed* affects the HCCI operation both directly and indirectly: the time available for any process to evolve within a certain crank angle interval is inversely proportional to the engine speed. This affects all rate controlled processes and particularly heat transfer, ignition and combustion. Increasing the engine speed decreases the available time for the pre-ignition reactions to occur, which tends to delay the ignition timing (in ^{o}CA) and reduce the peak heat release rates [13]. However, the speed increase decreases the time available for heat transfer to occur, and enhances turbulence [60]; the former tends to decrease the heat loss per cycle and the latter tends to increase it. Moreover, increasing the engine speed, affects the residual and/or EGR gas temperature and quantity, thereby altering the temperature history of the mixture. It is obvious that the engine speed effect on HCCI combustion is not straightforward. However, the predominant effect is the reduction of the time available for combustion with engine speed increase [13], which results to retarded ignition and increased combustion duration (in terms of ^{o}CA) [60]. This imposes a limit to the maximum allowable engine speed under HCCI operation.

The attainable *load* in HCCI combustion mode can be adjusted by modifying the fuel quantity. Too low a fueling rate, may lead to misfire – due to the slow reaction rates – or to engine halt, if the work produced is too low to overcome the engine friction. In addition, at low load operation the local in-cylinder combustion temperature is low, which results to high HC and CO emissions. On the other hand, high fueling rates tend to increase the reaction rates, thereby reducing the ignition delay. Moreover, high fueling rates lead to high heat release rates, high peak combustion temperatures and pressures and high pressure rise rates. This leads to unacceptable engine operation due to abnormal combustion (knock).

Due to these limiting factors – knock and misfire – the HCCI load operating range is limited. The attainable load depends upon various parameters, such as the specific method used for the implementation of HCCI combustion, the fuel used, the engine speed *etc.* Typical values of the maximum load achieved in naturally aspirated HCCI engines under normal operation lie close to 4-6 bar imep. This limits the use of HCCI combustion mode to part-load operation. However, the attainable load can be extended with the use of super- or turbo-charging. With supercharging, 14 bar imep has been achieved in a supercharged natural gas HCCI engine with 2 bar boost pressure [22].

IMPLEMENTATION OF HCCI COMBUSTION TO ENGINES

The realization of HCCI combustion in conventional engines can be achieved in two ways, *i.e.* by modifying spark ignition (SI) engines or diesel (CI) engines. Both of these approaches are described in the subsequent sections.

Gasoline HCCI Engines

Spark ignition engines usually operate on low compression ratios (typically about 8-12), to avoid pre-ignition and knocking combustion. At such low compression ratios HCCI combustion cannot occur under ambient inlet conditions, since the maximum compression temperature is too low for the mixture auto-ignition to occur. This is the main challenge in the implementation of HCCI combustion in gasoline engines, which can be overcome either by altering the auto-ignition characteristics of the mixture or by increasing the compression temperature. The former implies fuel modification and/or the use of fuel blends, in which the fuels mixed have different auto-ignition properties. The modification of the compression temperature history can be achieved by regulating the compression ratio or by increasing the inlet temperature. The latter can be implemented with an external heater or by utilizing the thermal energy, *i.e.* the high temperature, of the exhaust gases. The thermal energy of the exhaust gases can be exploited externally, *i.e.* using heat exchangers or mixing the hot EGR with the inlet air, or internally by retaining or reinducting the exhaust gases in the combustion chamber. In the internal EGR strategies the hot EGR located in the combustion chamber is mixed with the incoming charge, thereby raising the initial temperature of the charge. Such strategies involve the regulation of the gas exchange process and they are realized by sophisticated variable valve timing systems. These can produce various exhaust and/or inlet valve timing strategies, some of which are analyzed in subsequent sub-sections.

External EGR and Intake Air Heating

Intake mixture heating was used in the early studies of Thring [13] as was mentioned in section "Origins of the HCCI combustion concept". Oakley *et al.* [68] also used intake mixture heating in combination with external recycled exhaust gas (EGR). The heater was used to maintain the inlet temperature at 320 ^{o}C, while the EGR rate was adjusted between 0 to 70%. The fuels used were gasoline, methanol, ethanol and primary reference fuels (PRF). The researchers obtained experimental equivalence ratio-EGR rate maps of satisfactory HCCI operation. The limits of satisfactory HCCI operation were knocking combustion at high loads, partial fuel oxidation at low loads and misfire at high EGR rates. Alcohols were found to be more knocking-resistant extending the upper load limit to about 1.6 times the one obtained with gasoline.

Exhaust Gas Retention

Exhaust gas retention is commonly used for the implementation of the HCCI combustion mode in gasoline engines. It is usually achieved by modifying the valve timings: in normal SI operation (Fig. **9a**), the inlet valve opening (IVO) starts prior to TDC and the exhaust valve closing (EVC) event occurs after TDC, resulting to a (positive) valve overlap.

In the exhaust gas retention strategy, the valve overlap becomes negative, *i.e.* the exhaust valve closes before the piston reaches TDC and the inlet valve opens after TDC. This strategy is depicted in Fig. **9b**. With this strategy hot residual gases are recompressed due to the early EVC and are subsequently mixed with the incoming charge after the IVO event. The hot trapped residual gases provide the necessary heat to the incoming fresh charge, which is required to achieve auto-ignition of the fuel; without this heating effect, ignition is not likely to occur under naturally aspirated (NA) inlet conditions, due to the low compression ratio used in SI engines.

Ma *et al.* [69] applied the negative valve overlap strategy to achieve controlled Auto-ignition (CAI), *i.e.* HCCI, combustion in a multi-cylinder, four-stroke engine fueled with gasoline. The fuel was injected at the inlet port. Since a stoichiometric air-fuel mixture was used and the engine operated unthrottled during the CAI mode, the load was varied *via* the valve timings. The attainable operating range was 0.5 to 4 bar bmep and 1000-3500 rpm engine speed. At lower loads, misfire occurred due to low EGR thermal energy and high EGR percentage. The highest attainable load was restricted by the maximum amount of fresh charge that could be inducted. At high engine speeds this amount was reduced due to the limited time available for the gas exchange process. Compared to the SI operation within the same operating range, the CAI operation provided better fuel consumption due to the unthrottled operation and the faster combustion, which approximated constant volume combustion. NO_x emissions were reduced significantly (by about 90%) and CO emissions were also reduced by 10-40% but HC emissions increased.

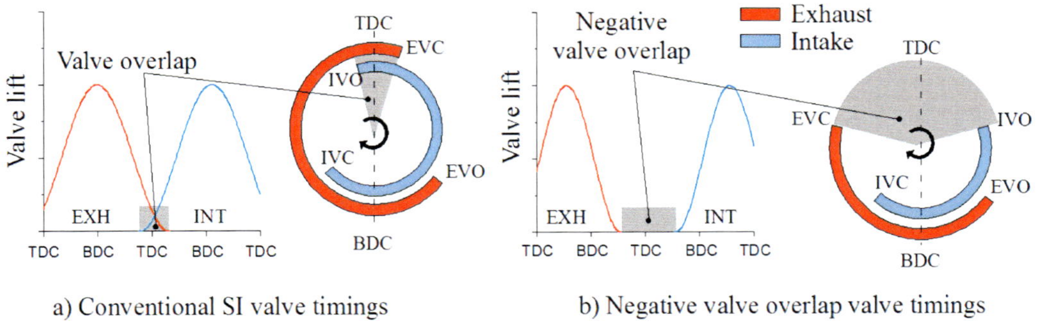

Figure 9: Valve timings in (a) conventional SI engines and (b) in negative valve overlap (NVO) strategies; EXH: exhaust stroke, INT: intake stroke.

In a similar study, Zhao *et al.* [70] found that the EVC timing affected significantly the internal EGR amount; earlier EVC resulted to higher EGR percentage and lower imep, the latter being inversely proportional to the EGR fraction. In addition, the researchers determined the work required for the recompression of the retained exhaust gases. It was found that the resulting pumping mean effective pressure (pmep) was about 0.4 bar regardless of the engine speed or the residual gas fraction (load). The IVO and IVC timing induced backflow effects, *i.e.* at early IVO or late IVC a portion of the residual gases escaped towards the inlet manifold due to the high in-cylinder pressure relative to the manifold pressure. These backflow effects tended to reduce the mixture temperature and resulted to retarded auto-ignition, slower combustion and lower NO_x emissions.

In the aforementioned studies, the fuel was introduced at the inlet port. In Ref. [71] an effort was made to apply the negative valve overlap concept to a gasoline direct injection (DI) engine. In this study, the fuel was injected in the cylinder during the residual gas recompression phase and even after the IVO event. The fuel injection induced a cooling effect to the recompressed residual gases due to the fuel vaporization. Moreover, depending on the injection timing, a small amount – up to 11% – of the total fuel chemical energy was released, due to the high temperature of the recompressed residuals. The fuel injection timing affected the pumping losses during the NVO period, since the cooling effect tends to lower the in-cylinder pressure, while the heat release tends to increase it. The pmep was 0.05 to 0.3 bar. Apart from the pmep effect, the fuel injection timing also affected the engine operation; early fuel injection resulted to earlier ignition of the charge, faster combustion and enhanced the fuel-air mixing, which further reduced NO_x emissions. The EVC timing was the main controlling variable for the residual gas fraction, thereby affecting load; earlier EVC resulted to higher amounts of residual gases and reduced the attainable load.

Law and Allen [72, 73] also conducted research on CAI combustion using variable valve timing. They used an active valve train configuration to control the valve timing on a single cylinder gasoline engine running on stoichiometric fuel-air mixture. The load regulation was achieved by the adjustment of the EVC and the IVO timings. With this configuration the load range was 0.5 to 4.5 bar bmep and the engine speed range of successful CAI operation was 1000-4000 rpm. The operating range was limited at low loads and engine speeds due to the insufficient thermal levels, which was the result of the low fuel quantity. At high speeds and loads, the limitation was imposed by abnormal combustion (uncontrolled detonation). Within the normal CAI operating range, the fuel consumption was reduced relative to the corresponding SI operation at the same operating region. This was attributed to the unthrottled CAI operation. NO_x emissions were greatly reduced (up to 97% reduction) and CO emissions were also reduced in the entire CAI operating range. HC emissions were found lower than in SI operation for a great portion of the load-engine speed operating range.

Koopmans *et al.* [74, 75] used an active valve train to regulate the valve timing events, in order to achieve CAI operation. In their study, the early EVC and the late IVO timing were symmetrically placed prior to and after TDC respectively. With this symmetrical EVC-IVO technique, the in-cylinder pressures at EVC and IVO were approximately the same. This minimized the residual gas backflow into the inlet manifold. Under symmetrical EVC-IVO timings, the researchers studied the effect of IVC, EVC and EVO on the performance and emissions of the CAI operation. The EVC timing produced the greatest effect on the operating characteristics; with early EVC more residual gases were retained and, due to the symmetrical IVO timing, less fresh charge was introduced within the combustion chamber. Using three valve timing events, the researches managed to operate the engine under CAI mode for a load range of 0.1 to 3.5 bar bmep and a speed range of 1300 to 3000 rpm. The operating region was limited at low speeds and low loads due to cycle-to-cycle variations, which occasionally led to misfire. The same problem of inconsistent combustion was observed at low load and high speeds. The highest attainable load at each engine speed was limited by abnormal (knocking) combustion.

Exhaust Gas Rebreathing

In the exhaust gas rebreathing or reinduction techniques, a portion of the exhaust gases temporarily escapes the combustion chamber and enters the inlet or exhaust manifold. These exhaust gases subsequently reenter the combustion chamber as a result of the suction induced by the descending piston. This EGR rebreathing technique can be implemented in various ways, three of which are depicted in Fig. **10a-c**.

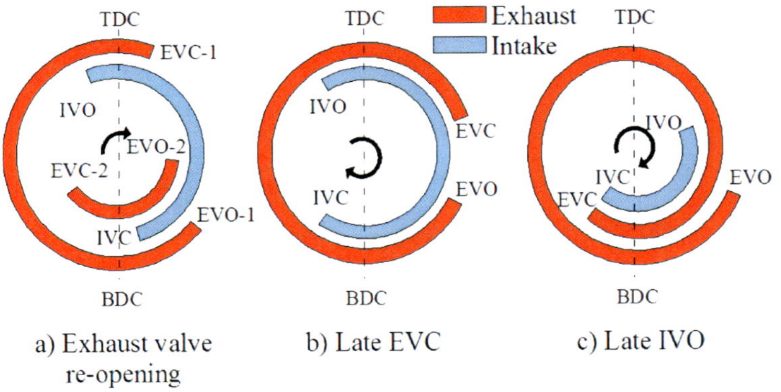

Figure 10: Valve timings for the exhaust gas rebreathing strategies.

Fig. **10a** describes an EGR rebreathing strategy, in which the exhaust valve reopens during the inlet stroke, allowing for the exhaust gases to re-enter into the combustion chamber through the exhaust port and mix with the incoming fresh charge. This method was used by Law *et al.* [76], in order to study the exact mechanism of CAI operation, *i.e.* to investigate the thermal and chemical effect of internal EGR. The single cylinder engine was equipped with four electrohydraulic valves, the timing of which could be independently regulated. With this configuration the researchers conducted three series of measurements: the first series involved the implementation of EGR rebreathing; the second and third series, involved the introduction of nitrogen or air respectively, which were heated to a temperature close to the EGR temperature of the first strategy prior to their induction through one of the exhaust valves. Therefore, the EGR used in the first series, was essentially replaced in the second and third series by nitrogen and air respectively. With the EGR rebreathing strategy, CAI (HCCI) combustion could be achieved, but when the EGR was replaced by hot nitrogen (400 ^{o}C) CAI did not occur. On the contrary, replacing the EGR with hot air (120 ^{o}C) resulted to uncontrolled auto-ignition of the mixture, *i.e.* to knocking combustion.

Kaahaaina *et al.* [77] conducted an extensive study using the valve timings depicted in Fig. **10b** and **c** under constant engine speed and stoichiometric propane-air mixture. In both of these configurations the EVC timing is placed well into the intake stroke. In the late EVC technique, (Fig. **10b**) the IVO timing was maintained to its conventional SI operation value, while the EVC timing was gradually retarded after TDC up to the point where the exhaust valve remained open throughout the intake stroke. The EVC timing controlled the EGR gas percentage; the later the EVC timing, the higher the EGR gas re-inducted into the combustion chamber. It was observed that HCCI combustion could not occur, unless the EVC timing was sufficiently retarded (after 110° aTDC), probably due to the insufficient thermal energy provided by the EGR at earlier EVC timings. Increasing EVC timing after this threshold reduced the fresh charge quantity, affected the heat transfer during the compression stroke and altered the ignition timing. These phenomena induced variations on the imep. Dynamic phenomena of the gas exchange process also affected the imep, especially when the EVC timing was set near the IVC timing. The imep values thus attained were about 5 bar or 55% of the imep of the unthrottled SI operation.

In order to increase the EGR percentage within the combustion chamber, the researchers proceeded to the late IVO technique illustrated in Fig. **10c**. In this method, the exhaust valve remains open throughout the intake stroke and the IVO timing is varied. When the IVO timing was placed near TDC the attainable imep was fairly constant. However, with an IVO of 10° aTDC or later, the imep was reduced due to advanced ignition, which increased the compression work, and also due to the reduction of the fresh charge. Overly retarded IVO (after 50° aTDC) produced too advanced ignition – thereby decreasing the imep – and this was one of the reasons, which prevented the further delay of the IVO event.

Other Control Methods

Apart from the EGR techniques presented in the previous sections, other methods have been used to regulate HCCI combustion in gasoline engines. *Variable compression ratio* has been used in HCCI engines by Haraldsson *et al.* [78] in order to control the combustion phasing. Using a control program, the researchers calculated engine performance characteristics in real time and adjusted manually the control variables, such as the fuel quantity, the inlet temperature and the compression ratio. High inlet temperatures required lower compression ratios to maintain the combustion phasing. With this method they achieved HCCI operation in an operating range of 0-4.4 bar bmep and 1000-5000 rpm.

In another study by Olsson *et al.* [79], the researchers constructed a *closed-loop control* system for the control of dual-fueled HCCI engine under varying load and engine speed. The engine was equipped with a heater to regulate the inlet temperature. The variables used to assess the engine status at any time where the engine speed, inlet temperature, inlet pressure, heater power, pressure traces, combustion phasing (defined *via* CA50 – the crank angle at which 50% of the total heat energy of the fuel energy has been released), amount of fuel, fuel mixing rate *etc.* The ignition timing was mainly controlled by the relative amount of the injected fuels – the easily ignitable n-heptane and the more auto-ignition-resistant isooctane. Load control was achieved by regulating the total fuel amount. The researchers studied the response of their control system over load and speed variations. They achieved satisfactory transient HCCI operation for an engine speed increase or decrease between 700 and 2000 rpm. The maximum load attained was 6.5 bar imep.

Diesel HCCI Engines

Although in gasoline HCCI engines the mixture preparation system is quite similar – if not identical – to the traditional fuel introduction system used during normal SI operation, the same does not hold for Diesel HCCI engine operation. The term *Diesel HCCI* refers to the HCCI combustion concept applied to conventional CI engines with appropriate modifications. The fuel can be introduced in the intake manifold or directly into the combustion chamber. In the case of port fuel injection, an additional injector must be mounted in the intake system. Moreover, the injected fuel must be easily atomized and vaporized, in order to achieve the formation of a homogeneous mixture and to avoid increases in HC and soot emissions. For port injected diesel fuel it is necessary to raise the intake temperature to promote the fuel vaporization [15, 46, 80].

For the implementation of direct injection HCCI combustion it may be necessary to modify the typical CI injection system and/or the combustion chamber design. This can be explained considering the traditional CI combustion process: CI combustion relies upon diffusion flame, which results from a highly non-homogeneous fuel-air mixture. This non-uniform mixture is formed by injecting the fuel into the combustion chamber towards the end of the compression stroke. The fuel is subsequently burned in a diffusion flame, while more fuel is injected into the combustion chamber. Since this process produces a highly in-homogeneous mixture, the CI fuel system requires modifications to establish a more uniform fuel distribution within the combustion chamber, which is necessary for Diesel HCCI operation. Due to the required homogeneity in fuel distribution, Diesel HCCI combustion is referred to as premixed charge compression ignition (PCCI), premixed compression ignition (PCI), partially premixed compression ignition (PPCI), premixed diesel combustion (PREDIC) *etc.* The term *premixed*, which is quite often used to describe the Diesel HCCI combustion, has a relative connotation: it is used to denote the formation of a fuel-air mixture that is homogeneous to a great extent, as is the case with the mixtures formed by port-fuel injection strategies; however, other researches use the term *premixed* to describe a mixture preparation technique, in which the injection duration is shorter than the ignition delay period, *i.e.* all the fuel has been injected into the combustion chamber prior to auto-ignition. In this case the mixture is relatively inhomogeneous and the term premixed is used in contrast to the typical CI combustion of conventional Diesel engines. In Diesel engines ignition usually occurs *during* the fuel injection, resulting to the formation of a diffusion flame.

Considering the homogeneous mixture requirement, one of the main concerns for the implementation of HCCI combustion in CI engines is the modifications required for the premixed mixture preparation. In order to achieve a mixture of sufficient homogeneity to avoid soot and NO_x formation, the injected fuel must be given enough time to mix with the air (and residual gases) in the combustion chamber. This can be achieved by advancing the injection timing – relative to conventional CI operation – to provide the necessary mixing time. However, the earlier fuel injection *via* the conventional injection system in combination with the low in-cylinder pressure and temperature encountered early in the compression stroke, usually result in poor fuel atomization, high spray penetration and eventually fuel impingement at the combustion chamber walls. This fuel spray-wall interaction is the major concern in Diesel HCCI combustion and it results to high HC emissions, high soot or particulate matter (PM), high fuel consumption and oil dilution. Fig. **11** shows the effect of spray angle on the aforementioned operating characteristics. The wide spray angle A results to fuel impingement on the cylinder liner, which induces lubricating oil dilution, increases soot formation, introduces fuel in the cold region outside the bowl – thereby increasing HC emissions – and results to an increase in fuel consumption. From the figure it is evident that the aforementioned phenomena are also affected by the injection timing, which is coupled to the position of the piston during the injection interval. Moreover, the spray-wall interaction becomes significant at high load, since the injected fuel quantity increases.

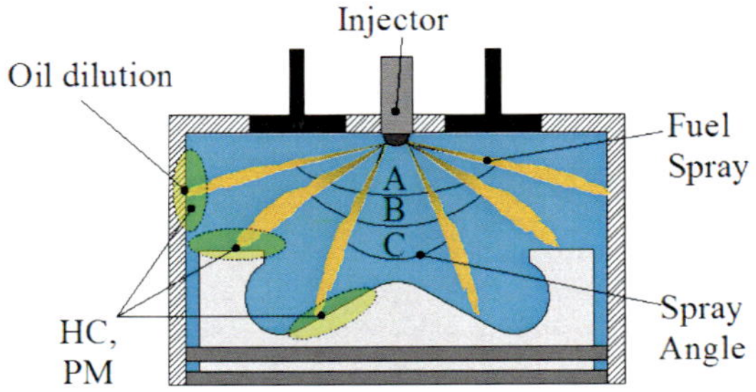

Figure 11: Major problems arising from fuel impingement on the combustion chamber walls of a Diesel HCCI engine, due to inappropriate selection of the spray angle. HC: hydrocarbon emissions, PM: particulate matter.

The fuel spray-wall impingement and the introduction of fuel in quench regions outside the piston bowl can be avoided by modifying the injection system or other operating parameters, *e.g.,* by:

- modifying the nozzle hole number and diameter to promote the fuel air mixing
- using low spray penetration-high dispersion nozzle geometries
- using multiple injectors
- adjusting the spray angle
- adjusting the injection pressure
- applying multiple injection pulses
- adjusting the initial gas temperature to promote fuel vaporization
- using volatile fuels

Apart from the mixture formation challenges, other concerns in Diesel HCCI engines is the ignition control, especially when the injection timing is too advanced to affect the ignition timing. Excessive combustion rate and noise are also issues to be resolved. In the aforementioned cases the usual means for control are:

- Adjustment of the compression ratio
- Adjustment of the initial temperature
- Use of heavy EGR
- Use of multiple injection pulses

The various methods used for the implementation of Diesel HCCI combustion follow the aforementioned generic principles. Specific methods are discussed in the following sub-sections.

Port Fuel Injection

Port fuel injection of diesel fuel was tested by the early researchers, namely Thring [13], Gray and Ryan [80] and Christensen *et al.* [46]. The main challenge in these studies is the poor fuel atomization and the low vaporization rate of diesel fuel under the low pressure and low temperature inlet conditions. Inlet preheating was required to enhance the diesel fuel vaporization and to avoid the fuel deposition on the intake system on combustion chamber walls. This tends to raise the smoke emissions and deteriorates the fuel conversion efficiency. Due to these issues the port injection of diesel fuel is not usually applied by itself, but rather in combination with other strategies [81].

Early Direct Injection

The general characteristics of Diesel HCCI combustion and its response to varying operating parameters depend largely upon the mixture formation strategy realized. In premixed diesel combustion (PREDIC) the early injection concept was employed [82]. In this study, the engine was equipped with three injectors: two side injectors and a central injector, as shown in Fig. **12**.

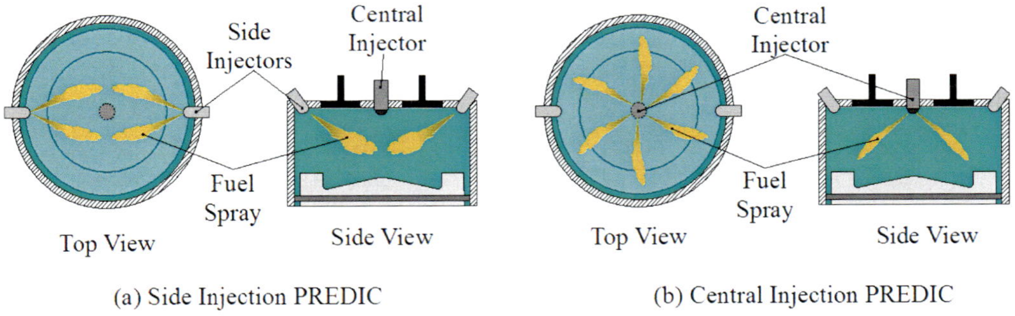

Figure 12: Injectors' configuration in the PREDIC combustion concept. (a) side injection PREDIC, (b) central injection PREDIC.

Each of the two side injectors produced two sprays and the central injector produced multiple sprays. The injection timings of the injectors could vary independently. In the side injection PREDIC (Fig. **12a**), the fuel was introduced simultaneously by the two side injectors at an injection pressure of 1000 bar. In order to avoid early ignition, the cetane number of the fuel was reduced (CN=19 or 40). With side injection PREDIC, the opposite fuel sprays collided with each other. This reduced the fuel penetration, improved the fuel-air mixing and confined the injected fuel within the central region of the combustion chamber. The fuel injection timing varied from about -120 to -30 °CA aTDC, for various equivalence ratios, *i.e.* load conditions. With proper injection timing, NO_x concentration was greatly reduced to about 1/10 of the conventional operation NO_x emissions, at low load cases. However, fuel consumption, HC and CO emissions were higher. The low- NO_x -emissions injection timing range was limited for the low CN fuel used. The PREDIC operation was sensitive to injection timing and air-fuel equivalence ratio: for constant load operation, the side PREDIC led to misfire when the injection timing was too advanced, due to overly lean mixture; at retarded injection timings knocking prevented normal operation. The injection timing affected ignition and NO_x emissions, which remained low compared to conventional operation. The attainment of high loads was a challenge, since the mixture preparation was more difficult at high fueling rates. The fuel cetane number and the relative injection timing of the two injectors also affected side PREDIC. Increasing the CN resulted to higher fuel consumption due to earlier ignition, but offered greater flexibility in the low- NO_x injection timing range. Asynchronous side injection PREDIC resulted to non-colliding fuel sprays, which produced overly lean mixtures and introduced fuel into quench regions. These effects reduced NO_x emissions but increased HC and CO emissions.

With central injection PREDIC (Fig. **12b**) the injection pressure was shifted to 1500 bar. Three types of nozzles were used, *i.e.* a conventional nozzle, a reduced diameter nozzle and a three-stage spray angle nozzle. All nozzle configurations produced similar results: early injection timings produced low NO_x and high HC and CO emissions. However, the three stage spray angle nozzle produced relatively low NO_x emissions and the lowest fuel consumption at advanced injection timings.

In a later study [83] of the same institute, the effect of fuel impingement during PREDIC operation was studied. With early injection timing (-130° aTDC) and side injection PREDIC a considerable amount of fuel impinged on the cylinder wall inducing oil dilution. To solve this problem, a pintle nozzle type with swirl grooves was used with a central injector PREDIC configuration. This provided a wider spray dispersion and shorter fuel penetration than the side injection PREDIC and reduced oil dilution. Since the ignition timing could not be controlled by the injection timing, the researchers used inlet air temperature regulation, EGR rate and CR adjustment to study their effect on the ignition timing. All the aforementioned factors affected significantly the ignition timing: inlet temperature and compression ratio increase, resulted in earlier ignition; high EGR rates retarded the ignition timing. The fuel CN and composition effect on the ignition timing was also studied. High CN fuels induced earlier ignition especially at relatively late injection timings. Apart from these parametric studies, the researchers also addressed the issue of extension to high load operation. Increasing the fuel quantity induced deterioration of combustion efficiency due to spray-wall interaction. For this reason, two approaches were implemented: supercharging and PREDIC multiple stage combustion. With supercharging the attainable load was extended to full load of conventional combustion. With multiple-stage diesel combustion (MULDIC) the fuel was injected in two equal amounts at two different timings. The first fuel amount was injected early from the side injectors and the second from the central injector near TDC. NO_x emissions and smoke were reduced relative to conventional operation. However, HC emissions were relatively higher but not as high as with the PREDIC type operation. When this concept was utilized in the Japanese emissions test cycle, higher PM emissions were produced relative to conventional operation, due to wall-wetting.

In premixed charge ignition (PCI) [17] the diesel HCCI combustion mode was also implemented *via* early injection during the compression stroke. A hole nozzle type was initially used for early fuel injection. The spray angle was reduced to 80° to produce a downward oriented spray and avoid spray-wall interaction. The compression ratio was also reduced from 18.5 to 12 to avoid ignition prior to TDC. The injection timing used with the hole nozzle type was limited by misfire and abnormal (knocking) combustion. Extremely early injection timings led to overly lean mixtures; retarded injection timings produced knock due to the existence of stoichiometric regions within the combustion chamber. With the hole nozzle and the aforementioned configuration, NO_x emissions were greatly reduced relative to conventional CI operation. However, fuel consumption, soot and HC emissions increased due to the fuel impingement on the combustion chamber wall. Other spray angles were also studied and the resulting sprays investigated numerically. It was concluded that with a hole nozzle the fuel-wall impingement had a significant influence on the mixture formation. Moreover, the spray angle affected the extent of fuel adhesion to the wall: too wide or too narrow a spray angle resulted to a greater portion of the fuel adhering to the combustion chamber walls.

In view of these findings, the researchers used an impinged type nozzle to avoid the fuel-wall interaction. In the impinged-type nozzles, the fuel is injected through a certain number of hole pairs, each pair consisting of a seat hole and a sac hole. Each pair of holes produces two intersecting sprays, as shown in Fig. **13**.

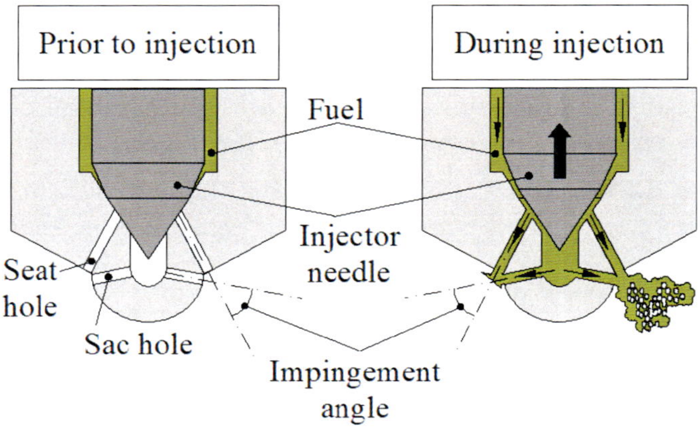

Figure 13: Impinged-spray nozzle configuration.

These conflicting sprays decreased the spray penetration and increased the fuel dispersion within the combustion chamber. With the impinged type nozzle the fuel consumption improved and soot emissions decreased relative to PCI operation with the hole nozzle type. HC emissions also decreased but remained high relative to conventional CI operation at the same air-fuel equivalence ratio. These HC emissions were reduced in PCI operation by using an oxidation catalyst, although its conversion rate deteriorated at low exhaust gas temperatures.

The formation of a premixed charge was greatly enhanced by the impinged nozzle type, but high fuel loading was difficult even with this wide-dispersion, low-penetration nozzle, since the injected fuel amount is too great to allow for sufficient fuel-air mixing. To attain higher loads, supercharging was utilized to increase the amount of air introduced into the cylinder and to allow for higher fueling rates. With a boost pressure of 0.8 bar, the output became comparable to the full load of conventional, naturally aspirated CI operation. NO_x emissions were significantly lower than the conventional CI operation and smoke was also reduced with supercharging.

Another issue of concern was the necessary reduction in the compression ratio, which was implemented to avoid early ignition. This reduction is thermodynamically unfavorable, since it tends to reduce the fuel conversion efficiency, which was already low for the naturally aspirated PCI operation. In order to minimize this loss, the researchers tried to increase the compression ratio from 12 to 13.5, while using heavy cooled EGR (up to 58%) to restrain early ignition. The cooled EGR allowed for richer mixtures to be used, reaching air-fuel equivalence ratios of 1.23. Moreover, cooled EGR was effective in delaying the ignition timing. However, its effect was small to allow for further compression ratio increase and was accompanied by problematic transient response and temperature instability.

Premixed charge compression ignition (PCCI) was another concept in which the early injection strategy was employed [84]. In this study the injection timing varied from -300 to -50 ^{o}CA aTDC. Gasoline fuel was used due to its higher volatility relative to diesel fuel, in order to enhance the homogeneous mixture preparation and avoid fuel impingement at the wall. For the same reason, a narrow spray angle was utilized and a pressure swirl atomizer nozzle was used. Moreover, the inlet temperature was regulated *via* an electric heater. These features allowed for smokeless operation over a wide range of injection timings (approximately -330° to -50° aTDC). The injection timing was varied to achieve different degrees of mixture homogeneity; retarding the injection timing increased the mixture stratification. This resulted in lower HC emissions, probably due to the lower dispersion of fuel towards quench regions. HC and CO emissions were reduced at higher load, *i.e.* at higher injection quantities, due to the resulting higher local combustion temperatures. However, NO_x emissions were found at extremely low concentration.

Late Injection - Ignition Delay Prolongation

This category includes strategies, in which the premixed charge preparation is based on the prolongation of the ignition delay. Strictly viewed, early direct injection is also a way of prolonging the ignition delay. However, in this sub-section the ignition timing is closer to TDC and the ignition delay is comparable to the injection duration. The goal is for the mixture to ignite only after all the fuel has been injected into the combustion chamber, in order to promote the fuel-air mixing. In the early direct injection strategies, the ignition delay is far greater than the injection duration, since the injection timing is placed early during compression. During this period, the in-cylinder gas temperature and pressure are too low for auto-ignition to occur. According to the aforementioned, late direct injection induces significant in-homogeneities in the mixture and the term *premixed* is used in contrast to conventional CI operation, in which the premixed combustion phase is merely a fraction of the total combustion event.

In Ref. [85] the premixed charge was formed by injecting diesel fuel prior to TDC (-40 to -5 ^{o}CA aTDC). Since the injection timing is retarded relative to the aforementioned premixed charge cases, the mixing time was too brief. In order to prolong the ignition delay of the mixture and enhance the fuel-air mixing, the researchers used a low cetane number fuel. They also used narrow spray angle and high injection pressure (1200 bar) to assist the formation of the premixed charge and avoid wall wetting. Using commercial diesel fuel with higher ignitability required the use of EGR to ensure the necessary fuel-air mixing time. With these modifications, NO_x and soot emissions were greatly reduced.

A distinct combustion concept, which implemented the late direct injection of fuel to achieve premixed charge combustion, is the modulated kinetics (MK) combustion concept [86, 87]. At first, this concept is counter-intuitive, since fuel injection after TDC reduces the time available for the fuel-air mixing process. In order to compensate for this apparent reduction in the available mixing time, the researchers employed high swirl in a toroidal combustion chamber and high injection pressure to enhance the fuel air mixing and reduce the injection duration. Moreover, they used heavy and/or cooled EGR, and reduced the compression ratio, to increase the ignition delay. The prolongation of the ignition delay and the formation of a relatively premixed mixture reduced soot emissions. The premixed nature of combustion and the use of heavy and/or cooled EGR also reduced NO_x emissions by at least 90%, due to the lower peak combustion temperatures (low temperature combustion). Heavy EGR also reduced the engine combustion noise.

A side-effect of this combustion concept was the reduction in peak heat fluxes relative to conventional CI operation. Moreover, it was found that increasing the swirl ratio in MK combustion mode decreased the peak heat loss fluxes, since the enhanced swirl confined the fuel within the central regions of the combustion chamber and prevented it from burning near the combustion chamber wall.

Multiple Injection Pulses

As already mentioned, the early direct injection of diesel fuel is problematic, since the diesel fuel is difficult to vaporize and over-penetration of the spray is likely. Moreover, the single injection strategy, relates to problems regarding the ignition timing control. To avoid this phenomenon, Kook and Bae [88] applied a two stage diesel fuel injection strategy. In this strategy, the diesel fuel was injected in two pulses: the first pulse occurred early in the compression stroke and the second one near TDC, which could act as an ignition promoter. The second injection pulse could be used to control ignition at low intake temperatures; at high intake temperatures, the ignition control *via* the second injection was not possible due to the advanced mixture auto-ignition. With this strategy, the researchers found a significant reduction in NO_x emissions (about 90%), but the HC and CO emissions increased relative to conventional DI-Diesel combustion.

Su *et al.* [89] conducted experiments in a Diesel HCCI engine using multiple injection pulses to introduce the fuel in the combustion chamber. The aim of the study was to determine the effect of various multiple injection patterns on the performance and emissions of the Diesel HCCI combustion process. The fuel was introduced in four, five or six consecutive pulses, of different pulse width and injected mass per pulse. The injection pattern, the mass ratio between pulses and the number of pulses was found to affect significantly virtually all engine parameters, *i.e.* the attainable imep, NO_x, soot, CO and HC emissions. The general conclusion of the study was that the early injection pulses must incorporate small and gradually increased injected fuel quantities to avoid wall wetting, since the early injection pulses introduce fuel into the cold, low pressure and low temperature combustion chamber. More fuel should be injected during the last pulses to take advantage of the high in-cylinder pressure and temperature, which promote the fuel dispersion and vaporization.

The various strategies for the implementation of Diesel HCCI combustion are, of course, not exhausted in the present text. More information on stratified HCCI combustion can be found in Yao *et al.* [90]. Moreover, the reader is referred to Ref. [81] for additional references and a comprehensive review on Diesel and Gasoline HCCI combustion.

CONCLUSIONS

Homogeneous charge compression ignition (HCCI) combustion is a promising combustion concept used in internal combustion engines. It provides simultaneous reduction of nitrogen oxide and soot emissions, an attribute which is quite attractive, in view of the ever stringent emissions standards. Moreover, HCCI combustion has the potential of high thermal efficiency due to the relatively high heat release rates, which approximate constant volume combustion, and also due to the unthrottled operation, which reduces the work consumed during the gas exchange process. However, HCCI combustion is not without problems, since carbon monoxide emissions and unburned hydrocarbons emitted from HCCI engines are usually higher than the corresponding ones of conventional engines. Moreover, the ignition timing and combustion rate cannot be easily controlled, rendering the potential gain in thermal efficiency quite dubious. The aforementioned characteristics relate to the nature of HCCI combustion, in which the ignition timing and combustion rate depend upon the autoignition and combustion chemistry of the fuel. The combustion progress does not depend upon flame propagation or diffusion flames; it is characterized by a gradual heat release which evolves almost simultaneously throughout the combustion chamber and is largely governed by chemical kinetics rather than a physical mechanism. In the absence of any external means to stimulate ignition or to control the combustion rates, various indirect methods are utilized to affect these operating parameters. These include altering the temperature history of the gas or altering the mixture reactivity. Specific methods used include inlet temperature or compression ratio regulation, use of internal or external EGR, use of various fuels and/or fuel blends *etc.* The selection of the appropriate method for the HCCI combustion control largely depends upon the type of engine, in which the HCCI concept is to be implemented: in Diesel HCCI engines different means are utilized for the ignition and combustion control than the ones used in gasoline (SI) HCCI engines. In any case, use is made of the existing equipment, which is also used during the conventional SI or CI operation. However, some modifications are required, either for the HCCI mixture preparation or for the HCCI ignition and combustion control. The main issues regarding the realization of HCCI combustion in actual engines is the ignition and combustion control over a wide load-engine speed range. Most of the current research is directed toward this ultimate goal.

ACKNOWLEDGEMENT

To the memory of my sister Dr. G.P. Komninou, Biologist.

REFERENCES

[1] F. Zhao, M-C. Laia and D.L. Harrington "Automotive spark-ignited direct-injection gasoline engines", *Prog. Energy. Combust. Sci.*, vol. 25, pp. 437-562, 1999.

[2] C.D. Rakopoulos, A.M. Dimaratos, E.G. Giakoumis and M.S. Peckham, "Experimental assessment of turbocharged diesel engine transient emissions during acceleration, load change and starting", SAE Paper No. 2010-01-1287.

[3] C.D. Rakopoulos and C.N. Michos, "Development and validation of a multi-zone combustion model for performance and nitric oxide formation in syngas fueled spark ignition engine", *Energy. Convers. Manage.*, vol. 49, pp. 2924-38, 2008.

[4] C.D. Rakopoulos, K.A. Antonopoulos, D.C. Rakopoulos and D.T. Hountalas, "Multi-zone modeling of combustion and emissions formation in DI diesel engine operating on ethanol-diesel fuel blends", *Energy. Convers. Manage.*, vol. 49, pp. 625-43, 2008.

[5] A.K. Agarwal, "Biofuels (alcohols and biodiesel) applications as fuels for internal combustion engines", *Prog. Energy. Combust. Sci.*, vol. 33, pp. 233-271, 2007.

[6] D.C. Rakopoulos, C.D. Rakopoulos, E.C. Kakaras and E.G. Giakoumis, "Effects of ethanol-diesel fuel blends on the performance and exhaust emissions of heavy duty DI diesel engine", *Energy. Convers. Manage.*, vol. 49, pp. 3155-62, 2008.

[7] D.C. Rakopoulos, C.D. Rakopoulos, E.G. Giakoumis, R.G. Papagiannakis and D.C. Kyritsis, "Experimental-stochastic investigation of the combustion cyclic variability in HSDI diesel engine using ethanol-diesel fuel blends", *Fuel*, vol. 87, pp. 1478-91, 2008.

[8] S. Verhelst and T. Wallner, "Hydrogen-fueled internal combustion engines", *Prog. Energy. Combust. Sci.*, vol. 35, pp. 490-527, 2009.

[9] C.D. Rakopoulos, G.M. Kosmadakis, E.G. Pariotis, "Evaluation of a new computational fluid dynamics model for internal combustion engines using hydrogen under motoring conditions", *Energy*, vol. 34, pp. 2158-2166, 2009.

[10] S. Onishi, S.H. Jo, K. Shoda, P.D. Jo and S. Kato, "Active thermo-atmosphere combustion (ATAC) - a new combustion process for internal combustion engines", SAE Paper No. 790501, 1979.

[11] M.Nogushi, Y. Tanaka, T. Tanaka and Y. Takeushi, "A study on gasoline engine combustion by observation of intermediate reactive products during combustion", SAE Paper No. 790840, 1979.

[12] P.M. Najt and D.E. Foster, "Compression-Ignited Homogeneous Charge Combustion", SAE Paper No 830264, 1983.

[13] R.H. Thring, "Homogeneous-Charge Compression-Ignition (HCCI) engines", SAE Paper No. 892068, 1989.

[14] R.H.Stanglemaier and C.E. Roberts, "Homogeneous Charge Compression Ignition (HCCI): benefits, compromises and future engine applications", SAE Paper No. 1999-01-3682, 1999.

[15] T.W. Ryan and T.J. Callahan, "Homogeneous charge compression ignition of diesel fuel", SAE Paper No. 961160, 1996.

[16] K. Tanaka, H. Endo, A. Imamichi, Y. Oda, Y. Takeda and T. Shimada, "Study of homogeneous charge compression ignition using a rapid compression machine", SAE Paper No. 2001-01-1033, 2001.

[17] Y. Iwabuchi, K. Kawai, T. Shoji and Y. Takeda, "Trial of new concept diesel combustion system -premixed compression-ignited combustion-", SAE Paper No. 1999-01-0185, 1999.

[18] M. Richter, A. Franke, M. Alden, A. Hultqvist and B. Johansson, "Optical diagnostics applied to a naturally aspirated homogeneous charge compression ignition engine", SAE Paper No. 1999-01-3649, 1999.

[19] A. Hultqvist, U. Endgar, B. Johansson and J. Klingmann, "Reacting boundary layers in a homogeneous charge compression ignition (HCCI) engine", SAE Paper No. 2001-01-1032, 2001.

[20] Z. Peng, H. Zhao and N. Ladommatos, "Effects of air/fuel ratios and egr rates on HCCI combustion of n-heptane, a diesel type fuel", SAE Paper No. 2003-01-0747, 2003.

[21] G. Shibata, K. Oyama, T. Urushihara and T. Nakano, "The effect of fuel properties on low and high temperature heat release and resulting performance of an HCCI engine", SAE Paper No. 2004-01-0553, 2004.

[22] M. Christensen, B. Johansson, P. Amneus and F. Mauss, "Supercharged homogeneous charge compression ignition", SAE Paper No. 980787, 1998.

[23] M. Christensen, B. Johansson and P. Einewall, "Homogeneous charge compression ignition (HCCI) using isooctane, ethanol and natural gas - a comparison with spark ignition operation", SAE Paper No. 972874, 1997.

[24] X. Lü, W. Chen and Z. Huang, "A fundamental study on the control of the HCCI combustion and emissions by fuel design concept combined with controllable EGR. Part 1. The basic characteristics of HCCI combustion", *Fuel*, vol. 84, pp. 1074-83, 2005.

[25] H. Machrafi, S. Cavadias and P. Gilbert, "An experimental and numerical analysis of the HCCI auto-ignition process of primary reference fuels, toluene reference fuels and diesel fuel in an engine, varying the engine parameters", *Fuel. Process. Technol.*, vol. 89, pp. 1007-1016, 2008.

[26] J.B. Heywood, *Internal Combustion Engine Fundamentals*. New York: McGraw Hill, 1988.

[27] M. Guenthner, W. Sauter, F. Schwarz, A. Velji and U. Spicher, "A study of the ignition and combustion process in a gasoline HCCI engine using port and direct fuel injection", The Sixth International Symposium on Diagnostics and Modelling of Combustion in Internal Combustion Engines, COMODIA 2004, August 2-5, Japan, Paper No. B2-2-021.

[28] Y. Aoyagi, Y. Asaumi, E. Kunishima, A. Harada, A. Morita and T. Seko, "Visualized analysis of a pre-mixed diesel combustion under the high boosting engine condition", The

Fifth International Symposium on Diagnostics and Modeling of Combustion in Internal Combustion Engines, COMODIA 2001, July 1-4, Japan, Paper No. 1-17.
[29] A. Hultqvist, M. Christensen, B. Johansson, A. Franke, M. Richter, and M. Alden, "A Study of the Homogeneous Charge Compression Ignition Combustion Process by Chemiluminescence Imaging", SAE Paper No. 1999-01-3680, 1999.
[30] L. Hildingsson, H. Persson, B. Johansson, R. Collin, J. Nygren, M. Richter, M. Alden, R. Hasegawa and H. Yanagihara", Optical diagnostics of HCCI and UNIBUS using 2-D PLIF of OH and formaldehyde", SAE Paper No. 2005-01-0175, 2005.
[31] Y. Ohta and M. Furutani, "Identification of cool and blue flames in compression ignition", Polish Academy of Sciences, *Archivum Combustionis*, vol. 11 (1-2), pp. 43-52, 1991.
[32] A. Hultqvist, M. Christensen, B. Johansson, M. Richter, J. Nygren, J. Hult and M. Alden, "The HCCI combustion process in a single-cycle - high-speed fuel tracer lif and chemiluminescence imaging", SAE Paper No. 2002-01-0424, 2002.
[33] N.P. Komninos, D.T. Hountalas and D.C. Rakopoulos, "A parametric investigation of hydrogen HCCI combustion using a multi-zone model approach" *Energy. Convers. Manage.*, vol. 48, pp. 2934-41, 2007.
[34] J.E. Dec and M. Sjöberg, "A parametric study of HCCI combustion--the sources of emissions at low loads and the effects of GDI fuel injection", SAE Paper No. 2003-01-0752, 2003.
[35] N.P. Komninos, "Assessing the effect of mass transfer on the formation of HC and CO emissions in HCCI engines, using a multi-zone model", *Energy. Convers. Manage.*, vol. 50, pp. 1192-201, 2009.
[36] N.P. Komninos, "Investigating the importance of mass transfer on the formation of HCCI engine emissions using a multi-zone model", *Appl Energy*, vol. 86, pp. 1335-1343, 2009.
[37] M. Jia and M. Xie. Numerical simulation of homogeneous charge compression ignition combustion using a multi-dimensional model. *Proc. Inst. Mech. Eng. Part D J. Automob. Eng.*, vol. 221, pp. 465-480, 2007.
[38] N.P. Komninos and D.T. Hountalas, "Improvement and validation of a multi-zone model for HCCI engine combustion concerning performance and emissions", *Energy. Convers. Manage.*, vol. 49, pp. 2530-2537, 2008.
[39] T. Shudo and Y. Ono, "HCCI combustion of hydrogen, carbon monoxide and dimethyl ether", SAE Paper No. 2002-01-0112
[40] G.R. Pucher, D.P. Gardiner, M.F. Bardon and V. Battista, "Alternative combustion systems for piston engines involving homogeneous charge compression ignition concepts--a review of studies using methanol, gasoline and diesel fuel", SAE Paper No. 962063, 1996.
[41] K. Yeom, J. Jang and C. Bae, "Homogeneous charge compression ignition of LPG and gasoline using variable valve timing in an engine", *Fuel*, vol. 86, pp. 494-503, 2007.
[42] N.K.M. Jothi, G. Nagarajan, andS Renganarayanan, "Experimental studies on homogeneous charge CI engine fueled with LPG using DEE as an ignition enhancer", *Renew Energy*, vol. 32, pp. 1581-93, 2007.
[43] A. Tsolakis, A. Megaritis and D. Yap, "Application of exhaust gas fuel reforming in diesel and homogeneous charge compression ignition (HCCI) engines fuelled with biofuels", *Energy*, vol. 33, pp. 462-70, 2008.
[44] J.P. Szybist, J. McFarlane and B.G. Bunting, "Comparison of simulated and experimental combustion of biodiesel blends in a single cylinder diesel HCCI engine", SAE paper 2007-01-4010, 2007.

[45] V. Manente, P. Tunestal, B. Johansson, "Influence of the wall temperature and combustion chamber geometry on the performance and emissions of a mini HCCI engine fuelled with DEE", SAE paper 2008-01-0008, 2008.

[46] M. Christensen, A. Hultqvist, B. Johansson, "Demonstrating the multi fuel capability of a homogeneous charge compression ignition engines with variable compression ratio", SAE Paper No. 1999-01-3679, 1999.

[47] M. Yao, Z. Chen, Z. Zheng, B. Zhang and Y. Xing, "Study on the controlling strategies of homogeneous charge compression ignition combustion with fuel of dimethyl ether and methanol", *Fuel*, vol. 85, pp. 2046-56, 2006.

[48] X. Lü, L. Ji, L. Zu, Y. Hou, C. Huang and Z. Huang, "Experimental study and chemical analysis of n-heptane homogeneous charge compression ignition combustion with port injection of reaction inhibitors", *Combust Flame*, vol. 149, pp. 261-70, 2007.

[49] G. Gnanam, A. Sobiesiak, G. Reader and C. Zhang, "An HCCI engine fueled with iso-octane and ethanol", SAE Paper No. 2006-01-3246, 2006.

[50] H. Xie, Z. Wei, B. He and H. Zhao, "Comparison of HCCI combustion respectively fueled with gasoline, ethanol and methanol through the trapped residual gas strategy", SAE Paper No. 2006-01-0635, 2006.

[51] M. Lemel, A. Hultqvist, A. Vressner, H. Nordgren, H. Persson and B. Johansson, "Quantification of the formaldehyde emissions from different HCCI engines running on a range of fuels", SAE Paper No. 2005-01-3724, 2005.

[52] C.D. Rakopoulos, D.C. Rakopoulos, D.T. Hountalas, E.G. Giakoumis, E.C. Andritsakis, "Performance and emissions of bus engine using blends of diesel fuel with bio-diesel of sunflower or cottonseed oils derived from Greek feedstock", *Fuel*, vol. 87, pp. 147-57, 2008.

[53] C.D. Rakopoulos, K.A. Antonopoulos and D.C. Rakopoulos, "Experimental heat release analysis and emissions of a HSDI diesel engine fueled with ethanol-diesel fuel blends", *Energy*, vol. 32, pp. 1791-808, 2007.

[54] C.D. Rakopoulos, K.A. Antonopoulos and D.C. Rakopoulos, "Development and application of multi-zone model for combustion and pollutants formation in direct injection diesel engine running with vegetable oil or its bio-diesel", *Energy. Convers. Manage.*, vol. 48, pp. 1881-901, 2007.

[55] A. Tsolakis and A. Megaritis, "Partially premixed charge compression ignition engine with on-board H2 production by exhaust gas fuel reforming of diesel and biodiesel", Int. J. Hydrogen Energ. vol. 30 (2005) 731-45, 2005.

[56] H. Zhao, Z. Peng, J. Williams and N. Ladommatos, "Understanding the effects of recycled burnt gases on the controlled autoignition (CAI) combustion in four-stroke gasoline engines", SAE Paper No. 2001-01-3607, 2001.

[57] D. Law, J. Allen and R. Chen, "On the mechanism of controlled auto ignition", SAE Paper No. 2002-01-0421, 2001.

[58] H. Machrafi, S. Cavadias and J. Amouroux, "A parametric study on the emissions from an HCCI alternative combustion engine resulting from the auto-ignition of primary reference fuels", *Applied Energy*, vol. 85, pp. 755-64, 2008.

[59] H. Machrafi, S. Cavadias and P. Guibert, "An experimental and numerical investigation on the influence of external gas recirculation on the HCCI autoignition process in an engine: Thermal, diluting, and chemical effects", *Combustion and Flame*, vol. 155, pp. 476-89, 2008.

[60] X. Lü, W. Chen and Z. Huang, "A fundamental study on the control of the HCCI combustion and emissions by fuel design concept combined with controllable EGR. Part 2. Effect of operating conditions and EGR on HCCI combustion", *Fuel*, vol. 84, pp. 1084-92, 2005.

[61] G. Haraldsson, P. Tunestål, B. Johansson and J. Hyvönen, "Transient control of a multi cylinder HCCI engine during a drive cycle", SAE Paper No. 2005-01-0153, 2005.

[62] Y. Iwashiro, T. Tsurushima, Y. Nishijima, Y. Asaumi and Y. Aoyagi, "Fuel consumption improvement and operation range expansion in HCCI by direct water injection", SAE Paper No. 2002-01-0105, 2002.

[63] N. Milovanovic, D. Blundell, R. Pearson, J. Turner and R. Chen, "Enlarging the operational range of a gasoline HCCI engine by controlling the coolant temperature", SAE Paper No. 2005-01-0157, 2005.

[64] M. Christensen, A. Hultqvist and B. Johansson, "The effect of combustion chamber geometry on HCCI operation", SAE Paper No. 2002-01-0425, 2002.

[65] S.M. Aceves, D.L. Flowers, J. Martinez-Frias, J.R. Smith, C.K. Westbrook, W.J. Pitz and R.W. Dibble, "A Sequential Fluid-Mechanic Chemical-Kinetic Model of Propane HCCI Combustion", SAE Paper No. 2001-01-1027, 2001.

[66] S-C. Kong, C. Marriott, R.D. Reitz and M. Christensen, "Modeling and experiments of HCCI engine combustion using detailed chemical kinetics with multidimensional CFD", SAE Paper No. 2001-01-1026, 2001.

[67] M Jia, M.Z. Xie, Z. Peng, "A comparative study of multi-zone combustion models for HCCI engines", SAE Paper No. 2008-01-0064, 2008.

[68] A. Oakley, H. Zhao, T. Ma and N. Ladommatos, "Dilution effects on the controlled auto-ignition (CAI) combustion of hydrocarbon and alcohol fuels", SAE Paper No. 2001-01-3606, 2001.

[69] T. Ma, H. Zhao, J. Li and N. Ladommatos, "Experimental investigation of controlled auto ignition (CAI) combustion in a 4-stroke multi-cylinder gasoline engine and drive cycle simulations", In: Proceedings of the IFP international congress, Rueil-Malmaison, France, November 26-27, 2001, pp. 115-124. Paris: Editions Technip, 2002.

[70] H. Zhao, J. Li, T. Ma and N. Ladommatos, "Performance and analysis of a 4-stroke multi-cylinder gasoline engine with CAI combustion", SAE Paper No. 2002-01-0420, 2002.

[71] R.H. Standing, N. Kalian, T. Ma, H. Zhao, M. Wirth and A.R. Schamel, "Effects of injection timing and valve timing on CAI operation in a multi-cylinder DI gasoline engine", SAE Paper No. 2005-01-0132, 2005.

[72] J. Allen and D. Law, "Advanced combustion using a LOTUS active valve train internal exhaust gas recirculation promoted auto-ignition", In: Proceedings of the IFP international congress, Rueil-Malmaison, France, November 26-27, 2001, pp. 85-100. Paris: Editions Technip, 2002.

[73] D. Law and J. Allen, "Variable valve actuated controlled auto-ignition: speed load maps and strategic regimes of operation", SAE Paper No. 2002-01-0422, 2002.

[74] L. Koopmans and I. Denbratt, "A four-stroke camless engine, operated in homogeneous charge compression ignition mode with commercial gasoline", SAE Paper No. 2001-01-3610, 2001.

[75] L. Koopmans, I. Denbratt and O. Backlund, "Cycle-to-cycle variations: their influence on cycle resolved gas temperature and unburned hydrocarbons from a camless gasoline compression ignition engine", SAE Paper No. 2002-01-0110, 2002.

[76] D. Law, J. Allen and R. Chen, "On the mechanism of controlled auto ignition", SAE Paper No. 2002-01-0421, 2002.

[77] N.B. Kaahaaina, A.J. Simon, P.A. Caton and C. Edwards, "Use of dynamic valving to achieve residual-affected combustion", SAE Paper No. 2001-01-0549
[78] G. Haraldsson, P. Tunestal, B. Johansson and J. Hyvonen, "HCCI combustion phasing in a multi-cylinder engine using variable compression ratio", SAE Paper No. 2002-01-2858, 2002.
[79] J-O. Olsson, P Tunestal and B Johansson, "Closed-loop control of an HCCI engine", SAE Paper No. 2001-01-1031, 2001.
[80] T.W. Ryan, A.W. Gray, "Homogeneous charge compression ignition (HCCI) of diesel fuel", SAE Paper No. 971676, 1997.
[81] M. Yao, Z. Zheng and H. Liu, "Progress and recent trends in homogeneous charge compression ignition (HCCI) engines", *Prog. Energy. Combust. Sci.*, vol. 35, pp. 398-437, 2009.
[82] Y. Takeda and N. Keiichi, "Emission characteristics of premixed lean diesel combustion with extremely early staged fuel injection", SAE Paper No. 961163, 1996.
[83] H. Akagawa, T. Miyamoto, A. Harada, S. Sasaki, N. Shimazaki and T. Hashizume, "Approaches to solve problems of the premixed lean diesel combustion", SAE Paper No. 1999-01-0183, 1999.
[84] C.D. Marriott, S-C. Kong and R.D. Reitz, "Investigation of hydrocarbon emissions from a direct injection-gasoline premixed charge compression-ignited engine", SAE Paper No. 2002-01-0419, 2002.
[85] N. Shimazaki, T. Tsurushima, T. Nishimura, "Dual-mode combustion concept with premixed diesel combustion by direct injection near top dead center", SAE Paper No. 2003-01-0742.
[86] S. Kimura, O. Aoki, H. Ogawa, S. Muranaka and Y. Enomoto, "New combustion concept for ultra-clean and high-efficiency small DI diesel engines", SAE Paper No. 1999-01-3681, 1999.
[87] S. Kimura, O. Aoki, Y. Kitahara and E. Aiyoshizawa, "Ultra-clean combustion technology combining a low-temperature and premixed combustion concept for meeting future emission standards", SAE Paper No. 2001-01-0200, 2001.
[88] S. Kook and C. Bae, "Combustion control using two-stage diesel fuel injection in a single-cylinder PCCI engine", SAE Paper No. 2004-01-0938, 2004.
[89] W. Su, H. Wang and B. Liu, "Injection mode modulation for HCCI diesel combustion", SAE Paper No. 2005-01-0117, 2005.
[90] M. Yao, H. Liu, Z. Zheng, "Fuel Chemistry and Mixture Stratification in HCCI Combustion Control", In: H. Machrafi (Ed.), Green Energy & Technology, Bentham Science Publishers.

APPENDICES

A. Nomenclature

q_{hv}	mixture heating value (J/kg)
q	heat released per unit mixture mass (J/kg)
m	mass (kg)

Abbreviations

a	after
ATAC	active thermo-atmosphere combustion
b	before
BDC	bottom dead center
bmep	brake mean effective pressure
CA	crank angle
CAI	controlled auto-ignition
CO	carbon monoxide
CFD	computational fluid dynamics
CI	compression ignition
CIHC	Compression Ignited Homogeneous Charge
CN	cetane number
CR	compression ratio
DEE	diethyl ether
DME	dimethyl ether
DI	direct injection
EGR	exhaust gas recirculation
EVC	exhaust valve closing
EVO	exhaust valve opening
HC	(unburned) hydrocarbons
HCCI	homogeneous charge compression ignition
HRR	heat release rate (J/deg)
HTHR	high temperature heat release
imep	indicated mean effective pressure (bar)

IVC	inlet valve closing
IVO	inlet valve opening
LTHR	low temperature heat release
LPG	liquefied petroleum gas
MRG	methanol reformed gas
MK	modulated kinetics
MULDIC	multiple stage diesel combustion
NA	naturally aspirated
NO	nitrogen monoxide
NO_x	nitrogen oxides
NTC	negative temperature coefficient
NVO	negative valve overlap
ON	octane number
PCCI	premixed charge compression ignition
PCI	premixed compression ignition
PLIF	planar laser-induced fluorescence
PM	particulate matter
pmep	pumping mean effective pressure
PPCI	partially premixed compression ignition
PREDIC	premixed diesel combustion
PRF	primary reference fuel
SI	spark ignition
TDC	top dead center
TS	Toyota-Soken
VCR	variable compression ratio

B. Terms

Bulk Gas Quenching: in HCCI combustion refers to the partial combustion of a great portion of the charge, resulting from low fueling rate.

Cool (or Blue) Flames: A distinct faint luminosity, which usually accompanies the low temperature heat release process in HCCI combustion.

Diesel HCCI: refers to the implementation of the HCCI concept in conventional Diesel engines.

Early Direct Injection: is a technique used for the implementation of HCCI combustion in Diesel engines. It corresponds to the direct fuel injection early during the intake or compression phases.

Exhaust Gas Rebreathing: is a technique used for the implementation of HCCI combustion in gasoline engines. With this strategy a portion of the exhaust gases, which have temporarily exited the combustion chamber, are reinducted *via* valve timing adjustment.

Exhaust Gas Retention: is a technique used for the implementation of HCCI combustion in gasoline engines. With this strategy a great portion of the exhaust gases are prevented from exiting the combustion chamber *via* valve timing adjustment.

Gasoline HCCI: refers to the implementation of the HCCI concept in conventional spark ignition engines.

Homogeneous Charge Compression Ignition Combustion (HCCI): is a distinct combustion mode implemented in internal combustion engines. It is used to describe the autoignition of a homogeneous fuel-air mixture, which has been formed early in the engine cycle, *i.e.* during the intake or compression stroke.

High Temperature Heat Release (HTHR): corresponds to the second energy release of the two stage ignition process encountered in HCCI combustion and takes place at a relatively high temperature.

Late Injection: is a technique used for the implementation of HCCI combustion in Diesel engines. It corresponds to the direct fuel injection at the latter stages or after the compression phase.

Low Temperature Heat Release (LTHR): refers to the first stage of the two stage ignition process in HCCI combustion. It corresponds to the first observable heat release and takes place at relatively low temperature.

Negative Temperature Coefficient (NTC): is the period between the two successive heat release stages of the two stage ignition process. During the NTC period the apparent chemical energy release ceases although the in-cylinder combustion chamber increases due to compression.

Two Stage Ignition: describes the heat release process in HCCI combustion, in which the fuel chemical energy is released in two successive stages.

CHAPTER 8

Fuel Chemistry and Mixture Stratification in HCCI Combustion Control

M. Yao*, H. Liu and Z. Zheng

State Key Laboratory of Engines, Tianjin University, China

Abstract: Homogeneous charge compression ignition (HCCI) is an autoignition combustion process with a lean or dilute fuel/air mixture. It can provide both good fuel economy and very low emissions of nitrogen oxides (NO_x) and particulates. Therefore, it is considered to be one of the most promising internal combustion engine concepts for the future. However, there are some obstacles that must be overcome before the potential benefits of HCCI combustion can be fully realized in commercialization, including combustion phasing control, operation range extending (high levels of noise, UHC and CO emissions), cold start, and homogeneous mixture preparation. All these HCCI characteristics have been summarized in Section 1. To overcome these obstacles, many effective technologies have been carried out and these technologies will be reviewed in Section 2 according to different fuel properties. HCCI can be applied to a variety of fuel types and the choice of fuel will have a significant impact on both engine design and control strategies. Some chemical components have the ability to inhibit or promote the heat release process associated with autoignition. Typical generalized diesel-fuelled HCCI combustion modes include: early direct injection HCCI, late direct injection HCCI, premixed/direct-injected HCCI combustion and low temperature combustion. Mixture control (mixture preparation), including charge components and temperature control in the whole combustion history and high pre-ignition mixing rate, is the key issue to achieve diesel HCCI combustion. The high octane numbers of gasoline fuels mean that such fuels need high ignition temperatures, which highlights the difficulty of autoignition. The main challenge for gasoline HCCI operation is focus on the obtaining sufficient thermal energy to trigger autoignition of mixtures late in the compression stroke, extending the operational range, and the transient control. In addition, alternative fuel can save the fossil fuel and reduce the CO_2 emission, therefore it has been got more attention in recent years. And to understand fundamental theory of HCCI combustion process, the primary reference fuel is the best choice due to the better understood chemical kinetics. All these fuels will be also introduced in the Section 2. Advanced control strategies of fuel/air mixture are more important than simple "homogeneous charge'' for the HCCI combustion control. Further, it is impossible to get an absolutely homogeneous mixture in the operation of practical HCCI engines. Modest inhomogeneity in fuel concentration or temperature appearing in mixing can affect the autoignition and combustion process. And stratification strategy also has the

***Address correspondence to M. Yao:** State Key Laboratory of Engines, Tianjin University, China; Tel.: +86 22 27406832; Fax: +86 22 27383362; E-mail: y_mingfa@tju.edu.cn

Hatim Machrafi (Ed)

potential to extend the HCCI operation range to higher loads. The thermal stratification can be caused by wall heat transfer and turbulent mixing during the compression stroke for a low-residual engine. This thermal stratification causes the combustion to occur as a sequential autoignition of progressively cooler regions, slowing the rate of pressure rise. For engines with high levels of retained residuals, incomplete mixing between the fresh charge and hot residuals could also contribute to the thermal stratification. Apart from the thermal stratification, more researches are about the charge or compositional stratification. The charge stratification is focus on the different injection strategies, while the compositional stratification means that all the EGR, internal or external, changes the composition of the charge therefore forming the different compositional stratification. These stratification combustion characteristics have been reviewed in Section 3. Finally, a summary for the progress of HCCI combustion and future research direction has been shown in Section 4.

Keywords: Fuel, chemistry, mixture stratification, homogeneous charge compression ignition (HCCI), combustion control, auto-ignition, operation range, mixture preparation, gasoline, diesel, fuel surrogate, natural thermal stratification, charge and compositional stratification, low temperature combustion (LTC), combustion mechanism, chemical kinetics.

INTRODUCTION

The internal combustion engine is one of the key drivers in modern industrial society, such as automobile, ship, power, propulsion, *etc.* There are two types of internal combustion engines for automobile: spark ignition (SI) and compression ignition (CI). The conventional SI combustion is characterized by flame propagation process. The onset of combustion in SI engines can be controlled by varying ignition timing from the spark discharge. Because the mixture is premixed and typically stoichiometric (λ=1), the emissions of soot are very low (except in gasoline direct injection engine). And with the three-way catalyst, other emissions, such as unburned hydrocarbon (UHC), carbon monoxide (CO), and nitrogen oxides (NO_x), are also very low. However, for a fixed air/fuel ratio, the throttle used for controlling the air mass flow gives rise to pumping losses and a reduction in efficiency. As a result, the major disadvantage of SI engines is its low efficiency at partial loads. The compression ratio in SI engines is limited by knock and can normally be limited in the range from 8 to 12 contributing to the low efficiency. Conventional diesel combustion, as a typical representation of CI combustion, operates at higher compression ratios (12–24) than SI engines. In this type of engine, the air–fuel mixture auto-ignites as a consequence of piston compression instead of ignition by a spark plug. The processes which occur between the two moments when the liquid fuel leaves the injector nozzles and when the fuel starts to burn, are complex and include droplet formation, collisions, breakup, evaporation, and vapor diffusion. The rate of combustion is effectively limited by these processes. A part of the air and fuel will be premixed and burn fast, but for the larger fraction of the fuel, the time scale of evaporation, diffusion, *etc.* is larger than the chemical time scale. Therefore, the mixture can be divided into high fuel concentration regions and high temperature flame regions. In the high fuel concentration regions, a large amount of soot is formed because of the absence of O_2. Some soot can be oxidized with the increase of in-cylinder temperature. The in-cylinder temperature in a conventional diesel engine is about 2700 K, which leads to a great deal of NO_x emissions. For diesel engines, a trade-off between these two emissions is observed, and their problem is how to break through the compromise between NO_x and particulate matter (PM) emissions.

Consequently, the obvious ideal combination would be to find an engine type with high efficiency of diesel engines and very low emissions of gasoline engines with catalytic converters, which saves more fuel and means lower greenhouse gases and other harmful emissions. One such candidate is the process known as homogeneous charge compression ignition, HCCI, on which we shall now focus attention. HCCI is characterized by the fact that the fuel and air are mixed before combustion starts and the mixture auto-ignites as a result of the temperature increase in the compression stroke. Thus HCCI is similar to SI in the sense that both engines use premixed charge and similar to CI as both rely on auto-ignition to initiate combustion.

The concept of HCCI was initially investigated for gasoline applications by Onishi *et al.* [1] in 1979, in order to increase combustion stability of two-stroke engines. They found that significant reductions in emissions and an improvement in fuel economy could be obtained by creating conditions that led to spontaneous ignition of the in-cylinder charge. Stable HCCI combustion could be achieved between low and high load limits with gasoline at a compression ratio of 7.5:1 over the engine speed range from 1000 to 4000 rpm. Noguchi *et al.* [2] performed a spectroscopic analysis on HCCI combustion by experimental work on an opposed piston two-stroke engine. From optical investigations they noted that ignition took place at numerous points throughout the cylinder and no discernable flame front was observed during combustion. Using spectroscopic methods to detect the intermediate species, they measured high levels of CH_2O, HO_2, and O radicals within the cylinder before auto-ignition. These species are characteristic of low temperature auto-ignition chemistry of larger paraffinic hydrocarbon fuels. After ignition, they observed high concentrations of CH, H, and OH radicals, which were indicative of high temperature chemistry during the bulk burn. Building on previous work on two-stroke engines, Najt and Foster [3] extended the work to four-stroke engines and attempted to gain additional understanding of the underlying physics of HCCI combustion. They concluded that HCCI auto-ignition is controlled by low temperature (below 1000 K) chemistry and the bulk energy release is controlled by the high temperature (above 1000 K) chemistry dominated by CO oxidation. Thring [4] further extended the work of Najt and Foster in four-stroke engines by examining the performance of an HCCI engine operated with a fully-blended gasoline. This is the first time that the name of this new combustion model, HCCI, was presented in his paper. It was also found that the operating regime was restricted to part load operation, and control of the auto-ignition timing was problematic.

As discussed above, initial efforts with HCCI involved gasoline-fuelled engines, and this technology continues to be strongly pursued today. In many researches, gasoline-fuelled HCCI combustion is also called controlled auto-ignition (CAI) combustion. However, the need to reduce emissions from diesel engines led to investigation into the potential of diesel-fuelled HCCI beginning in the mid-1990s. For diesel fuel, port fuel injection is perhaps the most straightforward approach to obtaining a premixed charge and this approach has been used in some of the earlier investigations of diesel-fuelled HCCI [5, 6]. Ryan *et al.* [5] used port fuel injection to supply diesel fuel into the intake air stream. An intake air heater upstream of the fuel injection point allowed preheating, with engine compression ratios varying between 7.5 and 17:1. This study and the following study on the same engine [6] found: for diesel-fuelled HCCI, very premature ignition and knocking occurred using normal diesel compression ratios; relatively high intake temperatures were required to minimize the accumulation of liquid fuel on surfaces in the intake system; and unburned HC emissions tended to be very high, but NO_x and smoke emissions were dramatically reduced.

The results from these pioneering investigations indicate the strong potential of HCCI to improve thermal efficiency of gasoline-fuelled engines and substantially reduce NO_x and soot emissions of diesel-fuelled engines. The mixture in HCCI is made very dilute either by being lean with fuel/air equivalence ratio (ϕ) typically less than 0.45, or through the use of high levels of exhaust gas recirculation (EGR) for equivalence ratios up to stoichiometric. Because of the high dilution, combustion temperatures are low, resulting in low NO_x emissions, and the charge is sufficiently well-mixed to prevent soot formation. Although these mixtures are typically too dilute to support flame-type combustion, they react and burn volumetrically as they are compressed to auto-ignition temperatures by the piston [7]. This ideal volumetric combustion process for HCCI results in a higher thermal efficiency, comparable with those of conventional diesel engines. Furthermore, these pioneering results confirmed the dominating role of chemical kinetics in HCCI combustion, which has significance to following studies.

However, they also foresaw some problems of this new combustion mode. There are some obstacles that must be overcome before the potential benefits of HCCI combustion can be fully realized in production applications became clear with more and more studies.

The First One is the Difficulty in Combustion Phasing Control

One of the principal challenges of HCCI combustion is control of the combustion phasing. Unlike conventional combustion, a direct method for controlling the start of combustion is not available. Instead, the start of combustion is established by the auto-ignition chemistry of the air-fuel mixture. Auto-ignition of mixture is influenced by the properties of the mixture and by the time-temperature history to which it is exposed. Hence, combustion phasing of HCCI engines is affected by the following factors [8-10]: auto-ignition properties of the fuel, latent heat of vaporization of the fuel, fuel concentration, mixture homogeneity, residual rate and, possibly, reactivity of the residual, compression ratio, intake temperature and pressure, engine temperature, heat transfer to the engine, and other engine-dependent parameters. Actually, these factors can be divided into two aspects: one is the fuel effect; the other is the engine geometry and operating parameters. Furthermore, fuel properties are fundamental to control the combustion phasing, which determine the engine geometry and operating parameters. Therefore, we will give a more detailed introduction in the Section 2 for the effects of fuel properties on the HCCI combustion.

The Second One is Operation Range or High Levels of Noise, UHC and CO Emissions

Another fundamental barrier in HCCI development is extending the operating load range whilst maintaining the full HCCI benefit. Extending the operating range is as important as the auto-ignition process. On the one hand, the HCCI operation needs to expand to higher load. At higher loads the rate of pressure rise can become so large that engine noise and knock increases significantly, and if left unchecked, engine damage may occur [11]. Therefore, the key to realize higher load is how to reduce the rate of pressure rise. On the other hand, the light load operation is also limited. For HCCI engines, as the load is reduced, either the mixture becomes leaner or the amount of dilution with EGR must be increased. Therefore, there is insufficient thermal energy to trigger auto-ignition of the mixture late in the compression stroke. Moreover, as all homogeneous combustion systems, a significant portion of the in-cylinder fuel is stored in crevices during the compression stroke and escapes combustion. But the burned gas temperature is too low to consume much of this unburned fuel when it re-enters the cylinder during the expansion stroke. This results in significant increase in both UHC and CO emissions relative to conventional combustion. In addition, the peak burned gas temperatures are too low (lower than 1400 K or 1500 K) to complete the CO to CO_2 oxidation at low loads [12]. Excess CO and UHC emissions in combination with low exhaust gas temperature at lower operating load makes this combustion mode less appealing from combustion efficiency and exhaust emissions perspectives. Therefore, the key to realize lower load is how to get sufficient in-cylinder temperature to make the combustion process more completely.

The Third One is Cold Start

Since temperatures are very low at cold start operations and the heat loss from the compressed charge to the cold combustion chamber walls is so high, the HCCI engine will encounter a major difficulty in firing during cold start operation. To overcome this difficulty, the engine may have to be started in a conventional mode and then switched to the HCCI mode after a short warm-up period. Therefore, maintaining a real homogeneous combustion after cold start will also be a real challenge. HCCI operation for cold starts is an area where much more developmental effort is needed.

The Last One is Homogeneous Mixture Preparation

Port injection is the simplest way of forming a homogenous mixture in the cylinder. This method takes advantage of the turbulence generated as the intake flow rushes past the intake valves to promote mixing. However, homogeneous mixture preparation is very difficult for fuels with low volatility, such as diesel, which requires elevated intake air temperatures for low-smoke operation when port-injected [6]. Furthermore, the fuel-air mixture is exposed to the entire time-temperature history of the cylinder, and thus injection timing cannot be used to control the start of reaction. Therefore, to control the combustion phasing, direct injection in the cylinder have been used in HCCI combustion. An early in-cylinder injection strategy, to some extent, can result in a quite homogeneous charge before ignition. However, due to lower charge density, in-cylinder pressure, and temperature, the liquid fuel impingement on the liner wall or piston wall is unavoidable, which leads to high UHC, CO, PM emissions, and oil dilution, even for moderately volatile fuels such as gasoline [13].

In fact, it is impossible to get an absolutely homogeneous mixture in the operation of practical HCCI engines. A little inhomogeneity in fuel concentration or temperature appearing in mixing can produce a modest effect on the auto-ignition and combustion process [14-18]. Furthermore, the chemiluminescence images have shown that the HCCI combustion is not homogeneous but has a strong turbulent structure even when the fuel and air are fully premixed prior to intake. The inhomogeneities are caused primarily by natural thermal stratification due to heat transfer during compression, combined with turbulent transport. This thermal stratification causes the combustion to occur as a sequential auto-ignition of progressively cooler regions, slowing the rate of pressure rise [19]. Utilizing charge inhomogeneity is an important path to achieve clean and high efficiency

HCCI combustion in engines, which can extend the operating range and control the combustion phasing. Therefore, we can say that advanced control strategies of fuel/air mixture are more important than simple "homogeneous charge'' in the control of the HCCI combustion process, such as using premixed in combination with direct injection. The key to stratification is finding the trade-off between combustion phasing control and harmful emissions increasing. A more detailed introduction will be indicated in the Section 3.

Finally, we should know that the meaning of HCCI has been greatly enriched with deeper and deeper research, and hence the broader subject area is often referred to as ''generalized HCCI combustion'' with many combustion concepts appearing under this heading, such as SCCI (Stratification Charge Compression Ignition), PCCI (Premixed Charge Compression Ignition) and LTC (low temperature combustion), *etc.* Although the combustion concepts are different, the final aim all is to realize the high efficiency and clean combustion in the complete engine operational range.

FUEL CHEMISTRY IN HCCI CONTROL

Due to the limitation caused by combustion mode and engine configuration, conventional combustion engines must use specific fuels. For spark-ignition engines, gasoline which has good volatility and anti-knock characteristic is appropriate; whilst for compression-ignition engines, diesel which has higher viscosity and lower resistance to auto-ignition is more appropriate. However, the HCCI combustion process can accept a variety of fuel types. Since ignition occurs in an HCCI engine by auto-ignition of the fuel/air mixture, the choice of fuel will have a significant impact on both engine design and control strategies.

HCCI combustion of diesel-like fuels displays a peculiar two-stage heat release, as shown in Fig. **1**. The first stage of the heat release curve is associated with low temperature kinetic reactions, and the time delay between the first and main heat releases is attributed to the "negative temperature coefficient (NTC) regime" which locates between the two heat releases stages [20]. In this NTC regime, the overall reaction rate decreases though the in-cylinder temperature increases, which leads to a lower reactivity of the system. Low temperature kinetics has been studied for some time, as this chemistry is responsible for knock in spark-ignition engines [21, 22]. Heat release from low temperature reaction relates to octane numbers of fuels. The lower the octane number is, the more obvious the heat release of low temperature reaction [23]. For gasoline-like fuels (high octane numbers), heat release from low temperature reaction (first-stage heat release) is less compared with diesel-like fuels at the same condition. Consequently, heat release from low temperature reaction is too little to obviously observe from the heat release profiles at most conditions for gasoline-like fuels [24].

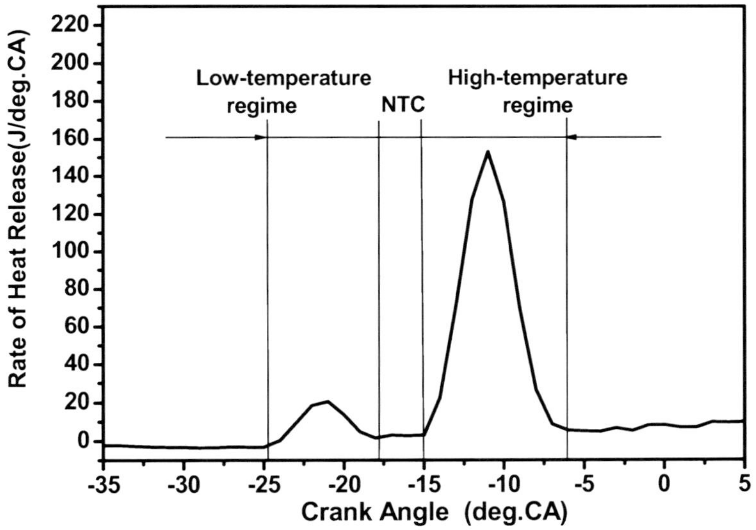

Figure 1: Typical heat release curve from HCCI combustion of n-heptane fuel.

In this part, we will introduce the gasoline-fuelled and diesel-fuelled HCCI engine. In addition, alternative fuel can save the fossil fuel and reduce the CO_2 emissions, thereby it has been got more attention in recent years. And to understand fundamental theory of HCCI combustion process, the primary reference fuel (PRF) is the best choice due to the better understood chemical kinetics. All of these fuels will be introduced in the Section 2.3: other fuels supply.

Gasoline HCCI Combustion

Gasoline fuel, with the feature of high volatility, is easily vaporized which is beneficial in forming a homogeneous charge. Therefore, there is a great benefit to use gasoline as fuel (based on diesel fuel) in HCCI engines. However, the high octane rating of this fuel means that it needs high ignition temperature, which brings the difficulty of auto-ignition by compression. This is the fundamental difficulty in achieving gasoline-fuelled HCCI combustion. Since gasoline engines are mainly used for light duty vehicles, the importance of gasoline-fuelled HCCI combustion research is focused on energy conservation and reducing emissions at medium and low loads. In addition, the ''CAI combustion'' was used broader than ''HCCI combustion'' in the literatures. But the essence of combustion is the same to them.

The main challenge for gasoline HCCI operation is obtaining sufficient thermal energy to trigger auto-ignition of the mixture late in the compression stroke. The most practical means to do this in a gasoline HCCI engine, where the compression ratio is limited, is through the use of large levels of residual exhaust gases. Ishibashi and Asai [25] used a variable exhaust port restriction to modulate residual exhaust in a single cylinder two-stroke engine, and in this way they were able to successfully control HCCI combustion phasing over one-third of the engine operating domain. However, they found that there was insufficient thermal energy to obtain repeatable HCCI combustion at near-idle loads, and there was too little charge dilution above medium loads to control the energy release process.

In four-stroke engines with flexible valve actuation, there are many strategies to obtain a high level of internal residuals. From an overall standpoint, all these valve strategies can be divided into two general modes. In the first mode, the exhaust gas re-enters the cylinder after leaving the engine through one of the ports. This is an exhaust re-breathing strategy. In the second mode, the exhaust valves are closed early in the exhaust stroke to trap the exhaust gas within the cylinder. This is an exhaust recompression strategy. There exists a wide range of valve timings that can be used with the re-breathing strategy. The intake valves can be opened for a short time during the exhaust stroke so that exhaust gas flows into the intake port, which then re-enters the cylinder with the remaining fresh charge during the intake stroke. Alternatively, residual gas dilution can be achieved by re-opening the exhaust valves during the intake stroke. Yet a third option is to hold the exhaust valve open through the exhaust stroke and well into the intake stroke (delay overall strategy). In comparison, the valve timings that can be used in the recompression strategy are simple. The exhaust valve is closed well before top dead center (TDC) of the exhaust stroke in order to trap the exhaust gas within the cylinder, and the intake valve opening location is symmetric with the exhaust valve-closing in order to minimize pumping losses during recompression.

All of these strategies have undergone experiments and been reported in the literature. For example, applying a VVA system, Kaahaaina *et al.* [26] achieved HCCI operation using the re-breathing strategy where the exhaust valve remained open throughout the intake stroke. Zhao *et al.* [27] used the exhaust recompression strategy in their HCCI experiments. Like the two stroke experiences of Ishibashi *et al.* these researchers were able to successfully modulate the levels of residuals to control HCCI combustion over a range of speed and loads. However, these dilution strategies were not able to provide enough thermal energy for satisfactory HCCI combustion at near-idle loads, and thus HCCI operation at near-idle loads remains problematic. Law *et al.* [28] compared the two dilution strategies and found no significant differences in the cylinder pressure profile or combustion characteristics between them. From a practical perspective, however, the exhaust recompression strategy appears to be easier to implement and has become the strategy favored in the literature.

Therefore, the most important and fundamental technology for gasoline-fuelled HCCI engines is obtain a high level of internal residuals with flexible valve actuation. It can be said that no matter what technologies are used in gasoline-fuelled HCCI engines, internal residuals trap is necessary for the engine. And the price of the flexible valve actuation will have a large effect on the development of HCCI engines.

Fuel Injection Strategies

To obtain the most homogeneous mixture, it is desirable to have a long mixing time between fresh air and fuel. Thus it seems that early injection using conventional port fuel injection would be more advantageous to obtain good homogeneous HCCI combustion. Successful HCCI operations have been achieved by many researchers using port fuel injection [29-33], but there are drawbacks to this operating mode. Port fuel injection offers no potential for additional combustion phasing control and limits the maximum usable compression ratio (10:1).

A switch to direct injection offers the potential for increasing compression ratios and thus extension of the HCCI light load limits. Gasoline direct injection (GDI) also offers the potential for combustion phasing control under conditions where the spark discharge is no longer effective. For the homogeneous charge preparing, a side-mounted or centrally located high-pressure fuel injector sprays fuel directly into the combustion chamber early enough in the cycle to promote homogeneous fuel-air mixing. Fuel wetting of the piston or cylinder walls is reduced by using a higher pressure (5-20 MPa) fuel injector compared to those in PFI engines, resulting in a more finely atomized spray (Sauter mean drop diameter of ~15 μm, *versus* ~120 μm for PFI) and faster vaporization. Advances in fuel-injector technology have been central in all GDI engines development, and many types of fuel injectors (*e.g.,* swirl, fan, multi-hole, and piezoelectric-pintle) have been used successfully in homogeneous charge preparing [34].

With different injection timing, the auto-ignition and combustion processes are quite different [35-38]. For example, early injection into the trapped burned gases of a lean-burn mixture during the negative valve overlap period has a large effect on combustion phasing, due to localized heat release and the production of chemically reactive species. As the injection is retarded to the intake stroke, the charge cooling effect tends to slow down the auto-ignition process. However, further retard of fuel injection to the compression stroke causes the earlier start of main combustion as fuel stratification was produced in the cylinder [35]. Combustion phasing changes are linked to changes in charge stratification resulting from changes in injection timing and can have a significant effect on emissions. Early injection created very homogeneous mixture, which leads to achieving extremely low NO_x and PM emissions, but HC emissions were quite high. By shifting to late injection timing, which resulted in significant charge stratification, they were able to cut the HC emissions, but NO_x and particulate emissions became unacceptably high. With optimum timing, however, they were able to achieve low levels of all three emissions [36]. In addition, the HCCI operational range could be expanded both to higher and lower load ranges through the adoption of direct fuel injection devices. The direct fuel injection extend the upper load range from an IMEP peak of about 400 kPa (homogeneous charge operation using port injection) to 650 kPa, but it also deteriorates the NO_x emissions. On the other hand, fuel injection during negative valve overlap expanded the operational range towards lower loads, because a reaction involving residual unburned hydrocarbons may have facilitated compression ignition [37]. A planar laser induced exciplex fluorescence technique also indicates that early injection led to homogeneous mixture but late injection resulted in serious stratification at the end of compression [38]. Furthermore, the HCCI engine performance and engine-out emissions can be optimized by double injection fueling and timing [35, 39]. In fact, the double injection can form a more intense stratification, which alters the combustion processes, emissions, and operating range. More information about stratification will be introduced in the Section 3.

Gasoline HCCI Applicable Technology

Combustion control of gasoline-fuelled HCCI engines can be divided into two areas. One covers heat release control, which could be of great benefit to enlarge the operation range. The other is auto-ignition timing control. The key factor for the ignition control is the in-cylinder gas temperature. Some approaches to control gasoline-fuelled HCCI combustion, such as charge boost, EGR, variable compression ratio, multi-mode, spark assisted, and ion sensor signal are reviewed here.

Boost

High dilution in HCCI combustion limits the amount of fuel that can be added for a given mass of charge, resulting in a low engine power. The necessity of running the engine with higher loads has prompted the investigation of supercharged or turbocharged HCCI. Some studies have applied charge boost methods to increase the speed-load window of HCCI include boosting with turbocharger on a PFI six cylinder engine [40] and boosting a single cylinder engine by standalone compressor [10, 41, 42]. Moreover, the SI, naturally aspirated (NA) HCCI and supercharged HCCI are investigated in [43], the results of which show that a smaller supercharger with a moderate boost is preferable to avoid an unacceptable penalty in fuel economy. All of these results are showing satisfactory improvement in terms of load and NO_x over aspirated HCCI. In addition, in combine with internal EGR, the IMEP can reach 7.6 bar (boost pressure is 1.4 bar), which is approximately 75% of the total engine load possible in this engine configuration with SI combustion [44], as shown in Fig. **2**. Therefore, boost should be the necessary for the HCCI engine.

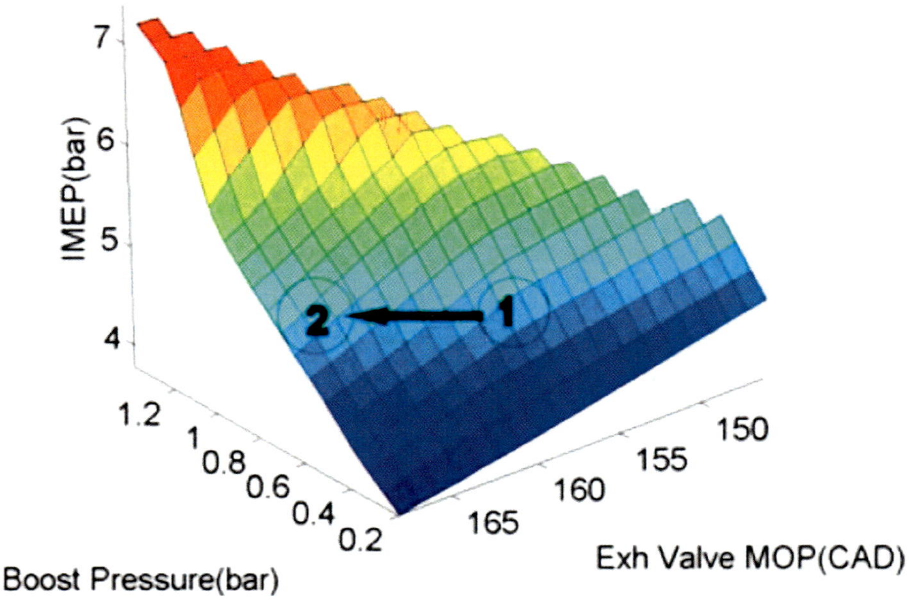

Figure 2: Engine load for various boost pressures and exhaust valve timing at 1500rpm. (Points "1" and "2" represent the same IMEP but very different NO_x emissions) [44].

EGR

A high level of internal residuals with flexible valve actuation is the most important and fundamental technology for gasoline-fuelled HCCI engines, which has been introduced in previously. In addition, the internal EGR can combine with external EGR to increase the attainable load [45]. The amount of residual gas trapped in the cylinder was adjusted *via* the NVO method (recompression). The flow of externally re-circulated exhaust gas was varied using a typical production level valve. Under stoichiometric fuelling conditions, the highest output achieved using internal exhaust gas was limited by excessive pressure rise and unacceptable levels of knock. Introducing additional external exhaust gas was found to retard ignition, reduce the rate of heat release and limit the peak knocking pressure. In Fig. **3**, it can be seen that addition of external EGR enabled significant increase in peak engine output, rising from 350 kPa to 580 kPa (~65% load). This supplementary cold diluting gas served to retard ignition, evident in the phasing of the 10% mass fraction burned. In addition, under conditions of combined EGR, NO_x values further decreased at high loads.

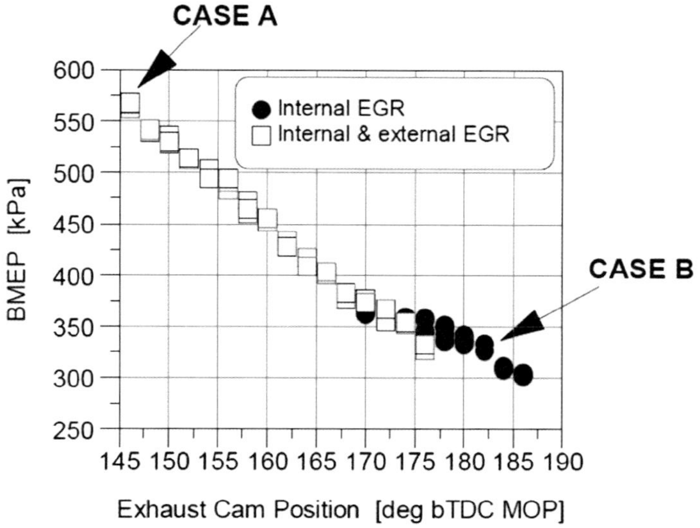

Figure 3: BMEP *versus* exhaust cam position (1500rpm, lambda=1.0) [45].

Variable Compression Ratio

HCCI combustion can be achieved by increasing the compression ratio to the point where the required temperature and pressure for auto-ignition are achieved through compression alone. The combustion phasing can also be controlled by variable compression ratio. However, it is difficult to apply in practice due to the complex engine design. Hyvonen *et al.* [46] investigated the operating range in terms of speed and load with naturally aspirated four-stroke multi-cylinder engine. HCCI combustion control is achieved by variable compression ratio and inlet air preheating with exhaust heat. The range of variable compression ratio is between 9:1 and 21:1. Operating ranges in Fig. **4** shows that lower compression ratios are used at intermediate loads as well as intermediate and high engine speeds to increase combustion efficiency, because higher inlet air temperature is available to keep the preferred CA50 (crank angle at 50% completion of heat release). However, compression ratios are close to maximum both at low load and high load. At low load and idle mode, maximum compression ratio and maximum available inlet air temperature are needed to auto-ignite and increase combustion efficiency. At high load the maximum compression ratio is used for reducing inlet air temperature to maximize volumetric efficiency when running NA. In addition, the results also show that brake efficiency is increased by 35% at 3 bars BMEP and 57% at 3.5 bar BMEP compared with conventional SI combustion with the same engine using a 14:1 compression ratio. In addition, almost any liquid fuel can be used in a HCCI engine using a variable compression ratio [47]. About more studies on the VCR can also be found by Christensen and Johansson [48] and Hyvonen *et al.* [49].

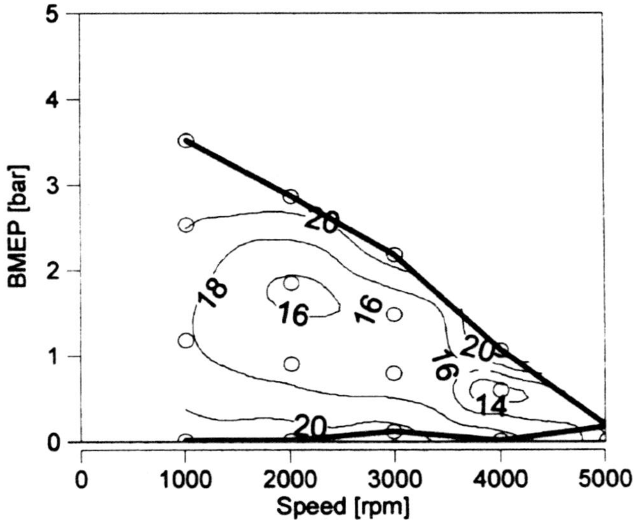

Figure 4: Operating range and compression ratio iso-lines with gasoline fuel [46].

Multimode

Multimode combustion is an idle combustion strategy to utilize HCCI for internal combustion engines. It combines HCCI combustion mode at low to medium loads with traditional SI mode at high speed and high loads. To fully utilize the potential of HCCI combustion for high fuel conversion efficiency and low NO_x emissions and to improve HCCI control for vehicle applications, a gasoline HCCI-SI dual mode engine system called the optimized kinetic process (OKP) was proposed by Yang *et al.* [50]. The OKP engine concept utilizes the waste thermal energy in an engine to heat the intake air, and mixes the heated and un-heated air streams to control the intake air temperature for HCCI combustion timing control. The concept also involves increasing engine geometric compression ratio to operating conditions about 15:1 for HCCI combustion, and reducing the effective compression ratio to about 10:1 for SI combustion. If the effective compression ratio is reduced by retarding the intake valve close timing to operate the engine in the Atkinson cycle mode, the volumetric efficiency will decrease. To compensate for the loss in volumetric efficiency, the intake air pressure has to be boosted at high-full loads. Using this strategy, a single-cylinder OKP engine has demonstrated a gross indicated mean effective pressure of 11.7 bar in SI mode [50].

The AVL-CSI system, as proposed, developed and demonstrated by Fuerhapter *et al.* [51], was used in an OPEL 2.0L 4-cylinder engine. At HCCI mode, the intake camshaft was shifted to small valve lift for early intake closing and the amount of fresh air was adjusted *via* a cam phaser and by setting the intake manifold pressure with an electronic throttle. SI combustion with early intake closing (Miller cycle), offers remarkable fuel reduction potential and the intake camshaft was switched to big valve lift in order to provide sufficient full load performance. Xie *et al.* [52] described a flexible mechanical VVA system named 4-variable valve actuating system (4VVAS). The 4VVAS system is capable of independent control of intake and exhaust valve lift and their timing. To investigate the potential of the 4VVAS system for HCCI engine operation and the switch between HCCI and SI operations on the same engine, an engine simulation program has been developed based on Matlab Simulink incorporating an engine block. And the experimental results also indicate that 4VVAS enables rapid and effective control over the in-cylinder residual gas, and therefore dynamic transitions between HCCI and SI can be stably achieved [53]. More references on the multi-mode can be found in Tian *et al.* [54], Koopmans *et al.* [55], Roelle *et al.* [56], and Santoso *et al.* [57], Shaver *et al.* [58], Zhao [59], and Wang [60].

The multi-mode is the developmental direction for the large scale production of gasoline-fuelled HCCI engines in the future. While the flexible valve actuation and injection strategies are the keystone in order to reach the combine HCCI combustion mode at low to medium loads with traditional SI mode at high speed and high loads.

Ion Sensor Signal

In general, expensive pressure transducers are used in laboratories to determine combustion characteristics in the engine. Ion sensors are significantly less expensive than pressure transducers. For example, an existing spark plug can often be used as an ion sensor. An ion sensor signal has been used for combustion diagnostics [61, 62], combustion stability control [63], peak pressure position control [64], predicting mass fraction burned [65] and the knock onset [66] in SI engines. The engines cited above operate at or near stoichiometric conditions where temperatures are high and the production of chemi-ions is significant. As engines operate leaner, the lower temperatures dramatically reduce the production of chemi-ions. In the case of HCCI engines, where equivalence ratio is seldom above 0.5, few researchers expected to find an ion signal. The paper written by Strandh *et al.* [67] has been the first to report an ion signal in a gasoline-fuelled HCCI engine. Mehresha *et al.* [68] investigated the ion current signal in HCCI engines. This research is to show that a measurable ion current exists even in the very lean combustion ($\phi = 0.35$) in an HCCI engine. The research shows that an inexpensive ion sensor may replace the expensive pressure transducers currently used in HCCI engines. It is equally capable of capturing cycle-to-cycle variations in HCCI combustion. The ion signal becomes stronger with increases in equivalence ratio and intake temperature. In their further researches [69], experiments were conducted to measure ion signals produced from the combustion of gasoline, ethanol, and n-heptane in a 4-cylinder HCCI engine with different equivalence ratios and intake pressures. It was found that the ion signal is reduced under several situations: with an increase in intake pressure, reducing equivalence ratio (under lean onditions), and decreasing the bias voltage source. The gasoline and ethanol produces more ions than n-heptane during combustion under the same operating conditions.

Spark Assisted

The spark ignition can play an important role in controlling HCCI combustion ignition in low load boundary region. Xie *et al.* [70] investigated the influence of spark ignition on HCCI combustion based on internal EGR strategy in a single engine. The low temperature chemical reaction process is shortened and the auto-ignition timing is advanced due to the spark discharge. Meantime, lower fuel consumption and cycle-to-cycle variations can be achieved. The spark discharge makes auto-ignition easier in low load boundary region, so that the HCCI operation can be expanded to lower load through assisted spark ignition. Therefore, it has the potential of spark ignition for control of auto-ignition timing and control of HCCI/SI mode switch. Hyvonen *et al.* [49] investigated the spark assisted HCCI combustion during combustion mode transfer to SI in a multi-cylinder VCR-HCCI engine. They found that spark assisted can be used for controlling the combustion phasing during a mode transfer between HCCI and SI combustion. However, the combustion fluctuations are large in the intermediate combustion region where some cycles have both spark ignited flame propagation and auto-ignition, but some cycles have only partially burnt flame propagation. More articles about spark assisted HCCI can be found by Persson *et al.* [71], Berntsson *et al.* [72], Bunting [73], and Felsch *et al.* [74].

Gasoline Fuel Surrogate Development on the Prediction of Auto-Ignition

Autoignition is of critical importance in HCCI engines. Auto-ignition is determined by the evolution of pressure and temperature in the fuel/air mixture with time and by the fuel. Practical fuels are complex mixtures of hydrocarbons and their auto-ignition characteristics must be described by empirical measures such as research and motor octane numbers, RON and MON, which are not always adequate [75]. Ideally, auto-ignition would be described by comprehensive chemical kinetics as the pressure and temperature evolve with time. However, such mechanisms applicable to HCCI engines have only been developed for a very few simple fuels, such as iso-octane [76, 77] or primary reference fuel [78, 79]. Nevertheless, the empirical knowledge gained about the auto-ignition of practical fuels needs to be understood in terms of fundamental chemical kinetics. This requires the development of chemical kinetic schemes for fuel mixtures that are more like practical fuels.

At first, fuels like toluene/n-heptane mixtures [80, 81] which behave more like practical fuels than isooctane/n-heptane mixtures. The fuel toluene/n-heptane mixtures have a significant improvement of the prediction of the general trend of auto-ignition phasing in HCCI combustion, in comparison to the PRF. The experimental data have been used to validate a new detailed chemical kinetics mechanism including cross-reactions [80].

And then, some studies use the PRF and toluene as a gasoline surrogate. Machrafi and Cavadias [82] carried out experimental and kinetic studies using a gasoline surrogate containing 11 vol% n-heptane, 59 vol% iso-octane, and 30 vol% toluene, PRF, and gasoline. The RON of these fuels is 95. Due to the additive of toluene in the surrogate, the auto-ignition timing and heat release rate trace is nearly the same to the gasoline fuel. But there is a large difference between PRF and gasoline fuel. Except the typical cool flame and final ignition stages are observed, another small heat release seems to be observed for gasoline and its three-component surrogate. In order to understand these phenomena, an experimentally validated kinetic chemical mechanism is used, which shows that the competition between the consumption of the fuel and the formation of an at that moment relatively stable benzyl radical from toluene cause the so-defined obstructed pre-ignition and delay the actual final ignition. Therefore, the kinetics of the auto-ignition of toluene plays an important role in the combustion process. Furthermore, a reduced surrogate mechanism has been composed out of reduced n-heptane, iso-octane and toluene mechanisms, containing 62 reactions and 49 species. This mechanism has been validated numerically in a 0D HCCI engine code against more detailed mechanisms (inlet temperature varying from 290 to 500 K, the equivalence ratio from 0.2 to 0.7 and the compression ratio from 8 to 18) [83]. Gauthier *et al.* [84] investigated the ignition-delay time of gasoline behind reflected shock waves and compared them to the results of gasoline surrogates comprised of isooctane/toluene/n-heptane by keeping the n-heptane fraction constant and varying the isooctane/toluene proportion. The optimized surrogates showed a good agreement to real gasoline in terms of ignition delay times. Fikri *et al.* [85] incorporate the diisobutylene (10%) with n-heptane (20%)/toluene (45%)/isooctane (25%) by liquid volume to investigate ignition times in high-pressure shock-tube experiments. The fuel has similar research octane numbers (RON 95) and motor octane numbers (MON 85) that correspond to the standard European gasoline. The data for this gasoline surrogate is in good agreement with the literature data for other gasoline surrogates, at high temperatures at a fixed pressure. The experimental findings can be used in future work to validate kinetic models for the auto-ignition of fuel blends containing diisobutylene.

In addition, due to the mixture of ethanol and gasoline are very common all over the world, some studies are carried out by fuels of PRF and ethanol. Fikri *et al.* [85] incorporate the ethanol (20%) with n-heptane (18%)/isooctane (62%) by liquid volume, which also was carried out in high-pressure shock-tube. Cancino *et al.* [86]investigate the ignition-delay times in shock-heated gases for a surrogate gasoline fuel comprised of ethanol/iso-octane/n-heptane/toluene at a composition of 40%/37.8%/10.2%/12% by liquid volume with a calculated octane number of

98.8. The experiments were carried out in stoichiometric mixtures in air behind reflected shock waves in a heated high-pressure shock tube. Initial reflected shock conditions were as follows: temperatures of 690–1200 K, and pressures of 10, 30 and 50 bar, respectively. Detailed kinetics modeling was used to simulate the auto-ignition of this surrogate fuel and two other gasoline surrogates (ethanol/isooctane/n-heptane: 20%/62%/18% by liquid volume, iso-octane/n-heptane/toluene 69%/17%/14% by liquid volume) based on experimental data from the literature [85, 87]. The model was developed by incorporating sub-mechanisms for ethanol and/or toluene oxidation into the primary reference fuel model of Curran *et al.* [78]. The resulting model shows the right trend but fails to capture absolute values of the ignition delay times, especially at low temperatures. All kinetics models over estimate the ignition delay times.

Finally, because the gasoline composition is very complicated, some other composition has been added to the fuels. Yahyaoui *et al.* studied the oxidation of iso-octane/toluene/1-hexane/ETBE mixtures in a jet stirred reactor [88]. The same group studied the auto-ignition of a binary mixture of 1-hexene and toluene in a shock tube in the 750–1860 K temperature and 2–10 bar pressure range [89]. All in all, the literature is rich in detailed kinetic models for the oxidation of mixtures of components of gasolines. These models can in most cases satisfactorily capture the evolution of auto-ignition delay times with temperature and pressure. Validated models for multicomponent mixtures representative of gasoline (*e.g.,* n-heptane/iso-octane/toluene, iso-octane/1-hexene/toluene, or n-heptane/iso-octane/ethanol blends) are starting to become available. The question about the optimum number of components, which should be considered in a surrogate for it to be well representative of an actual fuel, without increasing unreasonably the size of the model, is still open. It is certainly important that the components of a surrogate are carefully chosen so that they are well representative of the species present in the actual fuel. About more detailed reference about the gasoline surrogate can be found in the review by Battin-Leclerc [90].

Diesel HCCI Combustion and Low Temperature Combustion (LTC)

In diesel HCCI combustion, it is difficult to prepare homogeneous mixture because of the lower volatility, higher viscosity and lower resistance to auto-ignition of diesel fuel. First, elevated temperatures are required before significant vaporization occurs, making it easier to form a premixed homogeneous charge. Second, diesel fuel has significant cool-combustion chemistry, leading to rapid auto-ignition once compression temperatures exceed about 800 K [91]. This can lead to overly advanced combustion phasing and high combustion rates. Accordingly, the essential factor needed to achieve diesel HCCI combustion is mixture control, including both charge components and temperature control in the whole combustion history and high pre-ignition mixing rates.

Initial attempts to obtain diesel-fueled HCCI involved premixing using intake-port fuel-injection by Ryan and Callahan [5], Gray and Ryan [6], Christensen *et al.* [47]. However, significant intake heating (135-205 °C) was required to minimize the accumulation of liquid fuel in the intake system and to prevent soot and NO_x formation as a result of in-cylinder inhomogeneities. This heating, combined with the high cetane number of diesel fuel, required a lower CR, typically to values in the range of 9-11 to prevent knocking. In addition, combustion efficiency was poor due to high levels of unburned HC. As a result of the reduced CR and incomplete combustion, fuel consumption increased significantly compared to conventional diesel combustion, although NO_x and smoke levels were substantially reduced.

Early Direct Injection HCCI Combustion

Because of the disadvantage of port fuel injection, an early direct injection approach towards achieving diesel-fuelled HCCI has been investigated. Compared with premixing in the intake port, this approach offers three potential advantages [92]. 1) By injecting the fuel in the compression stroke, the higher in-cylinder temperatures and densities can help vaporize the diesel fuel and promote mixing. This allows cooler intake temperatures, reducing the propensity for early ignition. 2) With carefully designed fuel injectors, the possibility exists to minimize fuel wall wetting that can cause combustion inefficiency and oil dilution. 3) In principle, only one fuelling system is required for both HCCI and conventional diesel operation. The main disadvantage of early DI for HCCI is that it is easy to produce wall wetting due to over-penetration of the fuel. Finally, it should be noted that controlling combustion phasing is still a critical issue for early-DI

HCCI because injection timing does not provide an effective means of directly controlling combustion phasing as in conventional diesel combustion. Several different methods of early-DI HCCI have been investigated, including dual-injection techniques that combine early-DI HCCI with a conventional diesel injection. Many combustion systems of early direct injection HCCI have been developed such as PREDIC (premixed lean diesel combustion)/MULDIC (multiple stage diesel combustion), UNIBUS (uniform bulky combustion system), and MULINBUMP (multi-in-jection and BUMP combustion chamber).

One of the more significant efforts in early-DI HCCI is the work conducted at the New ACE Institute in Japan. Initial reports of this work by Takeda *et al.* [93] and Nakagome *et al.* [94]. To minimize the over-penetration and wall wetting that can occur when fuel is injected well before TDC, three injectors were used in their study (one at the centre, and two on the sides) to control the injection timing and quantity of each injector independently. When in-cylinder air densities and temperatures are low (during compression stroke), the double injectors mounted on either side of the cylinder, each with two sprays, were positioned so that the fuel sprays would collide in the middle of the chamber to limit wall wetting with very early injection. With PREDIC operation, self-ignition occurs near top dead center and NO_x emissions are drastically lower (20 ppm). When ignition occurs, the fuel-air mixture is thought to be nearly homogeneous, with only slight heterogeneity. A subsequent study [95] reported the development of a swirling-flow pintle-nozzle injector that produces a more uniform mixture and reduces wall wetting (as measured by reduced lube-oil dilution), thereby improving combustion efficiency. In an attempt to achieve higher power densities, this early-DI work was expanded to include a second fuel injection near TDC [96]. This was accomplished by using the dual side-mounted, impinging-spray injectors to produce the lean premixed charge, whilst a centre-mounted injector was used to inject additional fuel near TDC. Approximately half of the fuel was injected in each stage, which was near the maximum amount of premixed fuel that could be used without causing knock. Various combinations of injection timings (for both stages) and fuel CN were used to optimize performance and emissions. Since the second stage burns similarly to conventional diesel combustion, NO_x and PM emissions were much higher with this approach than those obtained with pure

HCCI. However, NO_x was reduced to about half that of conventional diesel combustion without a significant increase in fuel consumption. Smoke emissions also decreased, but HC emissions were higher. In order to solve the problem of high HC and CO emissions, the further research and development were carried out by Akagawa *et al.* [97], Morita *et al.* [98], and Nishijima *et al.* [99] using EGR, oxygenated fuel, water injection.

Yanagihara *et al.* [100] have also applied HCCI combustion with diesel fuel (Toyota UNIBUS). As described by Yanagihara, the engine operates in UNIBUS mode up to about half load and half speed using piezo-actuator injectors with pintle-type injector nozzles to reduce the spray penetration in an attempt to limit the over-leaning of the charge. The technique involves a combination of an early injection (around 50° BTDC) and a late injection (about 13° ATDC). The majority of data presented in the paper were taken with 12:1 compression ratio, presumably to avoid overly advanced combustion of the premixed fuel, or using the EGR to delay the HCCI combustion timing (a 60% EGR level can delay about 7 °CA). A luminous flame is not observed, due to the negligible PM available for oxidation, and confirmed by low PM emissions. The second injection significantly improves the combustion efficiency of the early-injected fuel, reducing HC emissions from 5000 ppm to 2000 ppm and CO emissions from about 8000 ppm to 2000 ppm. Therefore, fuel combustion from the second injection can be considered as a trigger for a portion of the heat release from the early-injected fuel. Although the second injection increases NO_x emissions above the near-zero levels for early injection alone, they are still very low for a diesel engine. For this particular injector nozzle, low levels of fuelling and low injection pressures were thought to be most appropriate to limit wall wetting and knocking. Further studies [101] were then carried out looking at the possibility of using a second injection closer to TDC to trigger the combustion, which reduced the NO_x benefits but also the fuel consumption penalty.

Su *et al.* [102-104] proposed a compound diesel HCCI combustion system – MULINBUMP, combining premixed combustion with "lean diffusion combustion". The premixed combustion was achieved by the technology of multi-pulse fuel injection (~6 times), as shown in Fig. **5**. The start of pulse injection, injection pulse number, injection period of each pulse and the dwell time between the injection pulses were controlled. The objective of controlling the pulse injection was to limit the spray penetration of the pulse injection so that the fuel will not impinge on the cylinder liner, and to enhance the mixing rate of each fuel parcel. The last or main injection pulse was set around TDC. A flash mixing technology was developed from the development of a so-called BUMP combustion chamber, which was designed with some special bump rings. The combustion of fuel injected in the main injection proceeds at much higher air/fuel mixing rate than in a conventional DI diesel engine under the effect of the BUMP combustion chamber, which leads to ''lean diffusion combustion''. By advancing the start timing of multi-pulse injection, noisy auto-ignition can be avoided, but over-advancing of the start timing will lead to over-mixing, resulting in higher total hydrocarbon emissions. The results also show that in the compound combustion mode, the fuel proportion of multi-pulse injection should be high, under the condition of no noisy auto-ignition, which is helpful in the reduction of NO_x emissions. Small amounts of fuel in multi-pulse injection offer no benefits to NO_x emissions reduction, but worsen smoke emissions. Compound combustion technology has potential in realizing HCCI combustion at rather wide operating conditions of a diesel engine. The basic idea of the MULINBUMP compound combustion system is: 1) at low loads of a diesel engine, there is control of the HCCI combustion process by the multi-pulse fuel injection strategy, which can achieve very low NO_x and PM emissions (<10 ppm); and 2) at medium and high loads, the strategy of combining premixed combustion with ''lean diffusion combustion'' is applied. The mixing rate of fuel and air is improved by combining a high mixing rate combustion chamber with a super high injection pressure, thus achieving rapid mixing. Over the full load range of a diesel engine, the best injection modes can be obtained at different conditions by controlling multi-pulse injection, thereby clean and high efficiency combustion of the diesel engine can be achieved over the full load range. To date, the IMEP of compound combustion has reached 0.93 MPa.

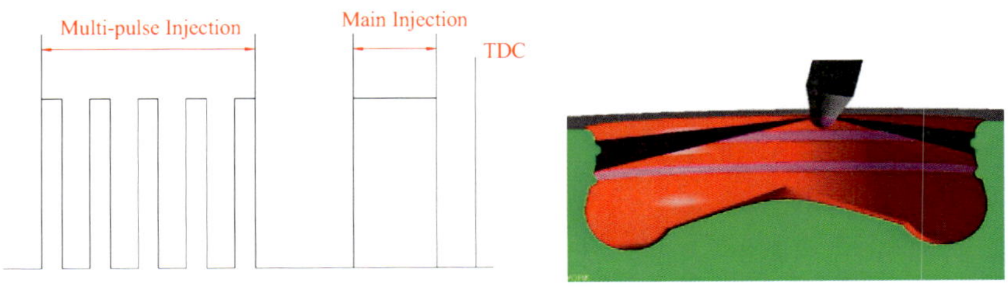

Figure 5: Multi-pulse injection and BUMP combustion chamber.

Many efforts have been made to improve mixture formation of these early-DI systems to reduce the fuel wetting on the piston head surface and cylinder wall, especially to use the special injection system, such as PREDIC, UNIBUS, MULINBUMP, or narrow angle [105-108]. However, for these early-DI HCCI combustion, maintaining the desired combustion-phasing with changes in speed and load remains a challenge for these early injection techniques in which the auto-ignition timing is only controlled by the chemical kinetics. The high cetane number of diesel fuel can lead to early ignition and knocking. Therefore, many studies try to use high levels of cooled EGR to slow the auto-ignition process [97, 100, 109], to reduce the negative work during compression stroke. Additionally, somewhat reduced compression ratios of 12:1~15:1 are employed in the diesel-HCCI concepts [100, 105], and some approaches also use late intake valve closing (IVC) to further reduce the effective compression ratio [110, 111]. Late IVC has the advantage of allowing the effective compression ratio be rapidly adjusted as part of the control system, and a high expansion ratio is preserved for good cycle efficiency. Early-injection approaches continue to be an area of research, but they are always considered as part of an overall combustion strategy, due to high load limitation.

Late Direct Injection HCCI Combustion

One of the most successful late injection DI HCCI systems for achieving diesel-fuelled HCCI is the MK (modulated kinetics) combustion system developed by the Nissan Motor Company. The principles of this late injection HCCI combustion process are described by Kawashima *et al.* [112] and Kimura *et al.* [113]. In order to achieve the diluted homogeneous mixture required for HCCI, a long ignition delay and rapid mixing are required for late injection DI HCCI combustion. In the MK system, the ignition delay is extended by retarding the injection timing from 7° BTDC to 3° ATDC and by using high levels of EGR, sufficient to reduce the oxygen concentration to 15–16%. Rapid mixing was achieved by combining high swirl with toroidal combustion-bowl geometry. The operating range for the first generation MK system was limited to about one-third of peak torque and half speed. In the MK mode, NO_x emissions were reduced substantially (to about 50 ppm) without an increase in PM. Combustion noise was also significantly reduced. In addition, it should be noted that with this late-DI HCCI technique, combustion phasing is controlled by injection timing to some extent, which is an advantage over port fuel injection and early-DI HCCI techniques.

In the development of a second-generation system, several modifications were made to expand the range of MK operation to higher loads and speeds [114]. Since more fuel must be injected at higher loads, a high pressure common-rail fuel system was used to provide high injection pressures at all speeds to reduce fuel injection duration. The ignition delay was increased by reducing the compression ratio to 16:1 and adding EGR cooling to reduce the intake temperature. To minimize the potential for liquid fuel impingement on the piston bowl wall, the piston bowl diameter was increased from 47 to 56 mm. This change significantly reduced HC emissions under cold-engine conditions. The results show that the second-generation MK system can be used over the entire range of everyday driving, indicating that the engine met the second-generation targets of about half load and three-quarters speed. NO_x emissions are stated to be reduced by 98% compared with conventional operation without EGR, whilst PM emissions are similar to the conventional engine. More recent work by Kawamoto *et al.* [115] proposed the use of higher CN fuel, which went against other HCCI operating requirements, but was recommended to reduce HC emissions under cold start conditions.

The major advantage of the MK combustion system is illustrated by its ease of implementation because it does not require additional or different hardware. Equally, its operation with conventional hardware did not affect negatively the engine's specific power output. However, the injection timing retard and lower compression ratio lead to a deterioration of cycle efficiency and higher unburned HC emissions.

Premixed/Direct-Injected HCCI Combustion

In premixed/direct-injected HCCI combustion, port injection was chosen for the main fuel supply to create a homogeneous charge, and direct fuel injection into the cylinder was used to change the concentration and position of local fuel-rich regions with the purpose of controlling HCCI combustion. A homogeneous charge diesel combustion (HCDC) system, which is the earliest premixed/direct-injected HCCI combustion, has been proposed by Odaka *et al.* [116-119]. In their system, most of the fuel (iso-octane [116, 117], diesel or mixture of diesel and MTBE [118, 119] was injected into the intake manifold to form a homogeneous pre-mixture in the combustion chamber and the premixture was ignited with a small amount of fuel directly injected into the cylinder. This system can reduce both NO_x and smoke emissions better than ordinary diesel engines. Smoke is reduced near-uniformly as the premixed fuel ratio is increased. However, a knocking due to uncontrolled self-ignition may occur under high premixed fuel ratio conditions with the diesel fuel. Thus, the maximum amount of premixed fuel was restricted by these knocking limits.

In addition, Midlam-Mohler *et al.* [120-122] have also developed a premixed/direct-injected HCCI system. Port/manifold fuel injection is achieved using a low pressure atomizer system and direct injection is achieved using a high pressure injection system. The atomizer system delivers fuel as a premixed lean homogeneous mixture into the cylinder. At low loads, the main torque is from the homogeneous charge fuel and the direct injection fuel is mainly for ignition; at high loads, the maximum homogeneous charge fuel is used and direct injection fuelling is increased to full load. The basic injection scheme is shown in Fig. **6**.

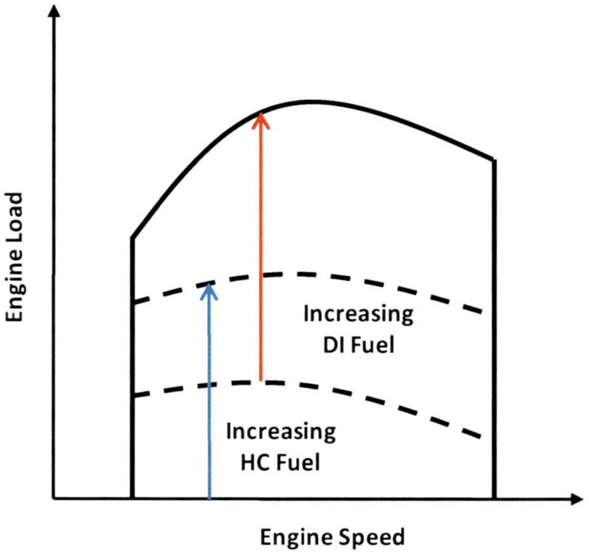

Figure 6: Basic injection scheme in a premixed/direct-injected HCCI mode. [120-122].

In fact, when the direct injection fueling rate is small, it can be seen a trigger to autoignite, but the DI fuel will form different stratification in the cylinder as the increase of DI fueling rate. Foster *et al.* [123] and Berntsson and Denbratt [124] presented results pertaining to expanding HCCI operations by charge stratification achieved by both port and direct fuel injection. In addition, in premixed/direct-injected HCCI combustion, these two fuels can also be different, and fuel characteristics can also be used to control combustion, for example the study by Inagaki *et al.* [125], gasoline was supplied to the intake air port and diesel fuel was injected directly into the engine cylinder to act as an ignition trigger. The combustion control through fuel characteristics will be introduced in the Section 2.3.

Low Temperature Combustion in Diesel Engines

Some approaches to implement HCCI combustion in diesel engines include port fuel injection, very early in-cylinder direct injection (single pulse or multiple pulses), late direct injection (single pulse or multiple pluses), or a combination of the two. Since the diesel fuel has low volatility, the port fuel injection is not a practical way without significant change of intake system. An early in-cylinder injection strategy, to some extent, can result in a quite homogeneous charge before ignition. But the liquid fuel impingement on the liner wall or piston wall is unavoidable due to lower charge density, in-cylinder pressure, and temperature, which leads to high HC and CO emissions as shown in upper literature. Therefore, some new types of injectors with low fuel penetration or narrow angle were developed to overcome these problems. However, because early injection HCCI combustion mode can only meet the requirement in low load conditions, in high load and high-speed region, the same injector will cause problems, for example serious fuel wall impingement and soot emission increase. Another issue for the early injection strategy is the ignition timing control. For early injection HCCI combustion, the ignition is mainly controlled by the chemical kinetics. The ignition is often advanced due to early injection timing and other measures have to be taken to delay the ignition by avoiding over-stress the engine resulted from high peak pressures like using heavy EGR, variable compression ratio, changing fuel property, *etc.* From the literature, it seems only MK combustion and UNIBUS combustion systems were successfully applied in production engines under part load regimes. Even though some new concepts were proposed by combining the HCCI mode and conventional diesel combustion, in the high load and high speed operating regimes, smoke and NO_x are still problems for diesel engines. The main concern of diesel engine emission reduction is to expand HCCI mode or improve the high load/speed diesel combustion. As stated in the paper [126], it is hard to get a real homogeneous mixture in diesel engines for a late injection strategy. Even for very early injection, there still exists partial charge heterogeneity [94]. Emission reduction in HCCI combustion is expected mainly due to low temperature reaction rather than homogeneity itself based on the previous studies. Therefore, low temperature combustion is the main reason for emission reduction.

Recently, the low-temperature combustion has gained tremendous attention due to its benefits of low NO_x and PM emissions [127-131]. For the LTC, the short times between the fuel injection event and the start of combustion preclude thorough pre-mixing, and significant regions exist where ϕ >1 at the start of combustion. Even though there is a locally rich region in the mixture of this strategy similar to conventional diesel combustion, the soot formation can still be suppressed. The main soot suppression mechanism is that the temperature is reduced by using large amounts of EGR, and this temperature reduction is sufficient to allow the combustion to avoid the soot formation region. This is the major reason why smokeless combustion can be accomplished with no adjustment required in the mixture formation by changing fuel spray system, combustion chamber geometry *etc.* under rich operating condition. In conventional combustion, this reduces temperatures just enough to decrease oxidation but not accumulation for the nucleation of soot, that is why the soot emission will go up with the increase of EGR rate. Whilst in the LTC regime, the addition reactions are also hindered and therefore the accumulation mode begins to slow down and this reduces soot with further increase in EGR [132]. Fig. **7** shows that soot incandescence can be negligible for fuel-rich local mixture stoichiometries that would result in intense soot incandescence under undiluted operating conditions, with a blue flame colour for LTC due to very low flame temperatures and a bright yellow colour for no EGR condition due to soot incandescence [133].

Simultaneously, the NO_x emissions can also be avoided due to the high EGR rates and thus low combustion temperature. Furthermore, the EGR rate influences the path not only through changes in the flame temperature, but also in ignition delay and the amount of ambient fluid that must be mixed with the fuel to attain a given equivalence ratio. In addition, the injection characteristic (including injection pressure, timing and multiple injections) influences the temperature (and density) during the ignition delay period, the peak flame temperature reached, and the premixing improvement. Finally, in order to keep the power density and the combustion efficiency of the engine at high EGR rates, high boost levels are required. Therefore, control and optimization of the EGR rate, injection characteristic and high boost are the keystones for the LTC.

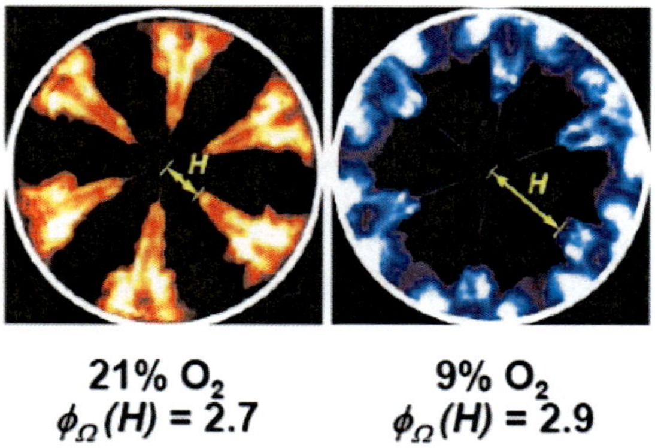

Figure 7: The imaging of conventional diesel and LTC (Engine speed=1200rpm, nitrogen dilution to simulate the EGR, using the DGE as fuel) [133].

For LTC, A large EGR rate results in a longer ignition delays and a lower ambient air temperature. And then, the air-fuel mixture starts combusting with a low flame temperature which inhibits the soot formation process at the beginning and also extends the time for air-fuel mixing without soot formation. Therefore, the LTC shows premixed combustion dominant heat release patterns [134]. But, a mixing-controlled combustion phase remains important under highly dilute operating conditions to achieve high fuel efficiency and low HC and CO emissions [128]. Many studies have been investigated to improve the mixing in LTC process through injection pressure [135, 136], swirl [135], piston bowl geometry [137], boosting [138, 139], and also including the different EGR rate.

However, due to the high level EGR rate, and thus the low temperature combustion processes, the HC and CO emissions are quite high and the fuel efficiency reduce. The optical diagnostic images show that the transient ramp-down at the end of fuel injection produces a low-momentum, fuel-lean mixture in the upstream region of the jet, which persists late in the cycle. At LTC conditions with long ignition delay, this mixture likely becomes too lean to achieve complete combustion, thereby contributing to UHC emissions. Therefore, mixtures near the injector after the end of injection are too lean to achieve complete combustion, thus contributing to UHC emissions [140, 141]. And increased dilution slows oxidation throughout the cylinder, but most noticeably within the squish volume, significantly increasing UHC and CO emissions from this region [142]. The HC and CO emission mechanisms in LTC at part load have been investigated using KIVA-3V Chemkin code by Opat *et al.* [143].

Due to the high HC and CO emissions, the "Trade-off" of soot-NO_x for conventional combustion changes to a "trade-off" of soot-fuel efficiency (HC and CO emissions). There are some methods can improve this new "trade-off" of soot-fuel efficiency by changing following: the injection timing, injection pressure, boost, and adding oxygenated fuel, *etc.* All these methods can shift the peak soot emissions towards lower oxygen concentration, thereby it can use a medium EGR, there is almost no fuel penalty, HC, CO and soot emissions are low, but NO_x emissions cannot meet Euro VI without aftertreatment [144].

Diesel Fuel Surrogate Development on the Prediction of Auto-Ignition

Since commercial engine fuels, gasoline and mainly diesel fuel, consist of a complex mixture of hundreds of medium-high molecular weight hydrocarbons, it is not feasible to consider the oxidation chemistry of all the compounds for modelling targets. Therefore, kinetic studies of commercial fuels can only be carried out by considering surrogate mixtures of a small number of very well known hydrocarbons that simulate properly some of the physical or chemical properties of the original fuel such as vapour-liquid equilibrium, viscosity, ignition delay, heat release rate, pollutant emissions, *etc.* The gasoline surrogate has been reviewed in the Section 2.3.1, this part will indicate the diesel surrogate.

However, due to the higher molecular weight hydrocarbons in diesel fuel, the research on the surrogate for diesel fuel is more complicate than the gasoline. Due to n-heptane has a cetane number (CN) equal to that of diesel fuel (CN = 56) and its oxidation chemistry is very well known, so it has been widely used as diesel fuel surrogate [145-147]. However, CN does not seem to be representative enough of the ignition characteristics of a fuel, because its experimental determination (in a cooperative fuel research) does not cover a real and wide range of engine conditions and because the physical delay time (involving the fuel atomization and evaporation processes and the mixture with air) is also affecting the ignition timing. And then, due to the very high aromatic content of diesel (around 30% in mass), many studies added the toluene to the n-heptane as a diesel surrogate and development the detailed chemical kinetic mechanism [148-150].

Therefore, on the one hand, the literature is relatively rich in detailed kinetic models for the oxidation of mixtures of components of gasolines, but far poorer for species included in diesel fuel. On the other hand, there is still a very important lack of experimental data to validate kinetic mechanisms of diesel surrogate fuels. Therefore, more experimental studies and detailed kinetic models are needed to be developed for the diesel surrogate.

Other Fuels Supply

Since ignition occurs in an HCCI engine by auto-ignition of the fuel/air mixture, the choice of fuel will have a significant impact on both engine design and control strategies. The HCCI combustion process can accept a variety of fuel types. A wide range of fuels have been tried in HCCI combustion except gasoline and diesel fuel, such as high octane number fuels: ethanol [41, 151, 152], methanol [153], liquefied petroleum gas [154], natural gas [155], iso-octane [23] *etc.*; and high cetane number fuels: dimethyl ether (DME) [153], bio-diesel [156], DEE [157], and n-heptane [23]. These investigations have been very useful for probing the effect of the fuel properties on HCCI combustion.

On the one hand, many researchers have carried out the study on the effect of octane number on the HCCI combustion by mixing iso-octane with octane number 100 and n-heptane with octane number 0 [23, 24, 47]. HCCI combustion can be controlled and HCCI operating range can be extended by burning different octane number fuel at different engine modes, *i.e.* burning low/high octane number fuel at low/high load mode respectively. There exists an optimum octane number that achieves the highest indicated thermal efficiency at different engine load. On the other hand, the effects of cetane number (CN) on HCCI auto-ignition, performance, and emissions were also investigated by some researchers [158-160]. High CN fuels exhibited a strong LTHR event and no LTHR was detected for fuels with $CN \leq 34$. Furthermore, at advanced combustion phasing, low CN fuels had significantly higher pressure rise rates and higher NO_x

emissions than the high CN fuels. Decreasing cetane number in fuels significantly reduces smoke emission due to an extension in ignition delay and the subsequent improvement in mixture formation. Smokeless combustion, ultra-low NO_x, and efficient operating range with regard to EGR and IMEP are significantly extended by decreasing fuel cetane number. And the gasoline was injected in a single cylinder diesel engine, that significantly higher loads could be attained while maintaining low smoke and high efficiency compared to a diesel fuel [161, 162]. NO_x could be controlled independently of smoke using EGR. When the same amount of gasoline was injected 15-25 crank angle degrees (CAD) before TDC, the engine ran smoothly presumably because there was not enough time for the fuel to mix fully with air and there were rich pockets where auto-ignition took place. However, the ignition delay is long in comparison with the diesel fuel and even these rich pockets were not rich enough to produce smoke because the engine could be run almost smoke free over a wide range of load conditions.

The primary conclusion from these experiments is that the best test fuel based on overall observations was medium octane fuel or cetane number. It appears that the timing of the main reaction and reaction rate can be controlled by optimizing the fuel composition, because some chemical components in the fuels have the ability to inhibit or promote the heat release process of auto-ignition. And this also suggests that for HCCI or LTC, we may need a new fuel that have different composition in comparison to conventional gasoline or diesel fuel. Therefore, HCCI auto-ignition can be controlled by modifying the fuel so that it is more chemically reactive or inhibitive by adding an ignition promoter or inhibitor. These effects can go back to the Leppard [163] and Westbrook [164] who used numerical additives to influence auto-ignition causing engine knock, which is nearly identical to the kinetics of HCCI ignition.

Ricklin *et al.* [165] reported that for a natural gas-fuelled HCCI engine, even a small amount of NO_x present in a trapped charge may be important in the HCCI operation. The large reactivity difference between methanol and DME can result in a wide range of HCCI timing control [166, 167]. This indicates that the dual-fuel can improve the HCCI operating range and control the auto-ignition. Aceves *et al.* [168] gave a numerical evaluation of fuels and additives for HCCI combustion. Their research included a ranking of many potential additives. Several additives were identified for advancing combustion by almost 11 crank angle degrees when added to the intake mixture at a concentration of 10 ppm. This is a big effect, equivalent to the increase of the intake temperature by more than 30 K. Shibata *et al.* [169] investigated the effects of fuel chemical composition on LTHR and HTHR. In their research, twelve fuels were tested. The base fuel is a mixture of equal parts of 11 pure hydrocarbons and another 11 test fuels include 15 vol% of one component and 8.5 vol% of each of the remaining ten components. The results indicated that the nparaffins, such as n-heptane, hasten the HTHR reaction and aromatics (except benzene), naphthenes and olefins have a mechanism that reduces the LTHR of their mixture counterparts. The structure of hydrocarbons strongly influences the tendency to exhibit LTHR. The similar results were also found by Farrell and Bunting [170]. They also found that all-paraffinic (normal, iso, and cycloparaffins) fuel exhibited distinctly earlier combustion phasing, increased rate of cylinder pressure rise, and increased rate of maximum heat release compared to the indolene reference fuel. Conversely, olefin-containing fuels exhibited retarded combustion phasing. It suggests that the effects on combustion phasing and engine operability range may need to be considered in the practical implementation of HCCI for fuels with large compositional variations.

MIXTURE STRATIFICATION IN HCCI CONTROL

A great deal of HCCI research has revealed that the process involved in mixing fuel and air has important effects on the HCCI combustion process. With HCCI, the start of combustion is dictated by auto-ignition chemical kinetics. Thus controlling the combustion phasing requires tuning of the auto-ignition kinetics, which is affected by the charge composition and the pressure and temperature histories of the reactants during the compression process. In fact, it is impossible to get an absolutely homogeneous mixture in the operation of practical HCCI engines. A little inhomogeneity in fuel concentration and temperature appearing in mixing can produce significant effects on the auto-ignition and combustion process. Therefore, solving the HCCI control problems has led to the investigation of various control strategies that may move away from homogeneous mixtures. The direction is the stratification that includes the following aspects: natural thermal stratification and charge/fuel-air stratification through split injection or EGR. We will introduce them in the following section.

Natural Thermal Stratification

High-load HCCI is typically limited by an excessive rate of pressure rise during combustion and the resulting engine knock. And knocking can cause excessive noise and can lead to engine damage, the maximum PRR must be kept to an acceptable level. The maximum allowable pressure rise rate (PRR) depends on the engine type and operating conditions such as engine speed and boost pressure. Fig. **8** indicates the PRR results using a single-zone CHEMKIN (Senkin application) computation and experiment [19]. In this figure, a limit of 9 bar/°CA was subjectively selected based on the perceived noise level. This limit results in a maximum allowable fueling rate of $\phi = 0.27$ for the experimental data controlling TDC combustion phasing. Although the PRR increases substantially with increasing ϕ, it does not increase nearly as rapidly as it would if the charge were fully homogeneous. In fact, it is not possible to produce a charge that is fully homogeneous in both mixture and temperature in a real engine. However, for a single-zone CHEMKIN (Senkin application) computation was used for the comparison in Fig. **8**, the increase in PRR with ϕ is much more rapid than observed experimentally and knocking becomes unacceptable for $\phi > 0.18$.

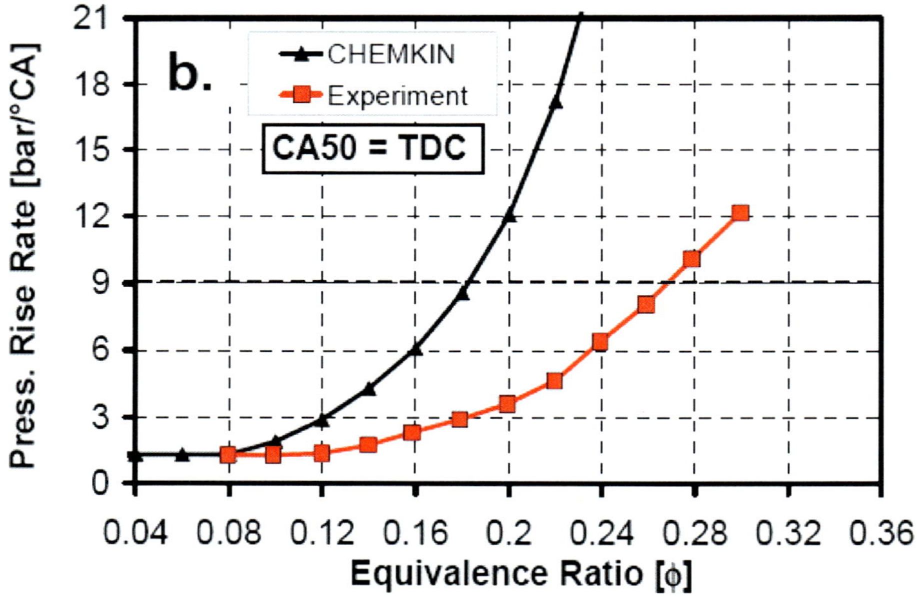

Figure 8: Experimental pressure rise rate traces as compared to a homogeneous charge simulated with CHEMKIN [19].

The reduced PRR of the real engine, and the corresponding increase in allowable ϕ, is due to naturally occurring thermal stratification. As shown in Ref. [19, 171], this thermal stratification is caused by wall heat transfer and turbulent mixing during the compression stroke for a low-residual engine. For engines with high levels of retained residuals, incomplete mixing between the fresh charge and hot residuals could also contribute to the thermal stratification [172]. As a result of thermal stratification, combustion does not occur simultaneously in all parts of the chamber, as evident in the chemiluminescence images presented in Fig. **9**. The HCCI combustion process is a sequential process from the hotter mixture to the cooler mixture around the combustion chamber walls [173] or other cooler regions. The sequential process of HCCI combustion is faster than a turbulent hydrocarbon flame [174], so it is not a flame-type propagation, which has been confirmed by Ref. [19, 175, 176]. The sequential auto-ignition slows the PRR considerably, allowing the fueling to be increased in comparison to the ideal homogeneous conditions.

-10° -8° -6° -4° -2°

Figure 9: Chemiluminescence images sequence obtained through the piston-crown window. The intake temperature is 95 °C, coolant temperature is 85 °C, and the port injection timing is -30° ATDC [173].

Since heat transfer occurs at the walls, thermal boundary layers are often considered to play a major role in the thermal stratification in HCCI engines. However, research based on chemiluminescence imaging [19, 177] and fuel-tracer planar laser induced fluorescence (PLIF) imaging [177] has shown that thermal stratification extends throughout the bulk gas. Since the majority of combustion occurs in the central part of the charge up past the time of the maximum PRR, it is mainly thermal stratification within the bulk gases that controls the maximum PRR and engine knock. The effect of thermal stratification between the bulk gases and the boundary layer has only a secondary effect on reducing the maximum PRR [19]. The researchers have taken many effective methods to increase the thermal stratification. For example, reducing the wall temperature (by lower coolant temperatures) [173, 178], combining swirl with a square piston bowl [179], and using different intake-air temperatures in each intake port [180, 181]. In addition, incomplete mixing of fresh charge with hot retained residuals has also been suggested as a method of increasing the thermal stratification [172, 182]. These techniques have shown promise, and research efforts continue, but significant enhancement of the naturally occurring thermal stratification remains a challenge.

Charge and Compositional Stratification

Except the thermal stratification, more researches are about the charge and compositional stratification. Firstly, the different premixing procedures were used in the intake port to obtain different degrees of homogeneity of the fuel/air charge. For example, Richter *et al.* [14] investigated the imaging of fuel and OH using PLIF in a HCCI engine. A standard port injection was used to create the charge and an additional preheated mixing tank of 20 liters was used to prepare a more homogenous charge. The PLIF confirmed that the fuel preparation strategy influences the fuel/air homogeneity at the time of combustion. Iida *et al.* [16, 17] investigated the effect of charge inhomogeneity by changing a device for mixing air and fuel in the intake manifold, and thus inhomogeneity in fuel distribution in the premixture was varied, see Fig. **10**. The results show that, for inhomogeneous charging in experiment, the maximum heat release rate and maximum pressure-rise rate decreased, the combustion duration got longer. Chemiluminescence, of which transition was identified from the side of intake valve to the side of exhaust valve, was observed, see Fig. **11**. Liu *et al.* [173] use the different port fuel injection strategies to form the different charge homogeneity, which can also result in different charge distributions in the combustion chamber, and thus affect the auto-ignition timing, chemiluminescence intensity, and combustion processes.

And then, the more research for charge stratification formation was focus on the injecting directly into the cylinder. Tian [39] investigated the effect of 2-injection on the HCCI combustion in a GDI engine. The first injection timing can choose the period of NVO or the intake stroke; the second injection timing will be in the compression process to form stratified charge. By combining various injection strategies, adjusting fuel quantity and injection timing of each pulse, the temperature, concentration and species of the mixture in cylinder can be controlled. As a result, the HCCI ignition timing and combustion rate can be controlled. For example, in low load region, the NVO pre-injection should be maintained at a higher mass to stable the combustion. The main injection can adjust load. When load is increased to bring light knocking, the NVO pre-injection should be decrease. The load can be extended to IMEP=0.43MPa. When extend the load region more, the NVO pre-injection should be reduce to zero, and IMEP can be increased to 0.53MPa. And the further computational study by Wang [183] had shown that the periphery of fuel-rich zone formed by the second injection ignited first, and then fuel-rich zone ignited and worked as an initiation to ignite the surrounding lean mixture zone formed by the first injection. The two injections can improve load level. Meanwhile, the periphery of fuel-rich zone leads to fierce burning, which results in slightly high NO_x emissions.

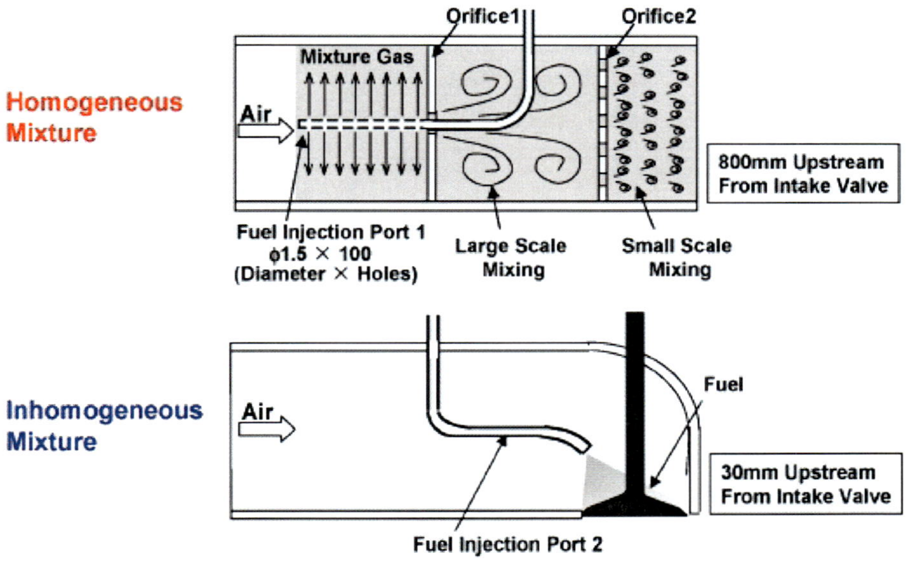

Figure 10: The different mixing process [16-17].

The UNIBUS is a representative charge stratification diesel-fuelled HCCI engine. Hasegawa *et al.* [101] investigated the ignition and combustion control of this system. Double injection technique was used by Common Rail injection system. The first injection was used as an early injection for fuel diffusion and to advance the changing of fuel to lower hydrocarbons (*i.e.* low temperature reaction). The second injection was used as an ignition trigger for all the fuel. It was found that the ignition of the premixed gas could be controlled by the second injection when the early injection was maintaining low temperature reaction. Luminous flame was not observed and many ignition points appeared in the combustion chamber. This means that the HCCI concept was realized. In addition, for the premixed/direct-injected HCCI combustion, it is also representative charge stratification. Port injection was chosen for the main fuel supply to create a homogeneous charge, and direct fuel injection into the cylinder was used to change the concentration and position of local fuel-rich regions with the purpose of controlling HCCI combustion.

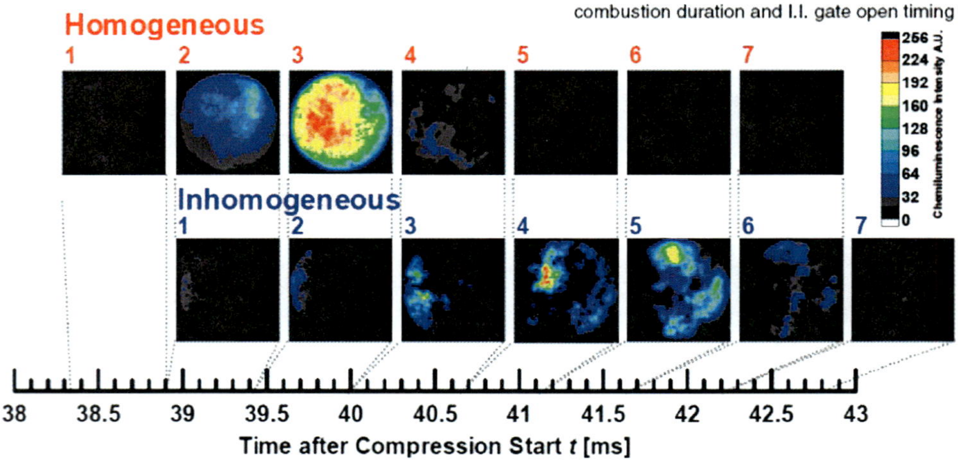

Figure 11: Chemiluminescence images with different mixing process [17].

The ideal charge stratification on combustion and emissions has also been investigated using coupled multi-dimensional CFD and a reduced n-heptane chemical kinetics model by Zheng and Yao [184]. Seven different kinds of imposed stratification have been introduced according that the position of maximal local equivalence ratio appears along the radial direction of the cylinder at IVC, as shown in Fig. **12**. In this study, the meshes are divided into 13 layers along the radial direction, and the "m" in Fig. **12** denotes the mesh layer where the maximal equivalence ratio appears ($1 \leq m \leq 13$). In addition, r_0 indicates the radius of the cylinder, and r indicates the distance between the cylinder centerline and the position where the maximal equivalence ratio appears in Fig. **12**. In the first type of imposed stratification, the concentration of fuel–air is the strongest in the centre of the cylinder and leanest along the cylinder wall; in the last type of imposed stratification, the concentration of fuel–air is the leanest in the centre of the cylinder and strongest near the cylinder wall (the concentration of fuel–air along the cylinder wall was assumed as 0 in order to accord with the operation condition of practical engine); in other types of imposed stratification, the concentration of fuel–air increases first then decrease and the positions of maximal local equivalence ratio are different in every type of imposed stratification. It is by different positions of maximal local equivalence ratio that the auto-ignition process and emission formation are affected. Consequently, the combustion process is controlled. The results show that the charge stratification results in stratification of the in-cylinder temperature. The former four kinds of stratification, for which the maximal local equivalence ratios are located between the cylinder centre and half the cylinder radius, advance ignition timing, reduce the pressure rise rate and retard combustion phasing. But three kinds of stratification, for which the maximal local equivalence ratios appear between half the cylinder radius and the cylinder wall, have little effect on the cylinder pressure. All seven kinds of stratification can reduce unburned fuel emissions; the stratification that the concentration of fuel–air is stronger in the centre of the cylinder increase the CO and NO_x emissions, whilst the last two kinds of stratification that the concentration is stronger near the wall can reduce unburned fuel, formaldehyde and CO emissions and maintain low NO_x emissions simultaneously.

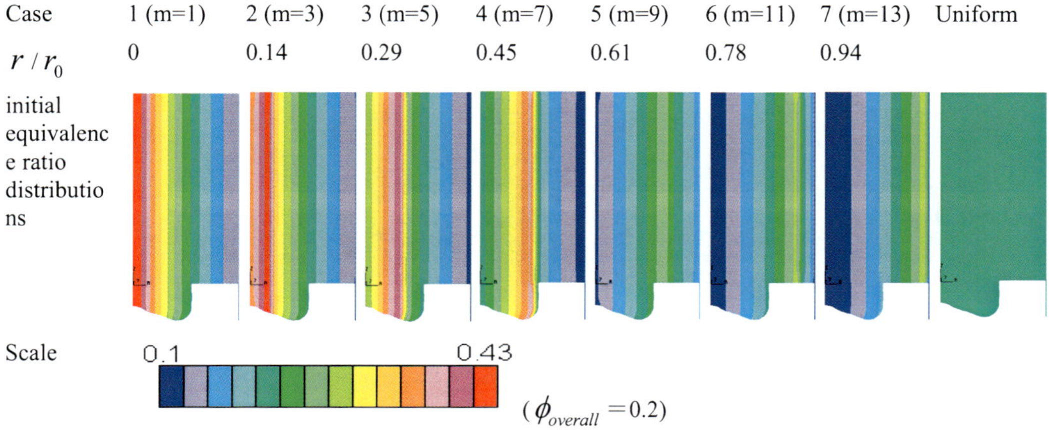

Figure 12: Imposed initial charge stratification at the two equivalence ratios [185].

For the compositional stratification, we can think that all the EGR, internal or external, change the composition of the charge, thereby forming the different compositional stratification. If the fuel additives change the properties of the fuel's auto-ignition, the compositional stratification by EGR changes the air composition and temperature, and thus has an effect on the auto-ignition. The Sjoberg *et al.* [109] has investigated the thermodynamic and chemical effects of real EGR, simulated EGR, dry EGR, and individual EGR constituents (N_2, CO_2, and H_2O) on the auto-ignition processes. This was done for gasoline and various PRF blends. The data show that addition of real EGR retards the auto-ignition timing for all fuels. However, the amount of retard is dependent on the specific fuel type. This can be explained by identifying and quantifying the various underlying mechanisms, which are: 1) Thermodynamic cooling effect due to increased specific-heat capacity, 2) O_2 reduction effect, 3) Enhancement of auto-ignition due to the presence of H_2O, 4) Enhancement or suppression of auto-ignition due to the presence of trace species such as unburned or partially oxidized hydrocarbons. The single-stage ignition fuels isooctane and gasoline are more sensitive to the cooling effect of EGR (without water), compared to the two-stage ignition fuels PRF80 and PRF60. On the other hand, the two-stage ignition fuels have much higher sensitivity to the reduction of O_2 concentration associated with the addition of EGR (N_2, CO_2, and H_2O). Furthermore, H_2O has a pronounced ignition-enhancing effect for the two-stage fuels that tends to counteract the strong cooling effect of H_2O. Yao *et al.* [130] also investigated the effects of CO_2 on combustion and emissions of diesel engine using a commercial diesel fuel.

In addition, Risberg *et al.* [185] investigated the influence of NO on combustion phasing with NO concentrations ranging from 4 up to 476 ppm in the fresh intake air. The combustion phasing, represented by CA50 was advanced up to 12.5 CAD by the influence of NO. Krasselt *et al.* [186] used the each side of the split intake port independently with different temperatures and reactant compositions. Therefore, two stratification approaches were enabled: thermal stratification and compositional stratification. Argon was used as a diluent to achieve higher temperatures and stratify the in-cylinder temperature indirectly *via* a stratification of the ratio of specific heats. Over all test conditions, the direct thermal stratification had no noticeable effect on the combustion duration or phasing within the limits of the experiment (a thermal difference of 45°C in the reactant gas mixtures, measured at the valve stems). However, the argon stratification case produced indirect thermal stratifications of sufficient magnitude to keep a constant combustion phasing with a mass-averaged intake temperature up to 14°C lower than the fully homogeneous case. For a given combustion phasing, it also increased the combustion duration by approximately 2 °CA (and thereby decreased the maximum pressure rise rate by 20~40%).

CONCLUSIONS

Four main results from the pioneering investigation of HCCI combustion have been established: 1), HCCI combustion demonstrates a strong potential to improve the thermal efficiency of gasoline-fuelled engines and substantially reduce NO_x and soot emissions of diesel-fuelled engines; 2), the chemical kinetics has a dominating role in HCCI combustion; 3), difficulties associated with the successful operation of HCCI engines need to be overcome including: i) combustion phasing control, ii) high HC and CO emissions, iii) extending the operation range, iv) cold start problems and homogeneous mixture preparation; 4), the HCCI concept could be expanded to all the new engine combustion modes achieving high efficiency and low emissions, such as PCCI, SCCI, LTC, *etc.* The main characteristics of HCCI combustion include: premixed mixture, compression ignition and low temperature combustion. After a great deal of research, the evolution in the investigation of HCCI can mainly be described as follows.

HCCI can be applied to a variety of fuel types and the choice of fuel will have a significant impact on both engine design and control strategies. Some chemical components have the ability to inhibit or promote the heat release process associated with auto-ignition. Therefore, HCCI auto-ignition can be controlled by modification of the fuel so that it is more chemically reactive or inhibitive by adding an ignition promoter or inhibitor. Typical generalized diesel-fuelled HCCI combustion modes include: early direct injection HCCI, late direct injection HCCI, premixed/direct-injected HCCI combustion and low temperature combustion. Mixture control (mixture preparation), including charge components and temperature control in the whole combustion history and high pre-ignition mixing rate, is the key issue to achieve diesel HCCI combustion. The high octane numbers of gasoline fuels means that such fuels need high ignition temperatures, which highlights the difficulty of auto-ignition. The main challenge for gasoline HCCI operation is focus on the obtaining sufficient thermal energy to trigger autoignition of mixtures late in the compression stroke, extending the operational range, and the transient control.

Utilizing charge inhomogeneity is an important path to achieve clean and high efficiency combustion in engines. Stratification strategy has the potential to extend the HCCI operation range to higher loads. The thermal stratification is caused by wall heat transfer and turbulent mixing during the compression stroke for a low-residual engine. For engines with high levels of retained residuals, incomplete mixing between the fresh charge and hot residuals could also contribute to the thermal stratification. Advanced control strategies of fuel/air mixture are more important than simple "homogeneous charge'' in the control of the HCCI combustion process. The charge stratification can control the auto-ignition timing and extend the operating range. EGR is the most important strategy to form the compositional stratification. EGR can change the temperature and composition in the cylinder; thereby it is a very important method to control the HCCI combustion process.

For the practical HCCI combustion system, some technologies should be developed further. The flexible injection strategies, EGR control and the closed-loop feedback control, will be the key issue for both the diesel-fuelled and gasoline-fuelled HCCI engines. For the diesel-fuelled HCCI engines, more attention should be paid to extend the LTC operation to higher loads, even the full load, while maintaining good fuel economy and low emissions (especially for the HC and CO emissions). For the gasoline-fuelled HCCI engines, more attention should be paid to individual cylinder control to ensure the same combustion phasing and reach HCCI/SI transition for multi-cylinders engines.

Finally, we think that the combustion mode design will be become one of the most interesting topics in the future. To realize the high efficiency and clean combustion in the complete engine operational range, the flexible multi-mode combustion processes must be necessary to various operating conditions and fuel properties by controlling the ϕ -T history. First of all, to the various operating conditions and fuel properties, the fuel/air/EGR mixing processes are different. Next, the combustion mode will be changed according to the different fuel/air/EGR mixing processes. And then, the combustion process should not be a single mode as the conventional CI or SI engines, but a multi-mode combustion process to the different fuel/air/EGR mixing processes. The fuel/air/EGR mixing processes can be controlled by some methods, such as multi-injection strategies, in-cylinder turbulence control, boosting, internal and external EGR rates, and flexible valve actuation. Therefore, the multi-mode combustion processes can be organized and optimized by the control of fuel/air/EGR mixing processes. Meantime, a closed-loop control system will construct a bridge between combustion mode and mixing processes. In the end, the optimization of fuel economy and emissions can be reached to the complete engine operational range through an optimum combustion mode design.

REFERENCES

[1] S. Onishi, S.H. Jo, K. Shoda, P.D. Jo, and S. Kato, "Active thermo-atmosphere combustion (ATAC) -a new combustion process for internal combustion engines", SAE Paper, 1979, no. 790501.

[2] M. Noguchi, T. Tanaka, and Y. Takeuchi, "A study on gasoline engine combustion by observation of intermediate reactive products during combustion", SAE Paper, 1979, no. 790840.

[3] P.M. Najt and D.E. Foster, "Compression ignited homogeneous charge combustion", SAE Paper, 1983, no. 830264.

[4] R.H. Thring, "Homogeneous-charge compression ignition (HCCI) engines", SAE Paper, 1989, no. 892068.

[5] T.W. Ryan III and T.J. Callahan, "Homogeneous charge compression ignition of diesel fuel", SAE Paper, 1996, no. 961160.

[6] A.W. Gray III and T.W. Ryan III, "Homogeneous charge compression ignition (HCCI) of diesel fuel", SAE Paper, 1997, no. 971676.

[7] J.E. Dec, "Advanced compression-ignition engines-understanding the in-cylinder processes", *Proc. Combust. Inst.*, vol. 32, pp. 2727–2742, 2009.

[8] R.H. Stanglmaier and C.E. Roberts, "Homogeneous charge compression ignition (HCCI): benefits, compromise, and future engine applications", SAE Paper, 1999, no. 1999-01-3682.

[9] H. Machrafi and S. Cavadias, "An experimental and numerical analysis of the influence of the inlet temperature, equivalence ratio and compression ratio on the HCCI auto-ignition process of primary reference fuels in an engine", *Fuel Process. Technol.*, vol. 89, pp. 1218–1226, Nov. 2008.

[10] H.F. Liu, M.F. Yao, B. Zhang, and Z.Q. Zheng, "Effects of inlet pressure and octane numbers on combustion and emissions of a homogeneous charge compression ignition (HCCI) engine", *Energy Fuel*, vol. 22, pp. 2207-2215, Jul.–Aug. 2008.

[11] J.A. Eng, "Characterization of pressure waves in HCCI combustion", SAE Paper, 2002, no. 2002-01-2859.

[12] J.E. Dec, "A computational study of the effects of low fuel loading", SAE Paper, 2002, no. 2002-01-1309.

[13] R.H. Stanglmaier, J. Li, and R.D. Matthews, "The effect of in-cylinder wall wetting location on the HC emissions from SI engines", SAE Paper, 1999, no. 1999-01-0502.

[14] M. Richter, J. Engström, A. Franke, M. Aldén, A. Hultqvist, and B. Johansson, "The influence of charge inhomogeneity on the HCCI combustion process", SAE Paper, 2000, no. 2000-01-2868.

[15] M. Richter, A. Franke, M. Aldén, A. Hultqvist, and B. Johansson, "Optical diagnostics applied to a naturally aspirated homogeneous charge compression ignition engine", SAE Paper, 1999, no. 1999-01-3649.

[16] K. Kumano and N. Iida, "Analysis of the effect of charge inhomogeneity on HCCI combustion by chemiluminescence measurement", SAE Paper, 2004, no. 2004-01-1902.

[17] J. Ozaki and N. Iida, "Effect of degree of unmixedness on HCCI combustion based on experiment and numerical analysis", SAE Paper, 2006, no. 2006-32-0046.

[18] H.F. Liu, Y.F. Wang, Z.Q. Zheng, and M.F. Yao, "Study on the homogeneous charge compression ignition under different ported injection strategies", *T. CSICE*, vol. 27, pp. 493-498, Dec. 2009.

[19] J.E. Dec, W. Hwang, and M. Sjöberg, "An Investigation of thermal stratification in HCCI engines using chemiluminescence imaging", SAE Paper, 2006, no.2006-01-1518.

[20] G.R. Pucher, D.P. Gardiner, M.F. Bardon, and V. Battista, "Alternative combustion systems for piston engines involving homogeneous charge compression ignition concepts-a review of studies using methanol, gasoline and diesel fuel", SAE Paper, 1996, no. 962063.

[21] R.M. Green, N.P. Cernansky, W.J. Pitz, and C.K.Westbrook, "The role of low temperature chemistry in the auto-ignition on n-butane", SAE Paper, 1987, no. 872108.

[22] S. Addargarla, Y. Henig, R.D. Wilk, D.L. Miller, and N.P. Cernansky, "Effect of fuel-air mixture stressing on pre-ignition heat release in a knock research engine", SAE Paper, 1989, no. 892082.

[23] M.F. Yao, B. Zhang, Z.Q. Zheng, and Z. Chen, "Experimental study on homogeneous charge compression ignition combustion with primary reference fuel", *Combust. Sci. Technol.*, vol. 179, pp. 2539-2559, Dec. 2007.

[24] H.F. Liu, M.F. Yao, B. Zhang, and Z.Q. Zheng, "Influence of fuel and operating conditions on combustion characteristics of a homogeneous charge compression ignition engine", *Energy Fuel*, ; vol.23, pp. 1422-1430, Mar.–Apr. 2009.

[25] Y. Ishibashi and M. Asai, "Improving the exhaust emissions of two-stroke engine by applying the activated radical combustion", SAE Paper, 1996, no. 960742.

[26] N.B. Kaahaaina, A.J. Simon, P.A. Caton, and C.F. Edwards, Use of dynamic valving to achieve residual-affected combustion, SAE Paper, 2001, no. 2001-01-0549.

[27] H. Zhao, J. Li, T. Ma, and N. Ladommatos, "Performance and analysis of a 4-stroke multi-cylinder gasoline engine with CAI combustion", SAE Paper, 2002, no. 2002-01-0420.

[28] D. Law, D. Kemp, J. Allen, G. Kirkpatrick, and T. Copland, "Controlled combustion in an IC engine with a fully variable valve train", SAE Paper, 2000, no. 2000-01-0251.

[29] J. Li, H. Zhao, N. Ladommatos, and T. Ma, "Research and development of controlled auto-ignition (cai) combustion in a 4-stroke multi-cylinder gasoline engine", SAE Paper, 2001, no. 2001-01-3608.

[30] G. Kontarakis, N. Collings, and T. Ma, "Demonstration of HCCI using a single cylinder four-stroke SI engine with modified valve timing", SAE Paper, 2000, no. 2000-01-2870.

[31] B.Q. He, H. Xie, Y. Zhang, J. Qin, and H. Zhao, "An experimental study on HCCI combustion in a four-stroke gasoline engine with reduced valve lift operations", SAE Paper, 2005, no. 2005-01-3736.

[32] J.L.C. Rocafort, M.M. Andreae, W.H. Green, W.K. Cheng, and J.S. Cowart, "A modeling investigation into the optimal intake and exhaust valve event duration and timing for a HCCI engine", SAE Paper, 2005, no. 2005-01-3746.
[33] M.F. Yao, H.F. Liu, B. Zhang, and Z.Q. Zheng, "An investigation on the effects of fuel chemistry and engine operating conditions on HCCI engine", SAE Paper, 2008, no.2008-01-1660.
[34] M.C. Drake and D.C. Haworth, "Advanced gasoline engine development using optical diagnostics and numerical modeling", *Proc. Combust. Inst.*, vol. 31, pp. 99-124, 2007.
[35] L. Cao, H. Zhao, and X. Jiang, "Analysis of controlled auto-ignition/HCCI combustion in a direct injection gasoline engine with single and split fuel injections", *Combust. Sci. Technol.*, vol. 180, pp. 176-205, Jan. 2008.
[36] C.D. Marriott and R.D. Reitz, "Experimental investigation of direct injection-gasoline for premixed compression ignited combustion phasing control", SAE Paper, 2002, no. 2002-01-0418.
[37] Y. Urata, M. Awasaka, J. Takanashi, T. Kakinuma, T. Hakozaki, and A. Umemoto, "A study of gasoline-fuelled HCCI engine equipped with electromagnetic valve train" SAE Paper, 2004, no. 2004-01-1898.
[38] Y.F. Li, H. Zhao, N. Brouzos, and T. Ma, "Effect of injection timing on mixture and CAI combustion in a GDI engine with an air-assisted injector", SAE Paper, 2006, no. 2006-01-0206.
[39] G.H. Tian, Z. Wang, and J.X. Wang, "HCCI combustion control by injection strategy with negative valve overlap in a GDI engine", SAE Paper, 2006, no. 2006-01-0415.
[40] J.O. Olsson, P. Tunestal, and B. Johansson, "Boosting for high load HCCI", SAE Paper, 2004, no. 2004-01-0940.
[41] D. Yap and A. Megaritis, "Applying forced induction to bioethanol HCCI operation with residual gas trapping", *Energy Fuel*, vol. 19, pp. 1812–1821, Sep.–Oct. 2005.
[42] M. Christensen, B. Johansson, P. Amneus, and F. Mauss, "Supercharged homogeneous charge compression ignition", SAE Paper, 1998, no. 980787.
[43] S. Gharahbaghi, T.S. Wilson, H. Xu, S. Cryan, S. Richardson, M.L.Wyszynski, and J. Misztal, "Modelling and experimental investigations of supercharged HCCI engines", SAE Paper, 2006, no. 2006-01-0634.
[44] D. Yap, M.L. Wyszynski, A. Megaritis, and H. Xu, "Applying boosting to gasoline HCCI operation with residual gas trapping", SAE Paper, 2005, no. 2005-01-2121.
[45] A. Cairns and H. Blaxill, "The effects of combined internal and external exhaust gas recirculation on gasoline controlled auto-ignition", SAE Paper, 2005, no. 2005-01-0133.
[46] J. Hyvonen, G. Haraldsson, and B. Johansson, "Operating range in a multi cylinder HCCI engine using variable compression ratio", SAE Paper, 2003, no. 2003-01-1829.
[47] M. Christensen, A. Hultqvist, and B. Johansson, "Demonstrating the multi fuel capability of a homogeneous charge compression ignition engine with variable compression ratio", SAE Paper, 1999, no. 1999-01-3679.
[48] M. Christensen and B. Johansson, "Supercharged homogeneous charge compression ignition (HCCI) with exhaust gas recirculation and pilot fuel", SAE Paper, 2000, no. 2000-01-1835.
[49] J. Hyvonen, G. Haraldsson, and B. Johansson, "Operating conditions using spark assisted HCCI combustion during combustion mode transfer to SI in a multi-cylinder VCR-HCCI engine", SAE Paper, 2005, no. 2005-01-0109.

[50] J.L. Yang, T. Culp, and T. Kenny, "Development of a gasoline engine system using HCCI technology-the concept and the test results", SAE Paper, 2002, no. 2002-01-2832.

[51] A.F. Furhapter, W. Piock, and G.K. Fraidl, "CSI-controlled auto ignition-the best solution for the fuel consumption -versus emission trade-off", SAE Paper, 2003, no. 2003-01-0754.

[52] H. Xie, S.Z. Hou, J. Qin, Y. Zhang, N. Li, and H. Zhao, "Control strategies for steady and transient operation of a 4-stroke gasoline engine with CAI combustion using a 4-variable valve actuating system (4VVAS)", SAE Paper, 2006, no. 2006-01-1083.

[53] Y. Zhang, H. Xie, T. Chen, N.H. Zhou, and H.Zhao, "Study of SI-HCCI-SI transition on a port fuel injection engine equipped with 4VVAS", SAE Paper, 2007, no. 2007-01-0199.

[54] G.H. Tian, Z. Wang, Q.Q. Ge, J.X. Wang, and S.J. Shuai, "Mode swith of SI-HCCI combustion on a GDI engine", SAE Paper, 2007, no. 2007-01-0195.

[55] L. Koopmans, I. Denbratt, H. Ström, S. Lundgren, and O. Backlund, "Demonstrating a SI-HCCI-SI mode change on a volvo 5-cylinder electronic valve control engine", SAE Paper, 2003, no. 2003-01-0753.

[56] M. Roelle, G. Shaver, and J.C.Gerdes, "Tackling the transition: a multi-mode combustion model of SI and HCCI for mode transition control", ASME International Mechanical Engineering Congress and Exposition, 2004, pp. 329–336.

[57] H.G. Santoso, J. Matthews, and W.K. Cheng, "Managing Si/HCCI dual-mode engine operation", SAE Paper, 2005, no. 2005-01-0162.

[58] G.M. Shaver, M.J. Roelle, and J.C. Gerdes, "Modeling cycle-to-cycle dynamics and mode transition in HCCI engines with variable valve actuation", *Control Eng. Practicevol.*, vol. 14, pp. 213-222, Mar. 2006.

[59] H. Zhao, *HCCI and CAI engines for the automotive industry*. Woodhead Publishing Ltd, 2007.

[60] Z. Wang, J.X. Wang, G.H. Tian, S.J. Shuai, Z.F. Zhang, and J.W. Yang, "Research on steady and transient operation of HCCI engine with gasoline direct injection", SAE Paper, 2008, 2008-01-1723.

[61] S. Yoshiyama, E. Tomita, and Y. Homamoto, "Fundamental study on combustion diagnostics using a spark plug as ion probe", SAE Paper, 2000, 2000-01-2828.

[62] S. Yoshiyama and E. Tomita, "Combustion diagnostics of a spark ignition engine using a spark plug as an ion probe", SAE Paper, 2002, 2002-01-2838.

[63] Andersson and L. Eriksson, "Ion sensing for combustion stability control of a spark ignited direct injected engine", SAE Paper, 2000, no. 2000-01-0552.

[64] M. Hellring and U. Holmberg, "A comparison of ion current based algorithms for peak pressure control", SAE Paper, 2001, no. 2001-01-1920.

[65] C.F. Daniels, G.G. Zhu, and J. Winkelman, "Inaudible knock and partial-burn detection using in-cylinder ionization signal", SAE Paper, 2003, no. 2003-01-3149.

[66] T. Suzuki, H. Ohara, A. Kakishima, K. Yoshida, and H. Shodi, "A study of knocking using ion current and light emission", SAE Paper, 2003, no. 2003-32-0038.

[67] P. Strandh, M. Christensen, J. Bengtsson, R. Johansson, A. Vressner, P. Tunestal, and B. Johansson, "Ion current sensing for HCCI combustion feedback", SAE Paper, 2003, no. 2003-01-3216.

[68] P. Mehresha, J. Soudera, D. Flowersb, U. Riedelc, and R.W. Dibble, "Combustion timing in HCCI engines determined by ion-sensor: experimental and kinetic modeling", *Proc. Combust. Inst.*, vol. 30, pp. 2701-2709, 2005.

[69] E.B. Jr Gregory, J. Chen, and R.W. Dibble, "The effects of intake pressure, fuel concentration, and bias voltage on the detection of ions in a homogeneous charge compression ignition (HCCI) engine", *Proc. Combust. Inst.*, vol. 32, pp. 2877-2884, 2009.
[70] H. Xie, L. Yang, J. Qin, R. Gao, H.G. Zhu, B.Q. He, and H. Zhao, "The effect of spark ignition on the CAI combustion operation", SAE Paper, 2005, no. 2005-01-3738.
[71] H. Persson, A. Hultqvist, B. Johansson, and A. Remon, "Investigation of the early flame development in spark assisted HCCI combustion using high speed chemiluminescence imaging", SAE Paper, 2007, no. 2007-01-0212.
[72] A.W. Berntsson, M. Andersson, D. Dahl, and I. Denbratt, "A LIF-study of OH in the negative valve overlap of a spark-assisted HCCI combustion engine", SAE Paper, 2008, no. 2008-01-0037.
[73] B.G. Bunting, "Combustion, control, and fuel effects in a spark assisted HCCI engine equipped with variable valve timing", SAE Paper, 2006, no. 2006-01-0872.
[74] C. Felsch, T. Sloane, J. Han, H. Barths, and A. Lippert, "Numerical investigation of recompression and fuel reforming in a SIDI-HCCI engine", SAE Paper, 2007, no. 2007-01-1878.
[75] G.T. Kalghatgi, "Auto-ignition quality of practical fuels and implications for fuel requirements of future SI and HCCI engines", SAE Paper, 2005, no. 2005-01-0239.
[76] H.J. Curran, P. Gaffuri, W.J. Pitz, and C.K. Westbrook, "A comprehensive modeling study of iso-octane oxidation", *Combust. Flame*, vol. 129, pp. 253-280, May 2002.
[77] M. Jia and M.Z. Xie, "A chemical kinetics model of iso-octane oxidation for HCCI engines", *Fuel*, vol. 85, pp. 2593–2604, Dec. 2006.
[78] H.J. Curran, W.J. Pitz, C.K. Westbrook, C.V. Callahan, and F.L. Dryer, "Oxidation of automotive primary reference fuels at elevated pressures", *Proc. Combust. Inst.*, vol. 27, pp. 379-387, 1998.
[79] C. Huang, X.C. Lu, and Z. Huang, "New reduced chemical mechanism for homogeneous charge combustion ignition combustion investigation of primary reference fuels" *Energy Fuel*, vol. 22, pp. 935-944, Feb. 2008.
[80] J. Andrae, D. Johansson, P. Bjornbom, P. Risberg, and G. Kalghatgi, "Co-oxidation in the auto-ignition of primary reference fuels and n-heptane/toluene blends", *Combust. Flame*, vol. 140, pp. 267–286, Mar. 2005.
[81] J. Andrae, P. Bjornbom, R.F. Cracknell, and G. Kalghatgi, "Autoignition of toluene reference fuels at high pressures modeled with detailed chemical kinetics", *Combust. Flame*, vol. 149, pp. 2–24, Apr. 2007.
[82] H. Machrafi and S. Cavadias, "Three-stage autoignition of gasoline in an HCCI engine: An experimental and chemical kinetic modeling investigation", *Combust. Flame*, vol. 155, pp. 557-570, Dec. 2008.
[83] H. Machrafi, S. Cavadias, and J. Amouroux, "The development and experimental validation of a reduced ternary kinetic mechanism for the auto-ignition at HCCI conditions, proposing a global reaction path for ternary gasoline surrogates", *Fuel Process. Technol.*, vol. 90, pp. 247–263, Feb. 2009.
[84] B.M. Gauthier, D.F. Davidson, and R.K. Hanson, "Shock tube determination of ignition delay times in full-blend and surrogate fuel mixtures", *Combust. Flame*, vol. 139, pp. 300–311, Dec. 2004.

[85] M. Fikri, J. Herzler, R. Starke, C. Schulz, P. Roth, and G.T. Kalghatgi, "Autoignition of gasoline surrogates mixtures at intermediate temperatures and high pressures", *Combust. Flame*, vol. 152, pp. 276–281, Jan. 2008.

[86] L.R. Cancino, M. Fikri, A.A.M. Oliveira, and C. Schulz, "Autoignition of gasoline surrogate mixtures at intermediate temperatures and high pressures: experimental and numerical approaches", *Proc. Combust. Inst.*, vol. 32, pp. 501-508, 2009.

[87] R.C.O.B. Delgado, A.S. Araujo, and V.J. Fernandes, "Properties of Brazilian gasoline mixed with hydrated ethanol for flex-fuel technology", *Fuel Process. Technol.*, vol. 88, pp. 365–368, Apr. 2007.

[88] M. Yahyaoui, N. Djebaili-Chaumeix, P. Dagaut, C.E. Paillard, and S. Gail, "Experimental and modelling study of gasoline surrogate mixtures oxidation in jet stirred reactor and shock tube", *Proc. Combust. Inst.*, vol. 31, pp. 385–391, 2007.

[89] M. Yahyaoui, N. Djebaili-Chaumeix, P. Dagaut, C.E. Paillard, B. Heyberger, and G. Pengloan, "Ignition and oxidation of 1-hexene/toluene mixtures in a shock tube and a jet-stirred reactor: Experimental and kinetic modeling study", *Int. J. Chem. Kinet.*, vol. 39, pp. 518–538, Jul. 2007.

[90] F. Battin-Leclerc, "Detailed chemical kinetic models for the low-temperature combustion of hydrocarbons with application to gasoline and diesel fuel surrogates", *Prog. Energy Combust. Sci.*, vol. 34, pp. 440-498, Aug. 2008.

[91] P.L. Kelly-Zion and J.E. Dec, "A computational study of the effect of fuel type on ignition time in homogeneous charge compression ignition engines", *Proc. Combust. Inst.*, vol. 28, pp. 1187–1194, 2000.

[92] F. Zhao, T.W. Asmus, D.N. Assanis, J.E. Dec, J.A. Eng, and P.M. Najt, *Homogenous charge compression ignition (HCCI) engine: key research and development issues.* Warrendale, PA: Society of Automotive Engineers, 2003.

[93] Y. Takeda, N. Keiichi, and N. Keiichi, "Emission characteristics of premixed lean diesel combustion with extremely early staged fuel injection", SAE Paper, 1996, no. 961163.

[94] K. Nakagome, N. Shimazaki, K. Miimura, and S. Kobayashi, "Combustion and emissions characteristics of premixed lean diesel combustion engine", SAE Paper, 1997, no. 970898.

[95] A. Harada, N. Shimazaki, S. Sator, T. Miyamoto, H. Akagawa, and K. Tsujimura, "The effects of mixture formation on premixed lean diesel combustion", SAE Paper, 1998, no. 980533.

[96] T. Hashizume, T. Miyamato, H. Akagawa, and K. Tsujimura, "Combustion and emission characteristics of multiple stage diesel combustion", SAE Paper, 1998, no. 980505.

[97] H. Akagawa, T. Miyamoto, A. Harada, S. Sasaki, N. Shimazaki, T. Hakeshizume, and K. Tsujimura, "Approaches to solve problems of the premixed lean diesel combustion", SAE Paper, 1999, no. 1999-01-0183.

[98] A. Morita, Y. Iwashiro, Y. Aoyagi, and T. Hashizume, "Observation of the diesel combustion process in multiple stage injection", Thermofluiddynamic Processes in Diesel Engines: Proceedings of the Thiesel 2000 Conference, 2000, pp. 447–452.

[99] Y. Nishijima, Y. Asaumi, and Y. Aoyagi, "Impingement spray system with direct water injection for premixed lean diesel combustion control" SAE Paper, 2002, no. 2002-01-0109.

[100] H. Yanagihara, "Ignition timing controlat Toyota "UNIBUS'' combustion system", Proceedings of the IFP International Congress, 2001, pp. 35-42.

[101] R. Hasegawa and H. Yanagihara, "HCCI combustion in DI diesel engine", SAE Paper, 2003, no. 2003-01-0745.

[102] W.H. Su, T.J. Lin, and Y.Q. Pei, "A compound technology for HCCI combustion in a DI diesel engine based on the multi-pulse injection and the BUMP combustion chamber", SAE Paper, 2003, no. 2003-01-0741.
[103] W.H., Su, X.Y. Zhang, and T.J. Lin, "Study of pulse spray, heat release, emissions and efficiencies in a compound diesel HCCI combustion engine" Proceedings of ASME-ICE ASME Internal Combustion Engine Division Fall Technical Conference, 2004, ICEF2004-927.
[104] W.H. Su, H. Wang, and B. Liu, "Injection mode modulation for HCCI diesel combustion", SAE Paper, 2005, no. 2005-01-0117.
[105] M.Y. Kim and C.S. Lee, "Effect of a narrow fuel spray angle and a dual injection configuration on the improvement of exhaust emissions in a HCCI diesel engine", *Fuel*, vol. 86, pp. 2871-2880, Dec. 2007.
[106] C.J. Mueller, G.C. Martin, T.E. Briggs, and K.P. Duffy, "An experimental investigation of in-cylinder processes under dual-injection conditions in a DI diesel engine", SAE Paper, 2004, no. 2004-01-1843.
[107] P. Duret, B. Gatellier, L. Monteiro, M. Miche, P. Zima, and D. Maroteaux, "Progress in diesel HCCI combustion within the european SPACE LIGHT project", SAE Paper, 2004, no. 2004-01-1904.
[108] B. Walter and B. Gatellier, "Near zero NOx emissions and high fuel efficiency diesel combustion: the NADITM concept using dual mode combustion", *Oil Gas Sci. Technol.*, vol. 58, pp. 101-114, Jan.-Feb. 2003.
[109] M. Sjoberg, J.E. Dec, and W. Hwang, "Thermodynamic and chemical effects of EGR and its constituents on HCCI autoignition", SAE Paper, 2007, no. 2007-01-0207.
[110] X. He, R.P. Durrett, and Z. Sun, "Late intake valve closing as an emissions control strategy at tier 2 bin 5 engine-out NOx level" SAE Paper, 2008, no. 2008-01-0637.
[111] Y. Murata, J. Kusaka, Y. Daisho, D. Kawano, H. Suzuki, H. Ishii, and Y. Goto, "Miller-PCCI combustion in an HSDI diesel engine with VVT", SAE Paper, 2008, no. 2008-01-0644.
[112] J.I. Kawashima, H. Ogawa, and Y. Tsuru, "Research on a variable swirl intake port for 4-valve high-speed DI Diesel engines", SAE Paper, 1998, no. 982680.
[113] S. Kimura, O. Aoki, H. Ogawa, S. Muranaka, and Y. Enomoto, "New combustion concept for ultra-clean and high-efficiency small DI diesel engines", SAE Paper, 1999, no. 1999-01-3681.
[114] S. Kimura, O. Aoki, Y. Kitahara, and E. Aiyoshizawa, "Ultra-clean combustion technology combining a low-temperature and premixed combustion concept for meeting future emission standards", SAE Paper, 2000, no. 2000-01-0200.
[115] K. Kawamoto, T. Araki, M. Shinzawa, S. Kimura, S. Koide, and M. Shibuya, "Combination of combustion concept and fuel property for ultra-clean DI diesel", SAE Paper, 2004, no. 2004-01-1868.
[116] H. Suzuki, N. Koike, and M. Odaka, "Exhaust purification of diesel engines by homogeneous charge with compression ignition. Part 1: experimental investigation of combustion and exhaust emission behavior under pre-mixed homogeneous charge compression ignition method", SAE Paper, 1997, no. 970313.
[117] H. Suzuki, N. Koike, and M. Odaka, "Exhaust purification of diesel engines by homogeneous charge with compression ignition. Part 2: analysis of combustion phenomena

and NOx formation by numerical simulation with experiment", SAE Paper, 1997, no. 970315.
[118] H. Suzuki, N. Koike, and M. Odaka, "Combustion control method of homogeneous charge diesel engines", SAE Paper, 1998, no. 980509.
[119] M. Odaka, H. Suzuki, N. Koike, and H. Ishii, "Search for optimizing method of homogeneous charge diesel combustion", SAE Paper, 1999, no. 1999-01-0184.
[120] Y. Guezennec, S. Midlam-Mohler, and G. Rizzoni, "A mixed mode HCCI/DI engine based on a novel heavy fuel atomizer", 8th Diesel Engine Emissions Reduction (DEER) Workshop, 2002.
[121] S. Midlam-Mohler, Y. Guezennec, and G. Rizzoni, "Mixed-mode diesel HCCI with external mixture formation: preliminary results", 9th Diesel Engine Emissions Reduction (DDER) Workshop, 2003.
[122] S. Midlam-Mohler. "Diesel HCCI with external mixture preparation", 10th Diesel Engine Emissions Reduction (DEER) Workshop, 2004.
[123] T. Aroonsrisopon, P. Werner, J.O. Waldman, V. Sohm, D.E. Foster, T. Morikawa, and M. Iida, "Expanding the HCCI operation with the charge stratification", SAE Paper, 2004, no. 2004-01-1756.
[124] A.W. Berntsson and I. Denbratt, "HCCI combustion using charge stratification for combustion control", SAE Paper, 2007, no. 2007-01-0210.
[125] K. Inagaki and T. Fuyuto, "Combustion system with premixture-controlled compression ignition", *R&D Rev Toyota CRDL*, vol. 41, pp. 35-46, 2006.
[126] D. Choi, P.C. Miles, H. Yun, and R.D. Reitz, "A parametric study of low temperature late injection combustion in an HSDI diesel engine", COMDIA, 2004.
[127] K. Akihama, Y. Takatori, K. Inagaki, S. Sasaki, and A.M. Dean, "Mechanism of the smokeless rich diesel combustion by reducing temperature", SAE Paper, 2001, no. 2001-01-0655.
[128] S. Kook, C. Bae, P.C. Miles, D. Choi, and L.M. Pickett, "The influence of charge dilution and injection timing on low-temperature diesel combustion and emissions", SAE Paper, 2005, no. 2005-01-3837.
[129] D. Han, A.M. Ickes, S.V. Bohac, Z. Huang, and D.N. Assanis, "Premixed low-temperature combustion of blends of diesel and gasoline in a high speed compression ignition engine" *Proc. Combust. Inst.*, vol. 33, pp. 3039-3046, 2011.
[130] M.F. Yao, Q.C. Zhang, Z.Q. Zheng, and P. Zhang, "Experimental study of effects of oxygen concentration on combustion and emissions of diesel engine", *Sci. China Ser. E.*, vol. 52, pp. 1527-1534, Jun. 2009.
[131] M.P.B. Musculus, "Multiple simultaneous optical diagnostic imaging of early-injection low-temperature combustion in a heavy-duty diesel engine", SAE Paper, 2006, no. 2006-01-0079.
[132] K.C. Natti, "PM characterization in an HSDI diesel engine under conventional and LTC regimes", SAE Paper, 2008, 2008-01-1086.
[133] C.J. Mueller and A. Upatnieks. "Dilute clean diesel combustion achieves low emissions and high efficiency while avoiding control problems of HCCI", 11th Annual Diesel Engine Emissions Reduction (DEER) Conference, 2005.
[134] T. Fang, "Low-temperature combustion within a small-bore high-speed direct injection optically accessible diesel engine", Ph.D. thesis, University of Illinois at Urbana-Champaign, Urbana, IL, 2007.

[135] N.A. Henein, B. Bhattacharyya, J. Schipper, A. Kastury, and W. Bryzik, "Effect of injection pressure and swirl motion on diesel engine-out emissions in conventional and advanced combustion regimes", SAE Paper, 2006, no. 2006-01-0076.
[136] D. Choi, P.C. Miles, H.H. Yun, and R.D. Reitz, "A parametric study of low-temperature, late-injection combustion in a HSDI diesel engine", *JSME Int. J. Ser. B*, vol. 48, pp. 656–664, 2005.
[137] C.L. Genzale, R.D. Reize, and M.P.B. Musculus, "Effects of piston bowl geometry on mixture development and late-injection low-temperature combustion in a heavy-duty diesel engine", SAE Paper, 2008, no. 2008-01-1330.
[138] Y. Aoyagi, H. Osada, M. Misawa, Y. Goto, and H. Ishii, "Advanced diesel combustion using of wide range, high boosted and cooled EGR system by single cylinder engine", SAE Paper, 2006, no. 2006-01-0077.
[139] W. Colban, P.C. Miles, and S. Oh, "Effect of intake pressure on performance and emissions in an automotive diesel engine operating in low temperature combustion regimes" SAE Paper, 2007, no. 2007-01-4063.
[140] M.P.B. Musculus, T. Lachaux, L.M. Pickett, and C.A. Idicheria, "End-of-injection overmixing and unburned hydrocarbon emissions in low-temperature-combustion diesel engines", SAE Paper, 2007, no. 2007-01-0907.
[141] T. Lachaux and M.P.B. Musculus, "In-cylinder unburned hydrocarbon visualization during low-temperature compression-ignition engine combustion using formaldehyde PLIF", *Proc. Combust. Inst.*, vol. 31, pp. 2921-2929, 2007.
[142] D. Kim, I. Ekoto, W.F. Colban, and P.C. Miles, "In-cylinder CO and UHC imaging in a light-duty diesel engine during PPCI low-temperature combustion", SAE Paper, 2008, no. 2008-01-1602.
[143] R. Opat, Y. Ra, D.M.A. Gonzalez, R. Krieger, R.D. Reitz, D.E. Foster, R.P. Durrett, and R.M. Siewert, "Investigation of mixing and temperature effects on HC/CO emissions for highly dilute low temperature combustion in a light duty diesel engine", SAE Paper, 2007, no. 2007-01-0193.
[144] M.F. Yao, Q.C. Zhang, H.F. Liu, Z.Q. Zheng, P. Zhang, Z.Q. Lin, T.J. Lin, and J. Shen, "Diesel engine combustion control: medium or heavy EGR", SAE Paper, 2010, no. 2010-01-1125.
[145] H.J. Curran, E.M. Fisher, P.A. Glaude, N.M. Marinov, W.J. Pitz, C.K. Westbrook, D.W. Layton, P.F. Flynn, R.P. Durrett, A.O. zur Loye, O.C. Akinyemi, F.L. Dryer, "Detailed chemical kinetic modeling of diesel combustion with oxygenated fuels", SAE Paper 2001, no. 2001-01-0653.
[146] L. Noel, F. Maroteaux, and A. Ahmed, "Numerical study of HCCI combustion in diesel engines using reduced chemical kinetics of n-heptane with multidimensional CFD code", SAE Paper, 2004, no. 2004-01-1909.
[147] H.J. Curran, "A comprehensive modeling study of n-heptane oxidation", *Combust. Flame*, vol. 114, pp. 149–177, Jul. 1998.
[148] J.M. Anderlohr, A. Piperel, A.P. Cruz, R. Bounaceur, F. Battin-Leclerc, P. Dagau, and X. Montagne, "Influence of EGR compounds on the oxidation of an HCCI-diesel surrogate", *Proc. Combust. Inst.*, vol. 32, pp. 2851-2859, 2009.
[149] H. Machrafi, S. Cavadias, and P. Gilbert, "An experimental and numerical analysis of the HCCI auto-ignition process of primary reference fuels, toluene reference fuels and diesel

fuel in an engine, varying the engine parameters", *Fuel Process. Technol.*, vol. 89, pp. 1007-1116, Nov. 2008.

[150] J.J. Hernandez, J. Sanz-Argent, J. Benajes, and S. Molina, "Selection of a diesel fuel surrogate for the prediction of auto-ignition under HCCI engine conditions", *Fuel*, vol. 87, pp. 655–665, May 2008.

[151] J.H. Mack, D.L. Flowers, S.M. Aceves, and R.W. Dibble, "Demonstrating direct use of wet ethanol in a homogeneous charge compression ignition (HCCI) engine", 2007 Fall Meeting of the Western States Section of the Combustion Institute, 2007, no. 07F-30.

[152] A. Megaritis, D. Yap, and M.L. Wyszynski, "Effect of inlet valve timing and water blending on bioethanol HCCI combustion using forced induction and residual gas trapping", *Fuel*, vol. 87, 732-739, May 2008.

[153] M.F. Yao, Z.Q. Zheng, Z. Chen, and B. Zhang, "Experimental study on homogeneous charge compression ignition operation by burning dimethyl ether and methanol", *Int. J. Green Energy*, vol. 4, pp. 283-300, May 2007.

[154] K. Yeom, J. Jang, and C. Bae, "Homogeneous charge compression ignition of LPG and gasoline using variable valve timing in an engine", *Fuel*, vol. 86, pp. 494-503, 2007.

[155] D. Yap, S.M. Peucheret, A. Megaritis, M.L. Wyszynski, and H. Xu, "Natural gas HCCI engine operation with exhaust gas fuel reforming", *Int. J. Hydrogen Energy*, vol. 31, pp. 587-595, Apr. 2006.

[156] A. Tsolakis, A. Megaritis, and D. Yap, "Application of exhaust gas fuel reforming in diesel and homogeneous charge compression ignition (HCCI) engines fuelled with biofuels", *Energy*, vol. 33, pp. 462–470, Apr. 2008.

[157] V. Manente, P. Tunestal, and B. Johansson, "Influence of the wall temperature and combustion chamber geometry on the performance and emissions of a mini HCCI engine fuelled with DEE", SAE Paper, 2008, no. 2008-01-0008.

[158] J.P. Szybist and B.G. Bunting, "Cetane number and engine speed effects on diesel HCCI performance and emissions", SAE Paper, 2005, no. 2005-01-3723.

[159] P. Risberg, G. Kalghatgi, H. Angstrom, and F. Wahlin, "Auto-ignition quality of diesel-like fuels in hcci engines", SAE Paper, 2005, no. 2005-01-2127.

[160] T. Li, Y. Okabe, H. Izumi, T. Shudo, and H. Ogawa, "Dependence of ultra-high EGR low temperature diesel combustion on fuel properties", SAE Paper, 2006, no. 2006-01-3387.

[161] G.T. Kalghatgi, P. Risberg, and H.E. Ångström, "Advantages of a fuel with high resistance to auto-ignition in late-injection, low-temperature, compression ignition combustion", SAE Paper, 2006, no. 2006-01-3385.

[162] G.T. Kalghatgi, P. Risberg, and H.E. Ångström, "Partially pre-mixed auto-ignition of gasoline to attain low smoke and low NOx at high load in a compression ignition engine and comparison with a diesel fuel", SAE Paper, 2007, no. 2007-01-0006.

[163] W.R. Leppard, "The chemical origin of fuel octane sensitivity", SAE Paper, 1990, no. 902137.

[164] C. Westbrook, W. Pitz, and R. Leppard, "The autoignition chemistry of paraffinic fuels and pro-knock and anti-knock additives: a detailed chemical kinetic study", SAE Paper, 1991, no. 912314.

[165] P.U. Ricklin, A. Kazakov, F.L. Dryer, S.C. Kong, and R.D. Reitz, "The effects of NOx addition on the auto ignition behavior of natural gas under HCCI conditions", SAE Paper, 2002, no. 2002-01-1746.

[166] M.F. Yao, Z. Chen, Z.Q. Zheng, B. Zhang, and Y. Xing, "Study on the controlling strategies of homogeneous charge compression ignition combustion with fuel of dimethyl ether and methanol", *Fuel*, vol. 85, 2046–2056, Oct. 2006.
[167] Z. Chen, M.F. Yao, Z.Q. Zheng, and Q.C. Zhang, "Experimental and numerical study of methanol/dimethyl ether dual-fuel compound combustion", *Energy Fuel*, vol. 23, pp. 2719-2730, May 2009.
[168] S.M. Aceves, D. Flowers, J. Martinez-Frias, F. Espinosa-Losa, W.J. Pitz, and R. Dibble, "Fuel and additive characterization for HCCI combustion", SAE Paper, 2003, no. 2003-01-1814.
[169] G. Shibata, K. Oyama, T. Urushihara, and T. Nakano, "Correlation of low temperature heat release with fuel composition and HCCI engine combustion", SAE Paper, 2005, no. 2005-01-0138.
[170] J.T. Farrell and B.G. Bunting, "Fuel composition effects at constant RON and MON in an HCCI engine operating with negative valve overlap", SAE Paper, 2006, no. 2006-01-3275.
[171] J.E. Dec and W. Hwang, "Characterizing the development of thermal stratification in an HCCI engine using planar-imaging thermometry", SAE Paper, 2009, no. 2009-01-0650.
[172] A. Babajimopoulos, G.A. Lavoie, and D.N. Assanis, "Modelling hcci combustion with high levels of residual gas fraction - a comparison of two vva strategies", SAE Paper, 2003, no. 2003-01-3220.
[173] H.F. Liu, M.F. Yao, Z.Q. Zheng, and Y.F. Wang, "An investigation of different ported fuel injection strategies and thermal stratification in HCCI engines using chemiluminescence imaging", SAE Paper, 2010, no. 2010-01-0163.
[174] J.B. Heywood, *Internal Combustion Engine Fundamentals*. New York: McGraw-Hill, 1988.
[175] A. Hultqvist, M. Christensen, B. Johansson, M. Richter, J. Nygren, H. Hult, and M. Aldén, "The HCCI combustion process in a single cycle-high-speed fuel tracer LIF and chemiluminescence imaging", SAE Paper, 2002, no. 2002-01-0424.
[176] P.G. Aleiferis, A.G. Charalambides, Y. Hardalupas, A.M.K.P. Taylor, and Y. Urata, "Modelling and experiments of HCCI engine combustion with charge stratification and internal EGR", SAE Paper, 2005, no. 2005-01-3725.
[177] D.L. Reuss and V. Sick, "Inhomogeneities in hcci combustion: an imaging study", SAE Paper, 2005, no. 2005-01-2122.
[178] M. Sjoberg, J.E. Dec, A. Babajimopoulos, and D.N. Assanis, "Comparing enhanced natural thermal stratification against retarded combustion phasing for smoothing of hcci heat-release rates", SAE Paper, 2004, no. 2004-01-2994.
[179] M. Christensen and B. Johansson, "The effect of in-cylinder flow and turbulence on hcci operation", SAE Paper, 2002, no. 2002-01-2864.
[180] A. Kakuho, M. Nagamine, Y. Ameromori, T. Urushihara, and T. Itoh, "In-cylinder temperature distribution measurement and its application to HCCI combustion", SAE Paper, 2006, no. 2006-01-1202.
[181] R.E. Herold and J. Ghandi, "Investigation of bulk in-cylinder stratification with split intake runners", SAE Paper, 2007, no. 2007-01-4044.
[182] B. Thirouard, J. Cherel, and V. Knop, "Investigation of mixture quality effect on cai combustion", SAE Paper, 2005, no. 2005-01-0141.
[183] Z. Wang, S.J. Shuai, J.X. Wang, and G.H. Tian, "A computational study of direct injection gasoline HCCI engine with secondary injection", *Fuel*, vol. 85, pp. 1831-1841, Sep. 2006.

[184] Z.L. Zheng and M.F. Yao, "Numerical simulation of the effects of charge stratification on combustion and emissions", *Energy Fuel*, vol. 21, pp. 2018-2026, May 2007.

[185] P. Risberg, D. Johansson, J. Andrae, G.T. Kalghatgi, P. Björnbom, and H.E. Ångström, "The influence of NO on the combustion phasing in an HCCI engine", SAE Paper, 2006, no. 2006-01-0416.

[186] J. Krasselt, J. Ghandhi, R. Herold, D. Reuss, and P. Najt, "Investigations into the effects of thermal and compositional stratification on HCCI combustion - part I: metal engine results", SAE Paper, 2009, no. 2009-01-1105.

CHAPTER 9

How Efficient are Hydrogen-Fueled Internal Combustion Engines?

S. Verhelst[a,*], R. Sierens[a] and T. Wallner[b]

[a]Department of Flow, Heat and Combustion Mechanics Ghent University, Belgium and [b]Energy Systems Division, Argonne National Laboratory, USA

Abstract: Hydrogen has long been recognized as an energy carrier for the transportation sector with a number of important advantages compared to the currently used fossil fuels. It can be produced from a variety of (renewable) energy sources and it can produce energy in an efficient and clean way. For powering vehicles, hydrogen can be used in two ways, either in a fuel cell (FC) producing electricity, or in an internal combustion engine (ICE) producing mechanical power. Converting an ICE to hydrogen operation is relatively straightforward and is interesting as it offers a bi-fuel possibility. What is less known is that a hydrogen-fueled ICE (H_2ICE) has a high efficiency potential, leading to a smaller gap in efficiency compared to a hydrogen-fueled FC than commonly assumed. This chapter describes the physical and chemical properties of hydrogen that theoretically allow a high engine efficiency, and presents experimental confirmation of these theoretical considerations. Published efficiency figures obtained by engine testing are reviewed and recent work on both port fuel injection (PFI) as direct injection (DI) H_2ICEs is discussed. Finally, an outlook is given on the potential for further increases in efficiency.

Keywords: Hydrogen, internal combustion engines, spark ignition, fuel, bi-fuel, NOx, combustion, emissions, efficiency, transportation, vehicles, injection, knock, mixture formation, injection strategies, injection strategies, power density.

INTRODUCTION

Research on the use of hydrogen as an energy carrier dates back to the first oil crisis and earlier. However, until about the year 2000 the space travel programs represented the only significant use of hydrogen as an energy carrier. Since then, several governments have recognized the potential importance of hydrogen and have launched research, development and demonstration plans to stimulate breakthroughs in hydrogen technology [1-4].

***Address correspondence to S. Verhelst:** Department of Flow, Heat and Combustion Mechanics Ghent University, Belgium; Tel: +32 9 264 33 06; Fax: + 32 9 264 35 90; E-mail: sebastian.verhelst@ugent.be

Hatim Machrafi (Ed)

For converting hydrogen to energy for propelling a vehicle, most attention has been focused on fuel cells. These are attractive for their promise of unrivalled efficiency, low noise signature and zero tailpipe emissions. However, there are a number of serious challenges for a large scale introduction of fuel cell powertrains, of which cost and durability are the most important ones [5].

An alternative to fuel cells is the well-known internal combustion engine (ICE). An ICE can be converted for dedicated hydrogen operation, or can be converted so that it is still able to run on the original fuel (*e.g.,* gasoline) [6]. In the latter case, the challenges in establishing a hydrogen infrastructure are less as these bi-fuel vehicles would not solely be dependent on the availability of hydrogen fueling stations. Also, converting an ICE for hydrogen combustion can be done significantly cheaper than producing fuel cells, and engines do not have the high hydrogen purity requirement that fuel cells have.

The authors have reviewed the necessary hardware changes in order to run an engine on hydrogen elsewhere [6, 7]. In the following, first the properties of hydrogen are compared to conventional fuels and the impact on the potential engine efficiency is discussed. Then the mixture formation strategies used for hydrogen engines are reviewed, as these affect the emissions (with oxides of nitrogen, or NO_X, being the only emission component to be considered) and attainable power output, thus leading to a number of load control strategies and ultimately the practically obtainable efficiency.

HYDROGEN PROPERTIES AFFECTING ENGINE EFFICIENCY

Table **1** lists some properties of hydrogen compared to methane and iso-octane [8-11], which are taken here as representing natural gas and gasoline, respectively, as it is easier to define properties for single-component fuels. The small and light hydrogen molecule is very mobile (high mass diffusivity) and leads to a very low density at atmospheric conditions.

The wide range of flammability limits, with flammable mixtures from as lean as $\lambda = 10$ to as rich as $\lambda = 0.14$ ($0.1 < \varphi < 7.1$) theoretically allows adjusting for a wide range of engine power output through changes in the mixture equivalence ratio. As will be discussed later, in practice, the lean limit of H_2ICEs is reached for lower air-to-fuel equivalence ratios than mentioned above, in the vicinity of λ =4 / φ = 0.25.

Table 1: Hydrogen properties compared with methane and iso-octane properties. Data given at 300 K and 1 atm.

Property	**Hydrogen**	**Methane**	**Iso-octane**
Molecular weight (g/mol)	2.016	16.043	114.236
Density (kg/m^3)	0.08	0.65	692
Mass diffusivity in air (cm^2/s)	0.61	0.16	~0.07
Minimum[a] ignition energy (mJ)	0.02	0.28	0.28
Minimum[a] quenching distance (mm)	0.64	2.03	3.5
Flammability limits in air (vol%)	4–75	5–15	1.1–6
Flammability limits (λ)	10–0.14	2–0.6	1.51–0.26
Flammability limits (φ)	0.1–7.1	0.5–1.67	0.66–3.85
Lower heating value (MJ/kg)	120	50	44.3
Higher heating value (MJ/kg)	142	55.5	47.8
Stoichiometric air-to-fuel ratio (kg/kg)	34.2	17.1	15.0
Stoichiometric air-to-fuel ratio (kmol/kmol)	2.387	9.547	59.666

[a] Corresponding equivalence ratios given in text.

The minimum ignition energy of a hydrogen–air mixture at atmospheric conditions is an order of magnitude lower than for methane–air and iso-octane–air mixtures. It is only 0.017 mJ, which is obtained for hydrogen concentrations of 22–26% (λ =1.2–1.5 / φ = 0.67–0.83) [12]. The quenching distance is minimal for mixtures around stoichiometry, and decreases with increasing pressure and temperature. As can be seen in Table **1**, it is about one-third that for methane and iso-octane. This affects crevice combustion and wall heat transfer, as will be discussed later.

Finally, note the large difference between the lower and higher heating values of hydrogen compared to methane and iso-octane, which is easily explained as H_2O is the sole combustion product of hydrogen. Also note the large difference in stoichiometric air-to-fuel ratio of hydrogen compared to methane and iso-octane, as well as the large difference in stoichiometric air-to-fuel ratio in mass terms *versus* in mole terms.

Table **2** lists the properties of hydrogen–air mixtures, at stoichiometric and at the lean limit mentioned above, compared to stoichiometric methane–air and iso-octane–air mixtures [8-11]. The volume fraction of fuel in the fuel–air mixture can be directly calculated from the molar stoichiometric air-to-fuel ratio listed in Table **1**. The large volume fraction occupied by hydrogen has consequences for the attainable engine power density (see further below). Combined with the wide flammability limits, it also has an important effect on mixture properties such as the kinematic viscosity, thermal conductivity, *etc.* These properties vary much more than in conventionally fueled engines.

Table 2: Mixture properties of hydrogen–air, methane–air and iso-octane–air. Data given at 300 K and 1 atm (with the exception of the laminar burning velocity, given at 360 K and 1 atm).

Property	**H_2–air**	**H_2–air**	**CH_4–air**	**C_8H_{18}–air**
	$\lambda = 1$ **$\varphi = 1$**	**$\lambda = 4$** **$\varphi = 0.25$**	**$\lambda = 1$** **$\varphi = 1$**	**$\lambda = 1$** **$\varphi = 1$**
Volume fraction fuel (%)	29.5	9.5	9.5	1.65
Mixture density (kg/m^3)	0.850	1.068	1.123	1.229
Kinematic viscosity (mm^2/s)	21.6	17.4	16	15.2
Autoignition temperature (K)	858[a]	>858[a]	813[a]	690[a]
Adiabatic flame temperature (K)	2390	1061	2226	2276
Thermal conductivity (10^{-2} W/mK)	4.97	3.17	2.42	2.36
Thermal diffusivity (mm^2/s)	42.1	26.8	20.1	18.3
Ratio of specific heats	1.401	1.400	1.354	1.389
Speed of sound (m/s)	408.6	364.3	353.9	334.0
Air-to-fuel ratio (kg/kg)	34.2	136.6	17.1	15.1
Mole ratio before/after combustion	0.86	0.95	1.01	1.07
Laminar burning velocity, ~360 K (cm/s)	290	12	48	45
Gravimetric energy content (kJ/kg)	3758	959	3028	3013
Volumetric energy content (kJ/m^3)	3189	1024	3041	3704

[a] See text.

There is some ambiguity concerning the autoignition temperature of fuels in general and hydrogen in particular. For instance, for hydrogen, values were found from 773 K [13] to 858 K [14]. Some sources list the autoignition temperature for hydrogen as lower than that for methane; other sources list the opposite. For spark-ignition engines, with a propagating flame front, autoignition of the unburned mixture ahead of the flame front is not desirable, as it can result in knocking combustion. The efficiency of a spark-ignition engine is influenced by the compression ratio and the ignition timing (among others), the choices of which are dependent on the autoignition temperature of the fuel–air mixture, so this is an important parameter. For liquid hydrocarbons, the octane rating is more commonly used as a measure of the propensity of a fuel–air mixture to undergo pre-flame reactions. For hydrogen, a research octane number (RON) in excess of 130 and a motor octane number (MON) of 60 have been reported [7, 15].

The laminar burning velocity of stoichiometric hydrogen–air mixtures is much higher than that of methane and iso-octane. However, if lean-burn strategies are used, the burning velocity can be much lower (see value for $\lambda = 4/\varphi = 0.25$). For mixtures around stoichiometry, the high burning velocity and high adiabatic flame temperature point to high nitrogen oxides (NO_X) emissions (see later).

Combining the lower heating value (LHV) of hydrogen, its density and the stoichiometric air demand, one can calculate the maximum theoretical power density of different engine concepts. Fig. **1** compares the theoretical power density of hydrogen-fueled engines with an iso-octane-fueled engine (as representative for gasoline) as the reference. Values for both port-fuel injection (PFI) engines and direct-injection (DI) engines are quoted. Note the large difference, with a (theoretical) power density increase of 38 % for hydrogen when switching from PFI to DI.

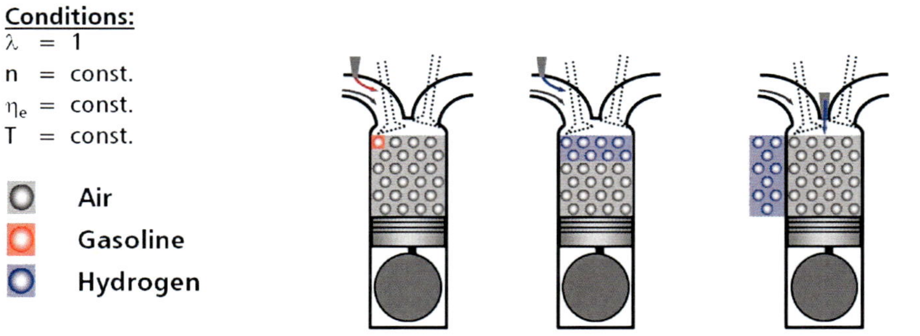

Fuel	Gasoline	Hydrogen	Hydrogen
Concept	Port Injection	Port Injection	Direct Injection
Mixture Calorific Value [MJ/m³]	3,9	3,2	4,5
Power Output [%] (Compared to gasoline)	**100**	**82**	**117**

Figure 1: Comparison of power densities of PFI and DI H_2ICEs compared to gasoline. For the DI illustration, the hydrogen is drawn outside the cylinder as it is only injected after intake valve closure.

Finally, it is noteworthy that the possibility of qualitative load control (changing the mixture richness at wide open throttle (WOT)), the tolerance for substantial mixture dilution (either through excess air or exhaust gas recirculation (EGR)), the high autoignition temperature (allowing high compression ratios) and the generally fast burn rate, all are factors contributing to potentially high engine efficiencies. As will be discussed, this has been experimentally confirmed. However, heat losses from cylinder gases to the combustion chamber walls can be higher with hydrogen compared to conventional fuels, negatively affecting efficiencies [16, 17]. Fig. **2** illustrates the efficiency potential by plotting the Otto cycle efficiency for hydrogen engines employing different equivalence ratios against compression ratio, compared to stoichiometric gasoline operation. Both, the potential for higher compression ratio and higher air-fuel ratio significantly increase the theoretical engine efficiency.

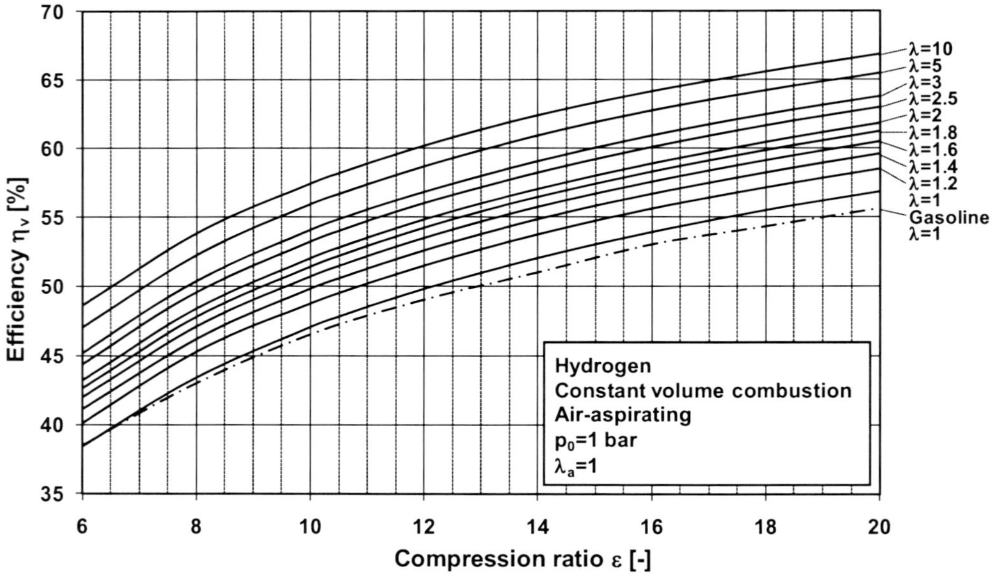

Figure 2: Theoretical engine efficiency (Otto cycle) for hydrogen operation as a function of equivalence ratio and compression ratio, compared to stoichiometric gasoline operation.

BASIC HYDROGEN MIXTURE FORMATION AND LOAD CONTROL STRATEGIES

The wide ignition limits of hydrogen allow an engine to be operated at extremely lean air–fuel ratios compared to conventional fuels. Concepts for part-load operation of hydrogen engines have been proposed using quantitative control (*i.e.* using a throttle valve to influence the mixture quantity entering the cylinders), qualitative control (*i.e.* changing the mixture's equivalence ratio) as well as combined strategies. Quantitative control refers to any approach that limits the amount of fresh charge introduced into the engine in order to limit the power output while keeping the ratio of air-to-fuel constant. On the other hand, qualitative control is used to describe systems that adjust the air-to-fuel ratio usually by adjusting the amount of fuel introduced to the engine while allowing maximum flow of air. The classification based on part-load operation is closely linked to the air–fuel ratio at which the engine is operated. Reports of hydrogen engines being operated at stoichiometric as well as lean air–fuel ratios have been published. The major advantages of operation at a stoichiometric air–fuel ratio are the increased power output compared to lean-burn strategies as well as the fact that a conventional aftertreatment system can be employed to reduce NO_X emissions. Lean-burn concepts, on the other hand, generally result in a significant increase in achievable engine efficiencies. More recently, strategies that employ both lean burn as well as stoichiometric operation have been proposed [18, 19].

Besides the above-mentioned classification that results from the specific properties of hydrogen as a fuel for internal combustion engines, the charging strategy is another important factor to influence the performance and efficiency of hydrogen internal combustion engines. Supercharged operation has been evaluated as a promising option to mitigate the significant reduction in power output related to hydrogen operation with external mixture formation. From an efficiency stand-point, turbocharging is considered the preferred option; however, reduced throttle response and reduced exhaust energy compared to conventional fuel operation make the implementation of a turbocharged hydrogen engine more challenging.

Although the self-ignition temperature of hydrogen is considerably higher than that of conventional fuels (H_2 = 858 K, diesel is approx. 523 K), which suggests using a spark plug as the ignition source, attempts have been made to operate hydrogen engines with compression ignition. Successful operation of compression ignition hydrogen engines has been reported for both large displacement stationary engines [20, 21] as well as automotive-size engines [22]. More recently, effort has been reported on hydrogen engines being operated in HCCI mode [23].

The proper design of the mixture formation process is crucial for achieving high engine efficiencies while meeting more and more stringent emissions targets. Similar to conventionally fueled engines, hydrogen engines have gone through continuing improvement and refinement in terms of mixture formation strategies.

A primary classification of mixture formation strategies can be done based on the location of mixture formation or the location of the hydrogen dosing devices. External mixture formation refers to concepts in which hydrogen and air are mixed outside the combustion chamber, whereas internal mixture formation refers to concepts with hydrogen being introduced directly into the combustion chamber. Some researchers have also proposed combined concepts with a combination of external and internal mixture formation [24-26]. As indicated in Fig. **1**, the mixture formation strategy, especially in hydrogen operation, has significant impact on the theoretical power output of the engine. The dramatic difference in theoretical power output is mainly caused by the low density of hydrogen, resulting in a significant decrease in mixture density when external mixture formation is being employed.

The two single most prominent mixture formation strategies for hydrogen engines are hydrogen port injection, which is being used in engine research as well as vehicle demonstrations, and hydrogen direct injection, which is still in a research stage. The following sections are intended to summarize the most typical characteristics and variations of these mixture formation concepts.

HYDROGEN PORT INJECTION

Hydrogen port injection is probably the most common hydrogen mixture formation strategy and is employed in a number of variants that differ in multiple aspects including part-load control, air–fuel ratio and charging strategy. Before evaluating the pros and cons of different mixture formation strategies, it is important to understand the principal correlation between the air–fuel ratio and oxide of nitrogen emissions that is applicable for all homogeneous mixture formation concepts. Fig. **3** shows a typical trace of oxide of nitrogen emissions as a function of the equivalence ratio for homogeneous port-injection operation. Combustion of lean hydrogen–air mixtures with fuel-to-air equivalence ratios of less than 0.5 ($\lambda > 2$) results in extremely low NO_X emissions. Due to the excess air available in the combustion chamber, the combustion temperatures do not exceed the NO_X critical value of approximately 1800 K [27]. Exceeding the NO_X critical equivalence ratio results in an exponential-like increase in oxides of nitrogen emissions, which peaks around a fuel-to-air equivalence ratio of 0.75 ($\lambda \sim 1.3$). At stoichiometric conditions, the NO_X emissions are at around 1/3 of the peak value. The highest burned gas temperatures in hydrogen operation occur around a fuel-to-air equivalence ratio near 1.1, but at this equivalence ratio, oxygen concentration is low, so the NO_X concentration does not peak there [28]. As the mixture gets leaner, increasing oxygen concentrations initially offset the falling gas temperatures, and NO_X emissions peak around a fuel-to-air equivalence ratio of 0.75 ($\lambda = 1.3$).

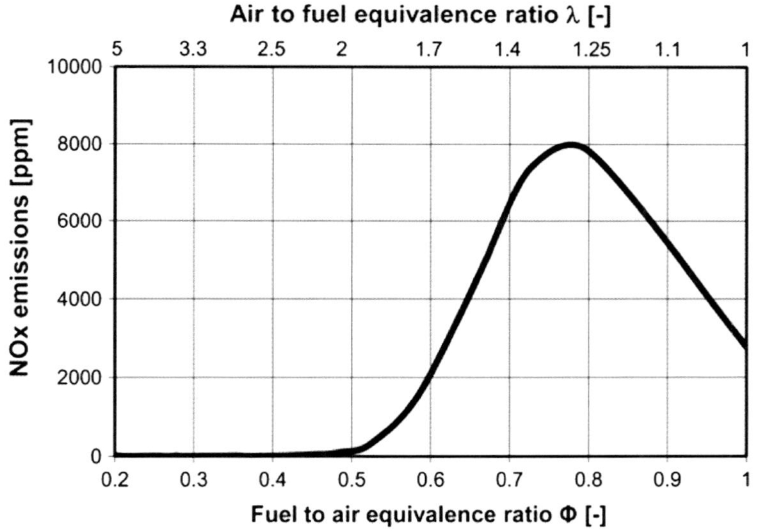

Figure 3: NO_X emissions as a function of air-fuel ratio in unthrottled hydrogen port fuel injection operation.

In light of these dependencies of oxide of nitrogen emissions as a function of air–fuel ratio, several operating strategies have been developed that mainly aim at achieving acceptable power densities while simultaneously avoiding excessive NO_X emissions. Conversion engines based on conventional gasoline engines have been operated on hydrogen employing a lean constant air–fuel ratio strategy. Using a conventional throttle and replacing the gasoline fuel system with hydrogen injectors, one can easily implement this strategy [29-31]. Selecting an equivalence ratio below the NO_X emissions critical limit of $\varphi \sim 0.5/\lambda \sim 2$ results in extremely low emissions signatures even without the use of any aftertreatment system. Due to the relatively low combustion temperatures, the reduced thermal load of the engine and the increase in required ignition energy for lean hydrogen–air mixtures, this lean operating strategy also effectively avoids combustion anomalies (*i.e.* knock, surface ignition or backfire).

A significant increase in engine efficiency (compared to gasoline operation) can be accomplished employing a constant lean air–fuel ratio operating strategy. The air–fuel ratio is generally set as high as possible to achieve acceptable power output while still meeting the emissions targets. However, increasing the air-fuel ratio (lowering power output) was shown to even further improve engine efficiencies with a peak at around $\lambda = 3.3/(\varphi = 0.30$. This peak results from a local minimum of the losses due to increased burn duration and heat transfer as well as the more favorable properties of the working fluid. Also a sharp decrease in indicated efficiency was observed at $\lambda = 4.5/(\varphi = 0.22$ due mostly to greater amounts of unburned hydrogen and a slower burn rate [32].

As a result, an operating strategy using a variable equivalence ratio as a function of engine load was evaluated both for naturally aspirated engines as well as supercharged engines and vehicles. Implementation of this variable equivalence ratio strategy in a range from $2 < \lambda < 5/0.2 < \varphi < 0.5$ on a GM 454 spark-ignited PFI engine (commonly known as the Chevrolet Big Block) showed a more than 20% increase in engine power compared to a carbureted version without increasing the danger of backfiring [33].

However, employing a constant lean air–fuel ratio strategy results in an even more significant loss in power density than shown in Fig. **1** for stoichiometric operation. Assuming a constant air-fuel equivalence ratio of 2 ($\phi = 0.5$) results in a theoretical maximum power output of the hydrogen engine that is only about 50% of a regular gasoline engine in stoichiometric operation. Thus researchers have investigated the potential of using supercharging in combination with constant lean air–fuel ratio operation to mitigate the significant power loss [34]. This concept has also been implemented in several versions of hydrogen-powered pickup trucks running at constant equivalence ratios around $\lambda \sim 2.5 / \varphi \sim 0.4$ with compression ratios of up to

12:1 and boost pressures of approximately 0.8 bar [35, 36]. The theoretical maximum power output based on a comparison of calculated mixture calorific values as a function of air–fuel ratio assuming constant efficiencies is in the range of about 80% compared to the naturally aspirated gasoline counterpart.

Another significant increase in power density can be achieved with supercharging in combination with a variable equivalence ratio strategy. Drive-cycle simulations based on steady-state engine efficiency and emissions maps of a supercharged 2.3 L hydrogen engine suggested a 3% increase in fuel economy when employing a variable equivalence ratio strategy compared to a constant $\lambda = 2/\ \varphi = 0.50$ strategy while simultaneously reducing NO_X emissions by more than 80%. Fig. **4** shows the brake thermal efficiency map obtained on a multi-cylinder supercharged engine, as a function of the torque and air-fuel equivalence ratio at 1500 RPM. At constant engine torque an increase in air-fuel equivalence ratio results in increased engine efficiency. The plot also shows the significant loss in peak torque with leaner air-fuel mixtures.

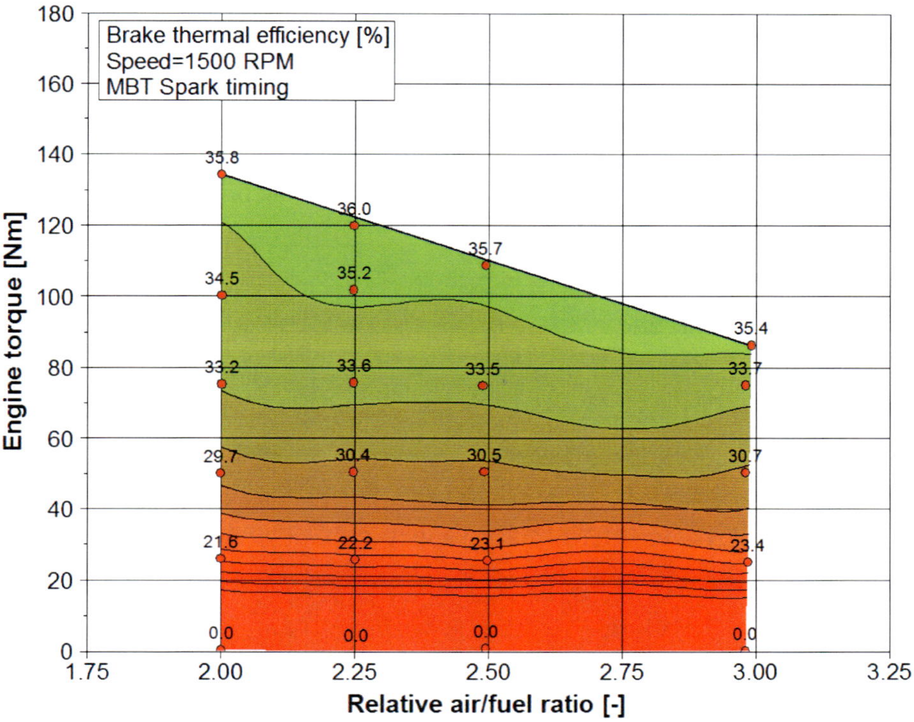

Figure 4: Influence of air-fuel ratio on engine efficiency on a supercharged multi-cylinder hydrogen engine.

Additional simulations with the hydrogen engine being operated in a full hybrid-electric vehicle with a pre-transmission parallel hybrid architecture suggest another almost 50% increase in fuel economy while reducing NO_X emissions by more than 99% compared to a conventional powertrain [37].

A similar engine control strategy was used for propelling a hydrogen shuttle bus using a 6.8 LV-10 engine. Due to the lean burn concept, the engine achieved a more than 12% better brake thermal efficiency compared to the gasoline counterpart at 1500 RPM and 2.62 bar brake mean effective pressure (BMEP) (Ford World Wide Mapping Point (WWMP)). In-vehicle tests confirmed an increase in power output enabling a more than 30% reduction in 0–35 mph acceleration time compared to the natural gas counterpart while achieving near-zero emissions [38].

An effective way to limit the power loss compared to gasoline or diesel engines is by running hydrogen port-injection engines at stoichiometric air–fuel ratios. However, stoichiometric port-fuel injection operation is prone to combustion anomalies and also requires an aftertreatment system to reduce the level of oxide of nitrogen emissions. As indicated in Fig. **1**, stoichiometric H_2 PFI operation results in a theoretical power density of approximately 86% compared to gasoline. A 6-cylinder 12 L displacement bus engine (MAN H 2866 UH01) converted to bi-fuel operation was shown to achieve 170 kW in gasoline operation and 140 kW in stoichiometric port-fuel injected hydrogen operation, which is approximately 82% of the gasoline power output, confirming the theoretical considerations. However, in order to avoid combustion anomalies, the compression ratio of the engine had to be reduced to as low as 7.5:1 [39]. With later engine conversions by the same manufacturer, significant improvements could be achieved by using solenoid-driven hydrogen injection valves instead of rotary hydrogen valves. The MAN H2876 UH01, a 12.8 L in-line 6-cylinder engine using these improved injectors with sequential injection, achieved a peak brake thermal efficiency of 31% in naturally aspirated stoichiometric hydrogen operation. The MAN H2876 UH01 engine uses a reducing catalyst with lambda control for emissions aftertreatment. Operated with slight hydrogen surplus, the engine can be operated well below Euro 5 emissions levels, which are mandatory since 2008 [39].

Fig. **5** shows results obtained on a single cylinder research engine, comparing two strategies for obtaining power outputs similar to or exceeding the power output of an equivalent, naturally aspirated, gasoline engine, without excessive tailpipe NO_X emissions [34]. The first option is to stay lean of the threshold equivalence ratio, the air-to-fuel equivalence ratio below which NO_X emissions rise exponentially, and make up for the power loss caused by the lean mixtures through supercharging. The second option is supercharging at stoichiometric mixtures, or in practice at slightly rich of stoichiometric so that a small amount of unburned hydrogen is present in the exhaust which is an effective reducing agent for NO_X, using a three way catalyst. For supercharging to be possible at stoichiometric operation, *i.e.* to avoid combustion anomalies, EGR had to be used to lower the peak temperatures.

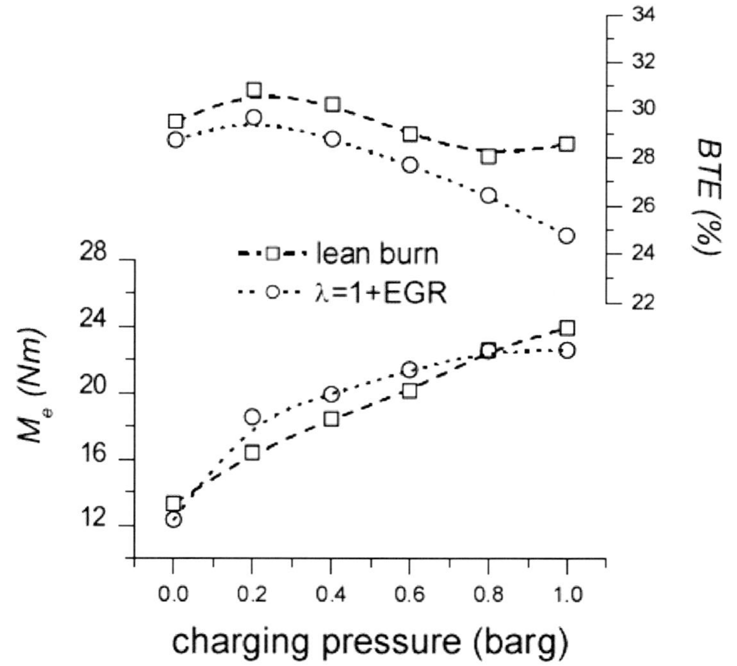

Figure 5: Comparison of brake torque and efficiency as a function of supercharging, between lean burn operation and stoichiometric + EGR operation, at the NO_X threshold, 2000 rpm.

The brake thermal efficiencies are lower for the stoichiometric + EGR strategy than for the supercharged lean-burn strategy, which has several causes. First of all, the higher heat-loss at stoichiometric operation leads to lower efficiency. In addition, to ensure the recirculation flow of exhaust gases, the inlet had to be throttled slightly (-0.1 barg), so operation at WOT was not possible. This causes additional pumping losses which decrease the efficiency. Another reason is the dilution of the mixture with EGR which decelerates the combustion resulting in lower efficiencies. On the other hand, the dilution of EGR decreases the cylinder temperatures resulting in lower heat loss. It should be noted that the brake thermal efficiencies are relatively low as these tests were conducted on a single cylinder research engine with relatively high friction losses.

The previous mixture formation concepts employing hydrogen port injection result in compromises, either in terms of power density with lean air–fuel ratio approaches, or engine efficiency with stoichiometric concepts. A potential solution to this trade-off is combining lean-burn and stoichiometric operating strategies. This concept has been proposed based on engine research results [25, 27, 40] and has also been implemented in hydrogen demonstration vehicles [18, 41]. At low engine loads, the engine is operated at variable lean air–fuel ratios, resulting in good engine efficiencies and extremely low engine-out emissions. Once a certain engine power demand is exceeded, the operating strategy is switched to throttled stoichiometric operation. The NO_X emissions critical operating regime at equivalence ratios $1 < \lambda < 2 / 0.5 < \varphi < 1$ is avoided, and a conventional aftertreatment system can be used to reduce oxide of nitrogen emissions in stoichiometric operation. Tests performed on a prototype demonstration vehicle with a 6.0 LV 12 engine employing this operating strategy showed that the emissions add up to only a fraction of the most stringent standards. The test results on a FTP-75 cycle revealed NO_X emissions as low as 0.0008 g/mi (3.9% of the Super Ultra Low Emission Vehicle (SULEV) standard) and hydrocarbon emissions that were lower than the ambient concentration, indicating that this vehicle actively reduces the concentration of certain emissions components [19].

HYDROGEN DIRECT INJECTION

Injection Strategies

Some of the downsides of hydrogen port injection, in particular the occurrence of backfiring as well as the low specific power density, can be mitigated when employing a direct injection strategy. Hydrogen direct injection operation strategies can best be classified according to the timing and number of injection events. In order to fully benefit from direct injection, the start of injection is typically timed after intake valve closing. This avoids loss in power density through displacement of intake air and also mitigates the risk of backfiring in the intake manifold.

Fig. **6** shows several basic injection strategies; various other strategies can be created by combining these basic strategies. Early direct injection typically occurs during the beginning of the compression stroke. At constant injection pressure the amount of fuel delivered to the cylinder is determined by the injection duration requiring a longer injection pulse for high engine load compared to low engine load. Due to the early start of injection for all load cases, the resulting mixture of hydrogen and air around spark timing is fairly homogeneous [42-45]. The wide flammability limits of hydrogen allow for unthrottled operation over the entire load regime of a combustion engine. When operating a hydrogen engine unthrottled, the overall air-fuel ratio increases with reduced engine load resulting in reduced flame speeds and longer combustion duration. Therefore lighter loads in unthrottled, lean hydrogen operation require more spark advance than higher loads.

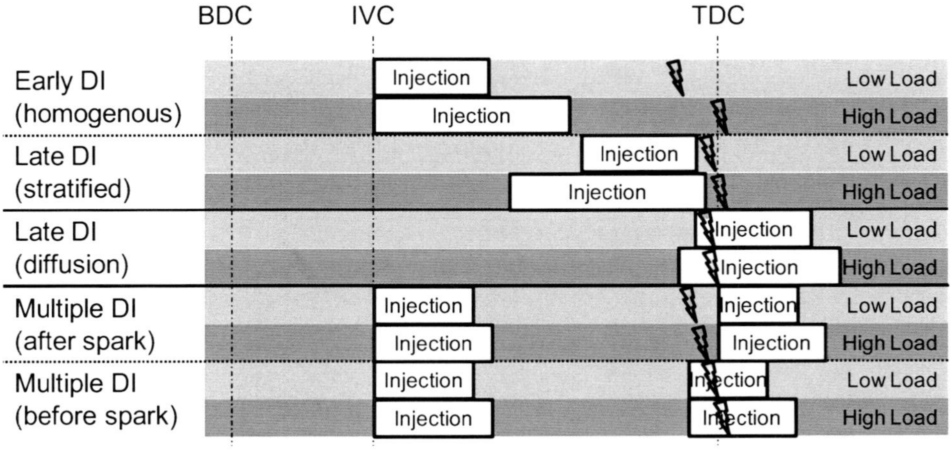

Figure 6: Injection strategies in hydrogen direct injection operation.

Late direct injection as another hydrogen direct injection operating strategy is characterized by an end of injection that occurs shortly before spark timing. The late injection does not provide sufficient time for the injected hydrogen to fully mix with the combustion air and therefore the resulting mixture remains stratified.

The range of early *versus* late injection strategies in terms of start and end of injection are not clearly defined; however, early injection usually refers to any direct injection strategy that results in homogeneous mixtures whereas late injection describes operating strategies employing stratified mixtures. Due to the limited time available between injection and ignition, the hydrogen is distributed unevenly throughout the combustion chamber. Since a combustible mixture has to be reliably delivered around the spark plug, the mixture in the proximity of the ignition source is typically richer than the average air-fuel ratio. These locally richer mixtures result in higher flame speeds and therefore shorter combustion duration compared to early injection with resulting homogeneous mixtures. The spread in spark timing between light and full load in stratified operation is typically less pronounced than with homogeneous early injection. So long as the injection event is completed before the spark occurs, the hydrogen combustion can be characterized as a premixed combustion with a flame front travelling through the combustion chamber. The flame front speed is mainly dependent on the local air-fuel ratio.

Experiments have also been performed with late direct injection occurring during the ignition event. The characteristic difference to the earlier concepts is that a certain amount of fuel is fed to the combustion chamber after spark timing and the subsequent start of combustion. The hydrogen delivered to the combustion chamber after spark timing is not burned in a premixed flame but rather a diffusion type flame best known from Diesel engines. Different variants of this strategy with ignition shortly after start of injection as well as ignition towards the end of the injection event have been tested.

Apart from single injection strategies with different start of injection several multiple injection strategies have been developed for hydrogen direct injection engines. All multiple injection strategies are characterized by a first injection pulse that generally occurs during the early compression stroke and thereby providing a fairly homogeneous lean background mixture for the combustion event. The second injection pulse can either be timed before or after ignition of the lean background mixture. A strategy with the second injection pulse occurring before spark timing actually resembles a combination of an early single injection with a late, diffusion controlled single injection.

Power Density

The higher mixture calorific value of hydrogen direct injection compared to hydrogen port injection and gasoline operation suggests hydrogen direct injection to be superior in power density.

Fig. 7 shows the indicated mean effective pressure as a function of air-fuel ratio for gasoline and hydrogen port injection as well as hydrogen direct injection operation. The data were measured on a single-cylinder 0.5 L research engine capable of operating on various fuels and injection modes [27]. The peak indicated mean effective pressure in hydrogen direct injection operation of 13 bar is approximately 15 % higher than the peak value in gasoline operation. The hydrogen direct injection strategy provides a more than 75 % higher IMEP than hydrogen port injection. The dramatic increase in IMEP with direct injection compared to port injection in hydrogen operation is partially due to the displacement of air during the injection of hydrogen into the intake pipe. Additionally, in this case, the occurrence of combustion anomalies in hydrogen port injection prevented the engine to be operated at stoichiometric fuel-air mixtures causing another significant loss in power density. The difference in engine load control in hydrogen operation compared to gasoline is also obvious from this graph. In gasoline operation the engine is operated at stoichiometric air-fuel ratio throughout the entire IMEP range, whereas the engine is operated at wide open throttle (WOT) conditions in hydrogen operation. The load is adjusted solely by changing the amount of fuel introduced to the engine. The pattern of the IMEP as a function of equivalence ratio also shows differences between the two H_2 mixture formation methods. Since during the direct injection of H_2, the air amount remains almost constant with a changing equivalence ratio (displacement effect is precluded because H_2 is not injected before the intake valves close), a leaner equivalence ratio under the same load (or the same IMEP) is established compared with the external mixture formation with H_2.

Injection Configurations

Hydrogen direct injection operation offers an additional degree of freedom through influencing the mixture homogeneity. Early injection generally results in a fairly homogeneous mixture, whereas late injection causes the air-fuel mixture to be rather stratified. Apart from the injection timing the injector configuration has the strongest influence on the mixture preparation process. Various injection concepts with different injector locations and injector nozzle designs have been tested. Nozzle designs vary in number and location of the injector holes ranging from simple single hole nozzles to complex multi-hole arrangements at various angles.

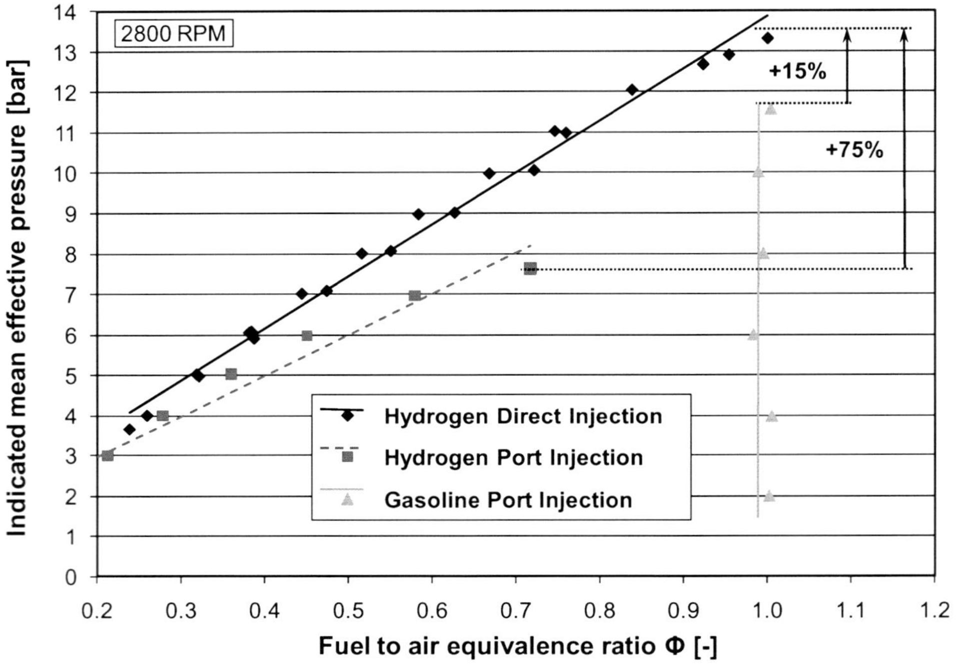

Figure 7: Indicated mean effective pressure as a function of air-fuel ratio for gasoline, hydrogen port injection and hydrogen direct injection operation.

The most typical injector locations in hydrogen direct injection operation are either to the side of the combustion chamber generally towards the cooler intake side of the engine or in central location adjacent to the spark plug. The location of the injector also influences the injector nozzle design. Configurations with the injector mounted in side location have been tested with symmetrical as well as asymmetrical nozzle design generally aiming at guiding the bulk of the injection jet towards the spark plug. The geometrical distance between the injector and the spark plug usually allows to create a fairly good stratification around the ignition source either by aiming the jets there resembling a spray-guided concept or by using the piston surface and combustion chamber walls to guide the jets creating a wall-guided approach. Concepts with injector in central and spark plug in side location have also been tested; the basic concept is similar to the inverse configuration. Central injector location adjacent to the ignition source on the other hand always results in the injection jets to move away from the spark plug. It takes considerable time for the injection jets to move away from the ignition source and return being guided by both the piston top and the combustion chamber walls. Therefore wall guided concepts with central injector location next to the spark plug generally require fairly advanced injection timings. However, concepts have also been developed where the injection jet is ignited when passing by the spark plug. Variations include ignition of the leading edge of the jet *versus* the end of the injection jet and whether or not the jets are directly aimed at the spark plug. In order to provide the spark within the ignitable air-fuel ratio limits, even strategies with extended electrodes have been suggested [46].

Fig. **8** shows the influence of the injection strategy on the indicated thermal efficiency [47]. A simple injector nozzle with a single injection hole was designed with the injector mounted in central location adjacent to the spark plug. The injection hole is located under an angle of 50° to the injector axis. The single-hole design was purposely chosen to simplify the setup and guarantee a single injection jet avoiding any jet-to-jet interaction inside the combustion chamber. An injection angle α was defined to distinguish between injection directions with α=0° referring to the injection direction towards the spark plug. Injection directions of 0, 45, 90, 180, 270 and 315° were investigated. For each injection direction a sweep of start of injection (SOI) was performed. This sweep was started out with early injection shortly after intake valve closing (SOI=140 °CA BTDC) and SOI was retarded with a step width of 20°CA until the end of injection coincided with spark timing or stable combustion could not be maintained.

As an example an injection away from the spark plug (α=180°) with an SOI of 80°CA BTDC yields an indicated thermal efficiency of 30 % at an emissions level of 5 ppm of NO_X. Changing the injection angle to α=0° (injection towards the spark plug) at the same start of injection results in an engine efficiency of 34 % at an identical NO_X emissions level.

At this low engine load the injection duration is short and late injection timings (SOI=20°CA BTDC) can be realized. However, as can be seen in Fig. **8**, these late injection cases cannot be realized for all injection angles. The dots indicating tested operating points clearly show that the late injection case with SOI=20°CA BTDC could only be run with injection towards the spark plug. The reason for the limitations in start of injection is likely due to the stratification resulting from late injection. Given that the overall air-fuel ratio for this particular operating point is very lean (φ ~ 0.25) an unfavorable stratification could result in zones with air-fuel ratio less than the lower flammability limit (φ ~ 0.1) in the vicinity of the spark plug. Only an injection towards the spark plug with very late start of injection ensures that there is sufficient hydrogen in the spark plug area to create an ignitable mixture.

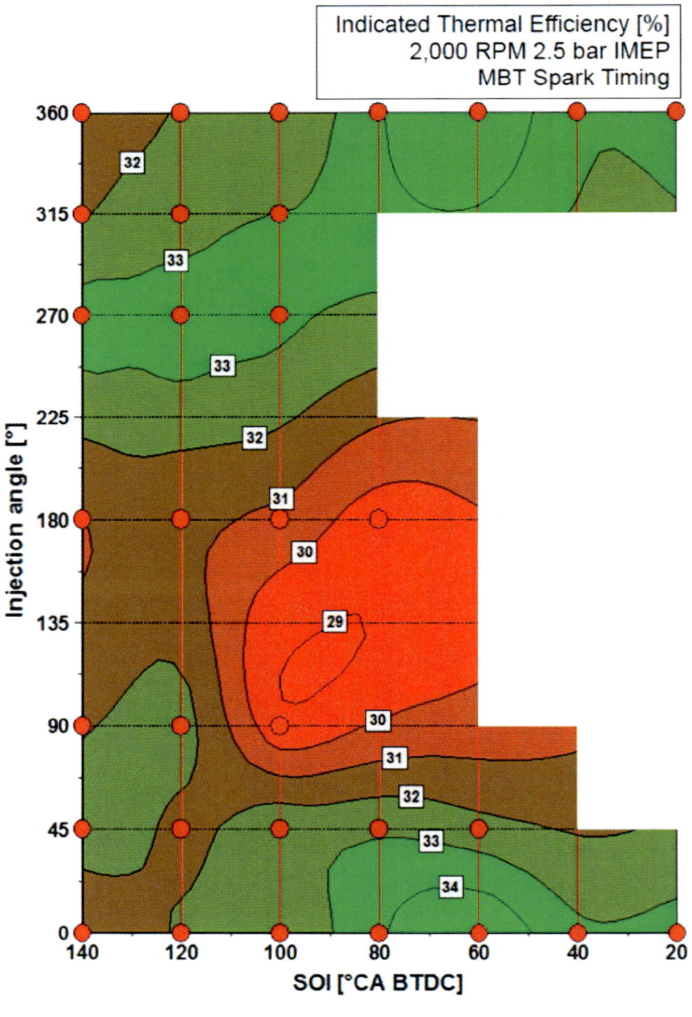

Figure 8: Indicated efficiency as a function of start of injection and injection angle.

All injection equipment for hydrogen direct injection operation is currently at a research and development stage. The pressures and temperatures inside the combustion chamber, the missing lubricity of hydrogen and the extreme tendency to leak pose significant challenges for the injection equipment. Hydrogen injectors for DI operation have to reliably operate at hydrogen supply pressures of up to 300 bar while the injector tip heats up to temperatures of 300–400 °C. It is estimated that a metering accuracy of approximately 2 % of the actual flow rate with minimum injection durations as short as 0.1 ms and leakage rates of less than 0.1 % of full flow are required for hydrogen DI injectors [48].

Over the past decades various actuation techniques for hydrogen direct injection systems have been developed ranging from hydraulic actuation to magnetostrictive and solenoid operation to piezo-electric systems in more recent prototype systems. Various types of systems with both, inward as well as outward opening mechanisms and flat seats as well as needle-nozzle seats have been tested. Despite the multitude of injection systems developed for hydrogen direct injection, none of the prototype systems would meet the durability requirements needed for automotive applications. A major development goal should be to substantially increase injector life from about 200 h to 1000 h. This would allow the development and demonstration of an early stage multi-cylinder engine with advanced features and material technology integrated in the DI fuel system. A longer term goal is to develop and prove 20, 000 h level durability, with a production-oriented design, through further fundamental materials research, accelerated testing on injector rigs and full-scale engine validation [49].

NO_X Emissions Trends with Direct Injection

The injection timing in hydrogen direct injection operation strongly influences the mixture homogeneity and ultimately the combustion characteristics. With early injection, the injected fuel has sufficient time to mix with the air inside the combustion chamber and form an almost homogeneous mixture. With late injection, only limited time for mixing is available, resulting in a stratified charge at spark timing. The impact of the above-mentioned characteristics on engine-out NO_X emissions is strongly dependent on the air-fuel ratio which itself is determined by the engine load. Fig. **9** shows the NO_X emissions results as a function of start of injection for various equivalence ratios at an engine speed of 2000 RPM collected on a single-cylinder hydrogen research engine [50, 51]. Because the engine is operated without throttling, the equivalence ratio corresponds to engine load. At low engine loads and high equivalence ratios respectively, early injection results in extremely low NO_X emissions because the mixture at ignition timing is very likely to be homogeneous. Thus, the lean homogeneous mixture burns without forming NO_X emissions. Late injection at low loads, on the other hand, results in a stratified mixture with hydrogen-rich zones, as well as zones with very lean mixtures or even pure air. Although the overall mixture is still lean, the combustion of rich zones causes a significant increase in NO_X emissions. At high engine loads, this trend appears to be inverted. Early injection results in homogeneous mixtures that approach stoichiometry and produce high NO_X emissions. Late injection is expected to result in stratification, with zones that are even richer than stoichiometric, along with lean zones. This kind of stratification avoids the NO_X critical equivalence ratio regime of λ~1.3 / φ~0.75 (see Fig. **3**) and thereby reduces overall NO_X emissions.

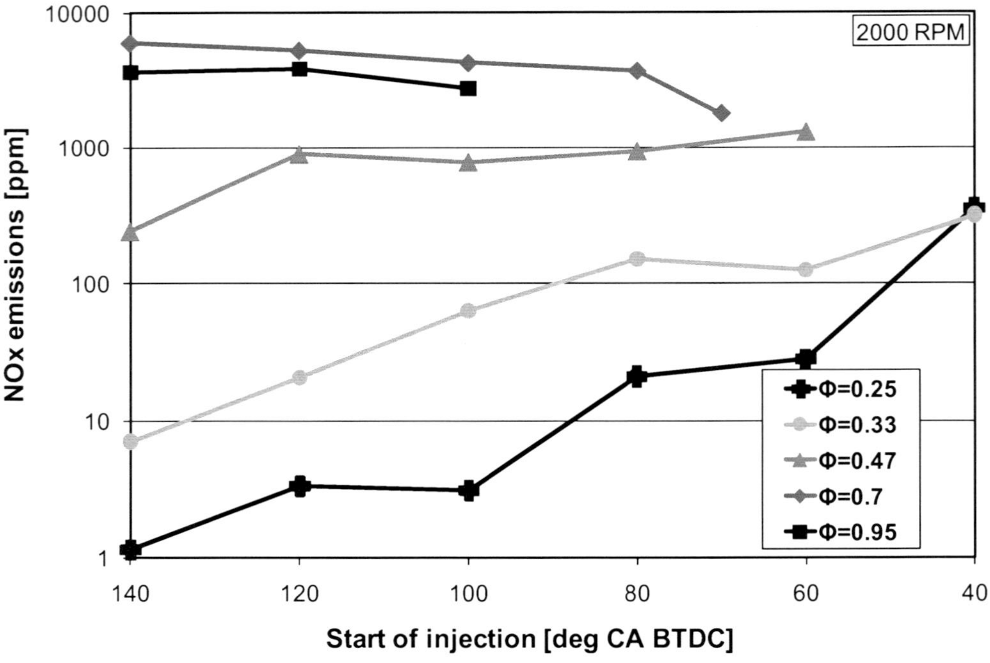

Figure 9: NO_X emissions as a function of air-fuel ratio with early and late premixed direct injection.

Measures for NO_X Emissions Reduction

As shown in Fig. **9**, the injection strategy also strongly influences oxides of nitrogen emissions. Hydrogen has proven to be a very efficient reductant for the oxidation of nitrogen oxide emissions in a 3-way catalyst [19]. However, since most hydrogen engines employ lean burn concepts for efficiency benefits, conventional 3-way catalysts do not provide a satisfying solution. One promising solution for an efficient NO_X emissions reduction without compromising engine efficiency is multiple injection. Fig. **10** shows the trade-off between indicated thermal efficiency and nitric oxide emissions for a medium load case (6 bar IMEP) at 1000 RPM measured on a single-cylinder engine [52]. The graph shows the results for several injection strategies including single injection only as well as multiple injections with various ratios of primary to secondary injection. A general trend of reduced emissions with decreased engine efficiency can be observed – however, certain combinations of multiple injections allow a significant reduction in nitric oxide emissions with a simultaneous increase in engine efficiency.

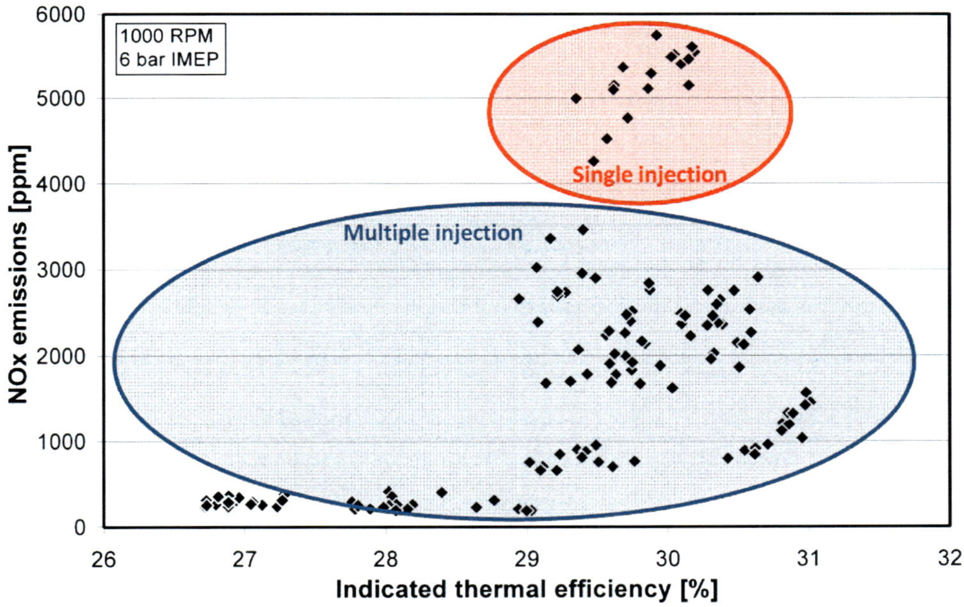

Figure 10: Trade-off between indicated efficiency and NO_X emissions with single and multiple injection strategies.

Reducing the duration of the primary injection such that the air-fuel ratio of the homogenous mixture from the first pulse is below the NO_X critical limit of φ~0.5

allows burning the homogeneous mixture without creating a significant amount of nitric oxide emissions. The second injection pulse required for achieving the requested engine load has to be considered to estimate the total amount of NO_X emissions. There is no time available for the second injection pulse to form a homogeneous mixture with the cylinder charge. The flammability limits of hydrogen are extremely wide, both for lean and rich mixtures ($0.1 < \varphi < 7$). The hydrogen jet from the second injection pulse is expected to burn at fairly rich conditions with lack of oxygen thereby reducing the formation of nitric oxide emissions. Overall a multiple injection strategy with an advantageous split of primary *versus* secondary injection allows avoiding the NO_X critical air-fuel ratio regimes by locally forming rich and lean mixtures.

Multiple injection strategies rely on avoiding the NO_X critical air-fuel ratio regime to reduce the engine-out oxides of nitrogen emissions. Additionally, reducing the peak in-cylinder temperatures was also found to reduce the raw emissions levels of hydrogen engines. The required temperature reduction can be accomplished using exhaust gas recirculation or even water injection. Fig. **11** compares the NO_X

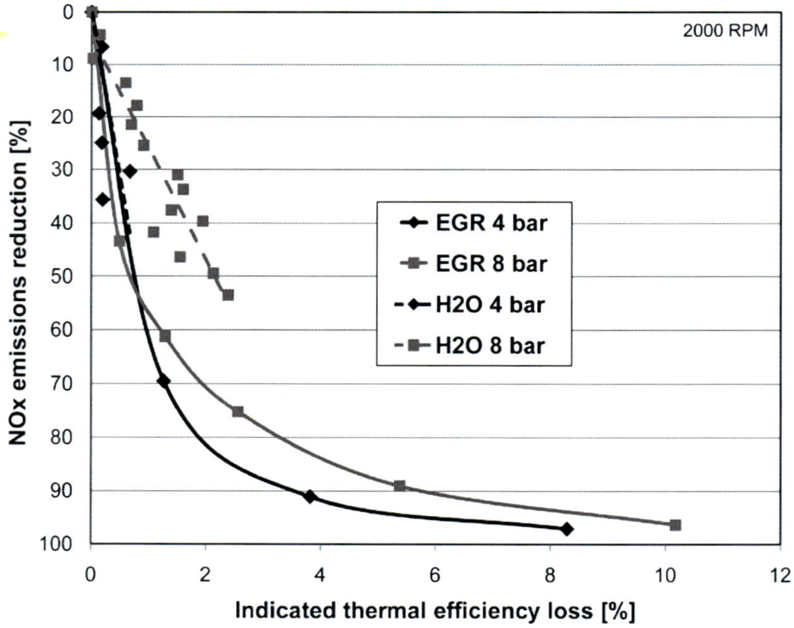

Figure 11: Comparison of EGR *versus* water injection in terms of NO_X emissions reduction potential and related efficiency loss for medium and high engine load.

emissions reduction potential of exhaust gas recirculation (EGR) to water injection [53]. Because the results are from different injection configurations and injection strategies the emissions reduction measures are compared on a relative basis. At the medium engine load a NO_X emissions reduction of up to 40 % compared to the base case could be achieved with water injection at a respective loss in indicated thermal efficiency below 1 %. These reductions were achieved at moderate amounts of water injection and similar results could be achieved with exhaust gas recirculation. At the high engine load of 8 bar IMEP exhaust gas recirculation proofs itself to be even more effective than water injection. A NO_X emissions reduction of 50 % with water injection resulted in a loss in indicated thermal efficiency of around 2 % whereas a NO_X emissions reduction in excess of 60 % could be achieved with exhaust gas recirculation at a respective efficiency loss of slightly more than 1 %.

ENGINE EFFICIENCY

Engine Efficiency Comparison Gasoline – H_2 PFI – H_2 DI

Fig. **12** shows the indicated efficiency as a function of engine speed and load for gasoline port injection, hydrogen port injection as well as hydrogen direct injection operation collected on a single-cylinder research engine. In gasoline operation the load control is quantitative by throttling the intake air. This negatively affects the engine efficiency especially at low engine load. The peak indicated mean effective pressure is around 11.5 bar. The maximum indicated engine efficiency is approximately 38 % and is achieved along the full load curve.

The wide flammability limits of hydrogen allow qualitative load control over the entire load and speed range. This positively influences the engine efficiencies in particular at part load. The maximum indicated engine efficiency reached at 2800 RPM and an IMEP of 6 bar is approximately 41 % which is significantly higher than in gasoline operation. The displacement of air in the intake manifold with hydrogen port injection and the occurrence of combustion anomalies at high engine loads with this configuration limit the maximum indicated mean effective pressure to approximately 7.5 bar which is 35 % less than in gasoline operation.

An injector with asymmetric injector nozzle design in side location between the intake valves is used for introducing the hydrogen in direct injection operation.

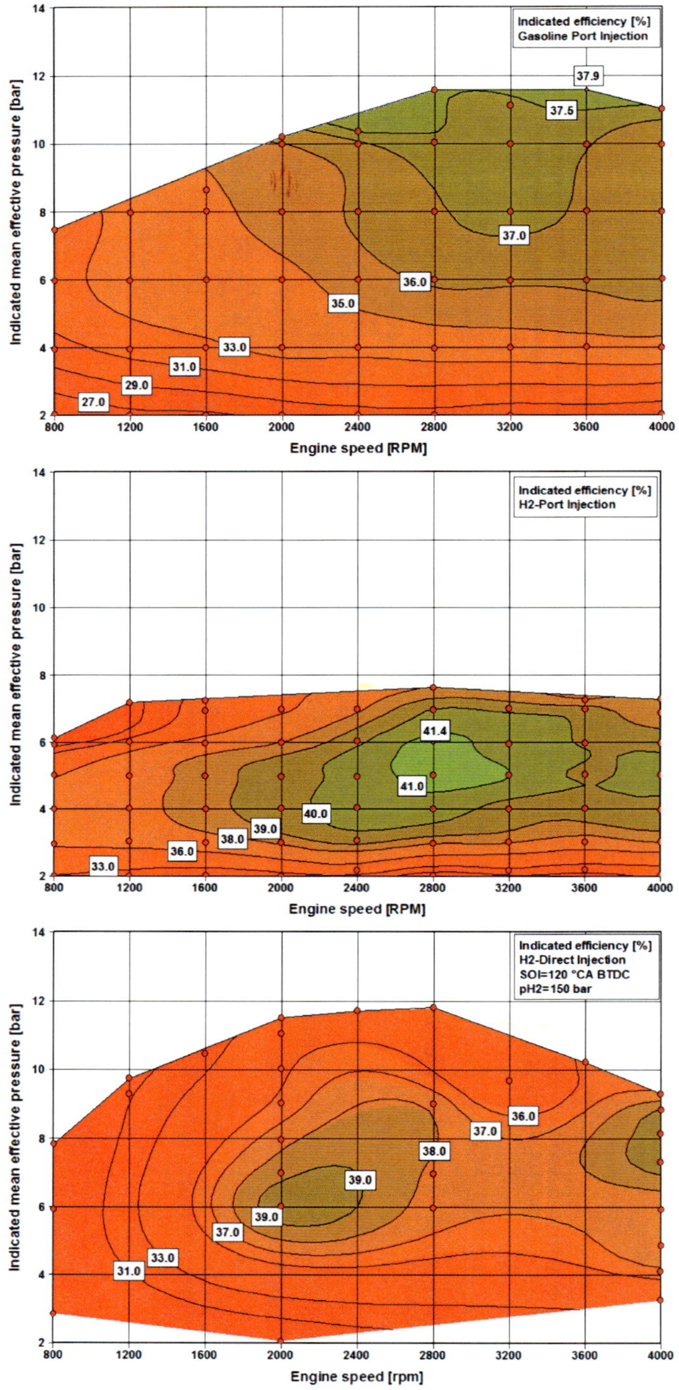

Figure 12: Indicated efficiency maps for gasoline port injection operation as well as hydrogen port and direct injection operation [54].

Since direct injection avoids air displacement in the intake manifold and reduces combustion anomalies, the peak indicated mean effective pressures are significantly higher than with hydrogen port injection. The engine efficiency characteristics in hydrogen direct injection are similar to hydrogen port injection, however, the absolute level of engine efficiency for this non-optimized configuration is slightly lower than with port injection.

Based on test data from different operating modes the engine efficiencies and losses of the working cycle can be estimated. The term loss is used to describe the difference in respective efficiencies that were calculated using actual engine test results. Starting from a theoretical efficiency of an engine without losses, the individual losses due to incomplete combustion, actual rate of heat release, wall heat losses and gas exchange can be calculated [9]. Fig. **13** shows the efficiencies and losses for gasoline and hydrogen operation with both port injection and direct injection. The data for all fuels represent the base case for each configuration with no efficiency improvement measures and were collected on a single-cylinder research engine at an engine speed of 2000 RPM and indicated mean effective pressures of 2 bar representing low engine load and 6 bar representing medium engine load. In hydrogen direct injection, the fuel is injected early during the compression stroke (with the start of injection at 120 °CA BTDC and an injection pressure of 150 bar), resulting in a fairly homogeneous mixture at spark timing. At the low load point of 2 bar indicated mean effective pressure the efficiency of the ideal engine calculated using actual gas properties in gasoline operation is significantly lower than in hydrogen operation. Since this value is mainly influenced by the compression ratio and the air-fuel ratio, hydrogen operation results in higher values due to the lean operation. Losses resulting from incomplete combustion can be determined by measuring unburned fuel components in the exhaust. These consist of hydrocarbons and carbon monoxide in gasoline operation and unburned hydrogen when fueled with hydrogen. The extremely lean conditions in hydrogen operation cause the losses due to incomplete combustion to be significantly higher than in stoichiometric gasoline operation. The loss due to actual combustion rather than an ideal constant-volume combustion at top dead center in gasoline operation is around 3%. The lean combustion in hydrogen operation results in the losses due to actual combustion to be slightly higher than in gasoline operation. Due to the higher pressure levels in hydrogen operation resulting from unthrottled operation, the wall heat losses are considerably higher than in gasoline operation. The higher in-cylinder charge motion in hydrogen direct-injection operation results in higher wall heat losses compared to port injection. Finally, the gas exchange losses in hydrogen operation are only a fraction compared to gasoline, since the engine is operated unthrottled. Overall this results in an advantage in indicated thermal efficiency for hydrogen port injection over hydrogen DI and gasoline operation.

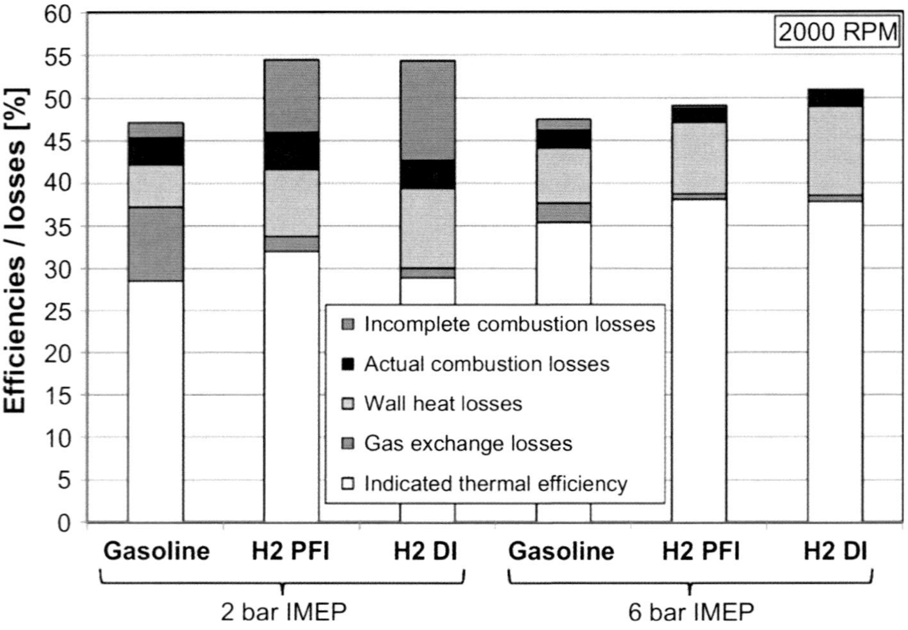

Figure 13: Analysis of losses in gasoline port injection operation as well as hydrogen port and direct injection operation for low and medium engine load.

Due to the lean operation the efficiency of the ideal engine in gasoline operation is also lower than in hydrogen operation at the medium load point of 6 bar IMEP. The air-fuel ratio, and hence the theoretical efficiency in hydrogen port injection, is lower than in hydrogen direct injection due to the air displacement effect with port-injection operation. For this operating point, the loss due to incomplete combustion in gasoline operation is slightly more than 1%. In hydrogen operation both with port and direct injection, the loss accounts for less than 0.5%. The very complete combustion in hydrogen operation is mainly due to the fast flame speed and small quenching distance at this air-fuel ratio. The loss due to actual combustion in gasoline operation is around 2%. Although the engine is operated unthrottled and therefore lean in hydrogen operation, the combustion is still faster than in gasoline operation. This results in lower losses due to actual combustion in hydrogen operation. However, the higher flame speeds and the smaller quenching distance, which positively affected previous losses, are the main reason for the increased wall heat losses in hydrogen operation compared to gasoline operation. Direct injection shows even higher wall heat losses compared to port injection, which is likely due to the higher level of in-cylinder charge motion and turbulence caused by the injection event. Finally, the gas exchange losses in hydrogen operation are significantly lower than in gasoline operation since the engine is operated unthrottled. At this operating point, this results in an overall advantage in indicated thermal efficiency of approximately 2.5% in hydrogen operation, both with port and direct injection, compared to gasoline operation.

Concepts for Engine Efficiency Optimization

Research on combustion system development for a heavy-duty truck operated on hydrogen was performed at Tokyo City University in collaboration with the National Traffic Safety and Environment Laboratory [46]. The experimental work was conducted on a Diesel-type single-cylinder engine with a displacement of approximately 1 L at a compression ratio of 13:1. A spark plug with an extended electrode was used to ignite the hydrogen plumes injected from a prototype DI injector. Three combinations of injection timing and ignition timing were classified as early injection and premixed ignition, late injection and plume head ignition and late injection and plume tail ignition. The early injection and premixed ignition regime with exceptionally low NO_X characteristics was applied in the low-load region and the late injection with plume tail ignition regime with high thermal efficiencies was combined with EGR in the high load range. As a result, a large NO_X reduction of as much as 1, 000 ppm was obtained without any decline in thermal efficiency over the entire operating range up to 45% indicated thermal efficiency. Based on these results the performance of a 6-cylinder DI hydrogen engine when operated under Japan's JE05 test mode was estimated. The simulation results showed that suitable management of a combination of combustion regimes together with the application of EGR has the potential to reduce NO_X emissions substantially and 0.35 g/kWh versus the limit of 0.5 g/kWh was achieved.

Research conducted at the Institute of Internal Combustion Engines and Thermodynamics at Graz University of Technology in collaboration with BMW focused on hydrogen direct injection and Diesel-type compression ignition. It was found that the high self-ignition temperature requires additional measures to ensure reliable combustion. A concept with glow-plug assisted combustion was developed that yielded calculated brake thermal efficiencies of up to 42 %. Fig. **14** shows the resulting shell diagram with each operating point optimized for engine efficiency. The diagram was derived from indicated data of a single-cylinder engine using the friction losses of a modern Diesel engine. At loads up to 6 bar IMEP the engine was operated naturally aspirated, above that load the charging pressure was gradually increased up to the maximum of 1 bar. Exhaust back pressure was adjusted simulating real turbocharger behavior. The fuel pressure was set at 200 bar for the majority of the operating regime, at full load a fuel pressure of 280 bar was used. Due to the high engine load in turbocharged operation the compression ratio was set to 18. The efficiency optimization included adjustment of injection parameters yielding a maximum indicated efficiency of the high pressure cycle in excess of 49 % [55].

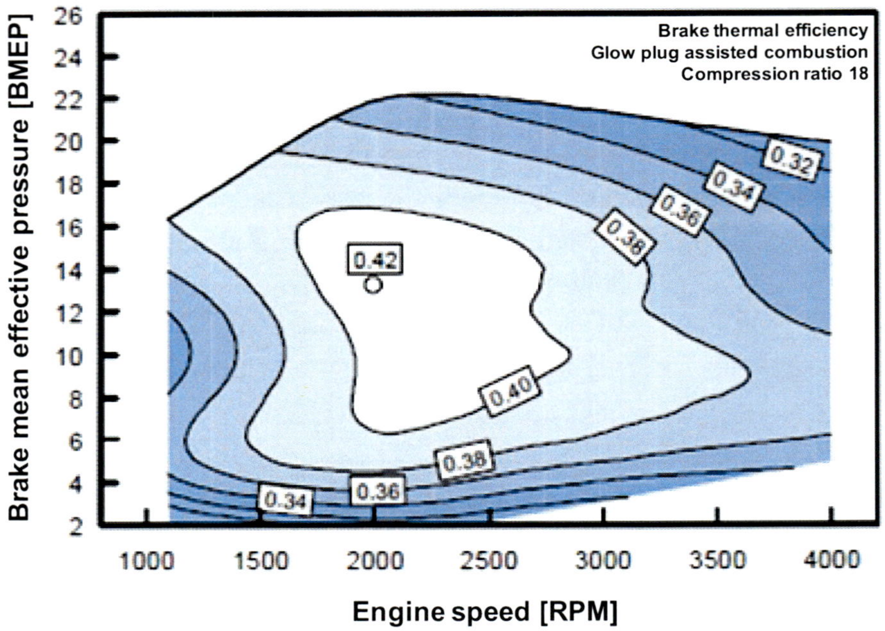

Figure 14: Efficiency potential of a hydrogen direct injection with glow plug assisted combustion [55].

Research performed at the engine test laboratories at Ford Motor Company, Argonne National Laboratory and Sandia National Laboratories moved the state of the art for peak brake thermal efficiencies of hydrogen powered internal combustion engines from less than 40% to over 45% brake thermal efficiency as demonstrated on a single cylinder test engine onsite at Ford. Fig. **15** shows a plot of brake thermal efficiencies for a wide range of test conditions during the development phase taken at Ford's hydrogen engine lab. Adjustments to the boost system, combustion and injection strategy as well as an optimized bore/stroke ratio resulted in a steady increase in calculated brake thermal efficiencies. For comparison the graph also shows peak brake thermal efficiency ranges of current technology gasoline and Diesel engines. With continuous improvement of combustion system and operation strategy a calculated brake thermal efficiency in excess of 45 % could be achieved at an engine speed of 3000 RPM and an indicated mean effective pressure of 14.56 bar employing a multiple injection strategy at an injection pressure of approximately 100 bar. The data was taken on a single cylinder engine with simulated boost conditions typical for a multi-cylinder engine. Assuming a friction mean effective pressure of 0.7 bar, which is consistent with multi-cylinder data, resulted in a calculated brake thermal efficiency of slightly above 45 % [56].

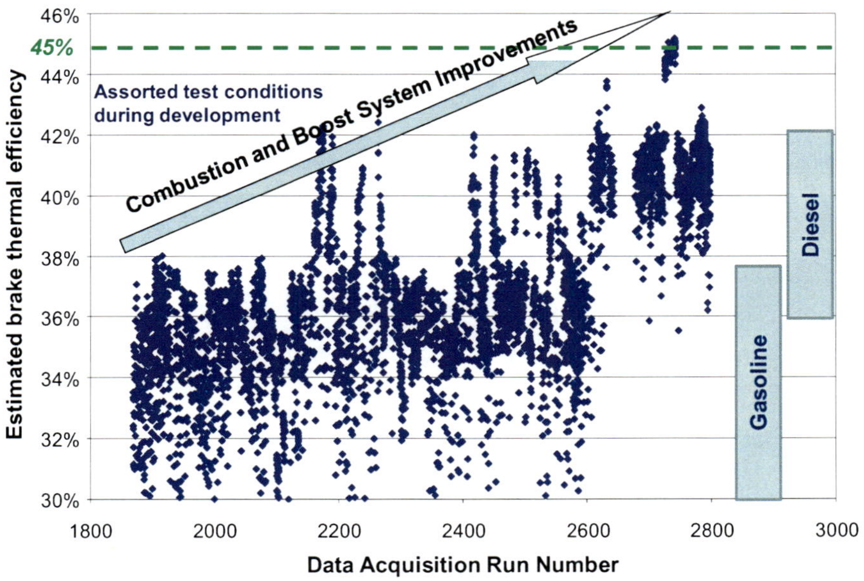

Figure 15: Evaluation of hydrogen internal combustion engine brake thermal efficiency in comparison to typical gasoline and diesel engines [56].

CONCLUSIONS

In the above, hydrogen has been shown to not only be suitable as a fuel for internal combustion engines, but to offer the potential to explore new concepts for mixture formation and load control.

Engines equipped with port fuel injection and operated on hydrogen are interesting as they are relatively easy to implement and offer a bi-fuel flexibility. Due to the properties of hydrogen, a hydrogen engine has more degrees of freedom that a conventionally fueled spark ignition engine. This enables load control strategies for part load that significantly increase efficiency compared to *e.g.,* gasoline engines, primarily through unthrottled operation and the corresponding favorable (lean) mixture properties. Using supercharging, either at stoichiometric operation (possibly with EGR) or with lean burn operation, allows power outputs up to or exceeding naturally aspirated operation on gasoline. In the lean burn case, this is possible with high efficiency and without the need for aftertreatment. In all cases, extremely low emissions can be reached.

With direct injection, extra degrees of freedom are possible in devising mixture formation and load control strategies. Using either stratification of the in-cylinder mixture, partly premixed combustion or compression ignition type of combustion, or combinations thereof, it is possible to obtain significantly higher engine efficiencies while still meeting emissions targets.

It has been shown that hydrogen fueled internal combustion engines thus have the potential for high engine efficiencies. This potential was deduced from the physical and chemical properties of hydrogen and has already been confirmed for some engine operating points. Clearly, because of the degrees of freedom offered by DI engines, there remains much to be investigated. This is also true for the hardware, in particular for increasing the flow rate and durability of DI injectors.

ACKNOWLEDGEMENTS

Parts of the submitted manuscript have been created by UChicago Argonne, LLC, Operator of Argonne National Laboratory (''Argonne''). Argonne, a U.S. Department of Energy Office of Science laboratory, is operated under Contract

No. DE-AC02-06CH11357. The U.S. Government retains for itself, and others acting on its behalf, a paid-up nonexclusive, irrevocable worldwide license in said article to reproduce, prepare derivative works, distribute copies to the public, and perform publicly and display publicly, by or on behalf of the Government.

Research referenced in this manuscript was partially funded by DOE's FreedomCAR and Vehicle Technologies Program, Office of Energy Efficiency and Renewable Energy. T. Wallner wishes to thank Gurpreet Singh and Lee Slezak, program managers at DOE, for their support.

A hydrogen engine used to run certain experiments presented in this manuscript was provided by Ford Motor Company. Special thanks to the team from Ford Motor Company for their support.

T. Wallner would also like to express his gratitude to all of the individuals from Graz University of Technology and BMW Research and Development involved in preparing, performing as well as analyzing hydrogen engine research referenced in this manuscript.

REFERENCES

[1] U.S. Department of Energy, "FreedomCAR and Vehicle Technologies Multi-Year Program Plan 2006-2011".

[2] European Commission, "The fuel cells and hydrogen joint technology initiative (FCH JTI)".

[3] Japanese Ministry of Economy, Trade and Industry, "Japan Hydrogen & Fuel Cell Demonstration Project".

[4] National Hydrogen Energy Board, Ministry of New and Renewable Energy, Government of India, "National hydrogen energy road map".

[5] U.S. Department of Energy, "2008 Annual Progress Report DOE Hydrogen Program", 2008.

[6] S. Verhelst, and T. Wallner, "Hydrogen-Fueled Internal Combustion Engines", *Prog Energ Combust*, vol. 35, pp. 490-527, Dec. 2009.

[7] S. Verhelst, R. Sierens, and S. Verstraeten, "A critical review of experimental research on hydrogen fueled SI engines", SAE Paper No. 2006-01-0430, 2006. Also in *SAE 2006 Transactions, Journal of Engines*, pp. 264-274, ISBN Number: 978-0-7680-1835-6, March 2007.

[8] I. Glassman, *Combustion*. Orlando, FL: Academic Press, Inc., 1987.

[9] J.B. Heywood, *Internal combustion engine fundamentals*. New York, NY: McGraw-Hill, 1988.

[10] C. Morley, *GASEQ, a chemical equilibrium program for Windows*.

[11] R.H. Perry, and D.W. Green, *Perry's chemical engineers' handbook*. New York, NY: McGraw-Hill, 1997.

[12] R. Ono, M. Nifuku, S. Fujiwara, S. Horiguchi, and T. Oda, "Minimum ignition energy of hydrogen-air mixture: effects of humidity and spark duration", *J Electrostat,* vol. 65, pp. 87-93, Feb. 2007.

[13] The Engineering Toolbox, "Fuels and chemicals - auto ignition temperatures". http://www.engineeringtoolbox.com

[14] U.S. Department of Energy, "Hydrogen analysis resource center". http://hydrogen.pnl.gov/cocoon/morf/hydrogen

[15] W. Peschka, *Liquid hydrogen - fuel of the future*. New York, NY: Springer, 1992.

[16] J. Demuynck, N. Raes, M. Zuliani, M. De Paepe, R. Sierens, and S. Verhelst, "Local heat flux measurements in a hydrogen and methane spark ignition engine with a thermopile sensor", *Int J Hydrogen Energy*, vol. 34, pp. 9857-9868, Nov. 2009

[17] A. Wimmer, T. Wallner, J. Ringler, and F. Gerbig, "H2-direct injection - a highly promising combustion concept", SAE Paper No. 2005-01-0108, 2005.

[18] B. Göschel, "DerWasserstoff-Verbrennungsmotor als Antrieb für den BMW der Zukunft", in Proceedings 24 Internationales Wiener Motorensymposium, Vienna, Austria, 2003.

[19] T. Wallner, H. Lohse-Busch, S. Gurski, M. Duoba, W. Thiel, D. Martin, and T. Korn, "Fuel economy and emissions evaluation of a BMW Hydrogen 7 mono-fuel demonstration vehicle", *Int J Hydrogen Energy*, vol. 33, pp. 7607-7618, Dec. 2008.

[20] H. Rottengruber, U. Wiebicke, G. Woschni, And K. Zeilinger, "Wasserstoff-Dieselmotor mit Direkteinspritzung, hoher Leistungsdichte und geringer Abgasemission - Part 3: Versuche und Berechnungen am Motor". *Motortechnische Zeitschrift,* vol. 61, pp. 122-128, Feb. 2000.

[21] S. Osafune, H. Akagawa, H. Ishida, H. Egashira, Y. Kuma, and W. Iwasaki, "Development of hydrogen injection clean engine", in CIMAC Congress, Paper No. 207, Kyoto, Japan, 2004.

[22] R. Heindl, H. Eichlseder, C. Spuller, F. Gerbig, and K. Heller, "New and innovative combustion systems for the H2-ICE: Compression ignition and combined processes", SAE Paper No. 2009-01-1421, 2009.

[23] P. Aleiferis, and M. Rosati, "Hydrogen SI and HCCI combustion in a direct-injection optical engine", SAE Paper No. 2009-01-1921, 2009.

[24] N. Wakayama, K. Morimoto, A. Kashiwagi, and T. Saito, "Development of hydrogen rotary engine vehicle", in 16th World Hydrogen Energy Conference, Lyon, France, 2006.

[25] H. Rottengruber, M. Berckmüller, G. Elsässer, N. Brehm, and C. Schwarz, "Operation strategies for hydrogen engines with high power density and high efficiency", in 15th Annual U.S. Hydrogen Conference, Los Angeles, California, 2004.

[26] H.S. Yi, K. Min, and E.S. Kim, "The optimised mixture formation for hydrogen fuelled Engines". *Int J Hydrogen Energy,* vol. 25, pp. 685-690, July 2000.

[27] H. Eichlseder, T. Wallner, R. Freymann, and J. Ringler, "The potential of hydrogen internal combustion engines in a future mobility scenario", SAE Paper No. 2003-01-2267, 2003.

[28] F. Salimi, A.H. Shamekhi, and A.M. Pourkhesalian, "Role of mixture richness, spark and valve timing in hydrogen-fuelled engine performance and emission", *Int J Hydrogen Energy,* vol. 34, pp. 3922-3929, May 2009.

[29] L.G. Olavson, N.R. Baker, F.E. Lynch, and L.C. Meija, "Hydrogen fuel for underground mining machinery", SAE Paper No. 840233, 1984.

[30] D. Davidson, M. Fairlie, and A. Stuart, "Development of a hydrogen-fuelled farm tractor", *Int J Hydrogen Energy,* vol. 11, pp. 39-42, 1986.

[31] H. Knorr, W. Held, W. Prümm, H. Rüdiger, "The MAN hydrogen propulsion system for city buses", in 11th World Hydrogen Energy Conference, pp. 1611-1620, Stuttgart, Germany, 1996.

[32] X. Tang, W.F. Stockhausen, D.M. Kabat, R.J. Natkin, and J.W. Heffel, "Ford P2000 hydrogen engine dynamometer development", SAE Paper No. 2002-01-0242, 2002.

[33] S. Verhelst, and R. Sierens, "Aspects concerning the optimisation of a hydrogen fueled engine", *Int J Hydrogen Energy,* vol. 26, pp. 981-985, Sept. 2001.

[34] S. Verhelst, P. Maesschalck, N. Rombaut, and R. Sierens, "Increasing the power output of hydrogen internal combustion engines by means of supercharging and exhaust gas recirculation", *Int J Hydrogen Energy,* vol. 34, pp. 4406-4412, May 2009.

[35] ETEC hydrogen internal combustion engine full-size pickup truck conversion. Hydrogen ICE truck brochure.

[36] J.E. Francfort, and D. Karner, "Hydrogen ICE vehicle testing activities". SAE Paper No. 2006-01-0433, 2006.

[37] T. Wallner, H. Lohse-Busch, and N. Shidore, "Operating strategy for a hydrogen engine for improved drive-cycle efficiency and emissions behavior", *Int J Hydrogen Energy*, vol. 34, pp. 4617-4625, May 2009.

[38] R. Natkin, A. Denlinger, M. Younkins, A. Weimer, S. Hashemi, and A. Vaught, "Ford 6.8l hydrogen IC engine for the E-450 shuttle van", SAE Paper No. 2007-01-4096, 2007.

[39] W. Prümm, "Hydrogen engines for city buses", in Proceedings 1st international symposium on hydrogen internal combustion engines. pp. 1-11, Graz, Austria, 2006.

[40] S. Verhelst, P. Maesschalck, N. Rombaut, and R. Sierens, "Efficiency comparison between hydrogen and gasoline, on a bi-fuel hydrogen/gasoline engine", *Int J Hydrogen Energy,* vol. 34, pp. 2504-2510, March 2009.

[41] G. Kiesgen, M. Klüting, C. Bock, and H. Fischer, "The new 12-cylinder hydrogen engine in the 7 series. The H_2 ICE age has begun", SAE Paper No. 2006-01-0431, 2006.

[42] R. Scarcelli, T. Wallner, V. Salazar, and S. Kaiser, "Modeling and Experiments on Mixture Formation in a Hydrogen Direct-Injection Research Engine", in 9th International Conference on Engines and Vehicles (ICE2009), Capri, Italy, 2009.

[43] W. Kirchweger, H. Eichlseder, F. Gerbig, and U. Gerke, "Optical measurement methods for the optimization of the hydrogen DI combustion", in 7th International Symposium on Internal Combustion Engines Diagnostics, Baden-Baden Germany, 2006.

[44] K. Kovac, A. Wimmer, M. Hallmannsegger, and A. Obieglo, "Mixture formation and combustion in a hydrogen engine - a challenge for the numerical simulation", in International Congress on Engine Combustion Processes - Current Problems and Modern Techniques, Munich, Germany, 2005.

[45] S.A. Kaiser, and C.M. White, "PIV and PLIF to evaluate mixture formation in a direct injection hydrogen-fueled engine", SAE Paper No. 2008-01-1034, 2008.

[46] K. Naganuma, K. Yamane, Y. Takagi, A. Kawamura, and T. Yanai, "Efficiency and Emissions-Optimized Operating Strategy of a High-pressure Direct Injection Hydrogen Engine for Heavy-duty Trucks", SAE Paper No. 2009-01-2683, 2009.

[47] T. Wallner, A. Nande, and J. Naber, "Study of Basic Injection Configurations using a Direct-Injection Hydrogen Research Engine", SAE Paper No. 2009-01-1418, 2009.

[48] P. Steinrück, and G. Ranegger, "Timed injection of hydrogen for fuel cells and internal combustion engines", in Proceedings 1st international symposium on hydrogen internal combustion engines, pp. 164-177, Graz, Austria, 2006.
[49] A. Welch, D. Mumford, S. Munshi, J. Holbery, B. Boyer, M. Younkins, and H. Jung, "Challenges in developing hydrogen direct injection technology for internal combustion engines", SAE Paper No. 2008-01-2379, 2008.
[50] T. Wallner, S. Ciatti, and B. Bihari, "Investigation of injection parameters in a hydrogen DI engine using an endoscopic access to the combustion chamber", SAE Paper No. 2007-01-1464, 2007.
[51] T. Wallner, S. Ciatti, B. Bihari, W. Stockhausen, and B. Boyer, "Endoscopic investigations in a hydrogen internal combustion engine", in Proceedings 1st international symposium on hydrogen internal combustion engines, pp. 107-117, Graz, Austria, 2006.
[52] T. Wallner, R. Scarcelli, A. Nande, and J. Naber, "Assessment of Multiple Injection Strategies in a Direct Injection Hydrogen Research Engine", SAE Paper No. 2009-01-1920, 2009.
[53] A. Nande, T. Wallner, and J. Naber, "Influence of Water Injection on Performance and Emissions of a Direct-Injection Hydrogen Research Engine", SAE Paper No. 2008-01-2377, 2008.
[54] T. Wallner, "Development of combustion concepts for a hydrogen powered internal combustion engine", PhD thesis, Graz University of Technology, Graz, Austria, 2004.
[55] C. Spuller, H. Eichlseder, F. Gerbig, and K. Heller, "Possibilities for Realising Diesel Engine Combustion Processes with Hydrogen", in Der Arbeitsprozess der Verbrennungskraftmaschine, Graz, Austria, 2009.
[56] FreedomCAR and Fuel Partnership, "Highlights of Technical Accomplishments". 2009.

APPENDIX: TERMS

Abbreviations

BMEP	brake mean effective pressure
BDC	bottom dead center
BTDC	before top dead center
CA	crank-angle
DI	direct-injection
DOE	(United States) Department of Energy
EGR	exhaust gas recirculation
FC	fuel cell
FTP	federal test procedure

HCCI	homogeneous charge compression ignition
H_2	hydrogen
H_2ICE	hydrogen-fueled internal combustion engine
IMEP	indicated mean effective pressure
LHV	lower heating value
MBT	minimum spark advance for best torque
MON	motor octane number
NO_X	nitrogen oxides
PFI	port-fuel injection
RON	research octane number
RPM	revolutions per minute
SI	spark-ignition
SOI	start of injection
SULEV	Super Ultra Low Emission Vehicle
TDC	top dead center
WOT	wide open throttle
WWMP	World Wide Mapping Point

Greek symbols

λ	air-to-fuel equivalence ratio
φ	fuel-to-air equivalence ratio

CHAPTER 10

Commercialization and Public Acceptance of Fuel Cell Vehicles

A. Kazim*

Department of Mechanical Engineering, United Arab Emirates University, Al-Ain, United Arab Emirates

Abstract: Currently, major automotive companies are involved intensively in the development of hydrogen-fuelled FCV's in order to be globally commercialized by 2005. However, the current cost of FCV's and lack of commercials and information addressing environmental, economical and technological advantages associated with such vehicles leave the general public to be completely unaware. This paper presents a general assessment of commercialization and public acceptance of utilizing FCV's in terms of their costs in comparison to ICV's, and safety and dependability at various scenarios over the next twenty years. A significant improvement in the cost fraction of FCV's was achieved in the 1990's in comparison with the cost fraction of ICV's in the same period. Moreover, the calculated results demonstrated a lower cost of FCV's and a higher safety and dependability could lead to a higher rate of commercialization and public acceptance.

Keywords: Hydrogen, energy, PEM fuel cell, fuel cell vehicle, internal combustion vehicle, public acceptance, commercialization, safety, dependability, cost fraction, fuel cell cost, pollution, emission, higher demand, innovative technology.

INTRODUCTION

Recently, the demand on internal combustion vehicles (ICV's) in the world has been on the rise as a result of booming population, higher demand on better livelihood and innovative technologies and a relatively lower cost of fossil fuels. With estimated emissions of an average-sized ICV of 213g of pollutants and greenhouse gases such as NO_x, CO, CO_2 per km on the road, a catastrophic global environmental pollution could take place in the near future [1]. Thus, there is a serious global need for an urgent, environmental friendly and economical alternative mode of transportation to replace such environmental damaging vehicles.

***Address correspondence to A. Kazim:** Department of Mechanical Engineering, United Arab Emirates University, Al-Ain, United Arab Emirates; Tel.: +971 4 390 1111 or + 971 4 391 4677; Fax: + 971 4 390 1110; E-mail: akazim@uaeu.ac.ae

Hydrogen-fueled proton exchange membrane fuel cell vehicles (PEM FCV's) are considered by many scientists, energy policy makers and automotive manufacturers to be the key technology to reduce the environmental pollution caused by ICV's. With their attractive characteristics such as being clean energy converters and operate at low temperatures and achieve a quick response, FCV's are at least 30% more efficient than ICV's since they are not limited by the Carnot Cycle and they have lesser hardware parts and do not generate noise like ICV's [2].

The economics of FCV's depend heavily on key variables such as the price of natural gas, electricity prices, fuel cells and reformers and their durability. Furthermore, policies were proposed such as the one in California mandating car manufacturers to include zero-emission vehicles (ZEV) among the vehicles that are to be delivered to the market in 2003 [3]. Consequently, auto manufacturers are now feeling obligated to offer consumers a viable alternative, especially in terms of cost, performance, and durability to survive in the ultra-competitive automobile market and to globally commercialize the vehicles by 2005 [3, 4]. For instance, Toyota and Honda started selling and leasing FCV's in the U.S. and Japan last year. In addition, 60 Ballard-powered Daimler-Chrysler FCV's appeared in the market in 2003, based on small Mercedes A-Class vehicles. Furthermore, Ford's FCV, which is based on the Focus, has a range of 200 miles and it is expected to reach the public by 2008 [1, 5].

Previously, I proposed a scheme through which FCV's are introduced to co-exist with ICV's in the transportation sector and to gradually increase the number of FCV's until they completely replace ICV's by 2025 [6]. However, general public awareness and acceptance of such technology if it is introduced in the market was not taken into consideration. Moreover, the impact of durability and safety of such technology on the commercialization and public acceptance were not addressed. Thus, the objectives of this paper are to compare FCV's with ICV's in terms of the cost fraction and to predict the projected cost of FCV's and its effects on commercialization and public acceptance in the next twenty years period. Moreover, effects of safety and dependability of FCV's on the commercialization and public acceptance at various scenarios are to be considered as well.

COSTS OF FCV's AND ICV'S

In the current study, the cost fraction of FCV's and ICV's can be defined as the ratio of the initial cost of the vehicle in 1990, C_{1990}, over the cost in the years after 1990, C_{1990+i}. Cost fraction at any year t, between 1990 and 2000, can be determined using the following equation:

$$CF = \sum_{i=1}^{10} \frac{C_{1990}}{C_{1990+i}} = (t-1989)^{a} \quad \textbf{(1)}$$

where, a is the annual growth rate and it is set to equal 1/10 for FCV and -1/17 for ICV. These values are used based on the cost trends of these two types of vehicles in the ten years period and they were taken from the available literature [7-10]. The initial costs of FCV's and ICV's in 1990 were estimated to be $30, 400 and $17, 300, respectively [7, 8].

The projected cost of the fuel cell vehicle C_{FCV}, at any year after the year 2000 is expressed as:

$$C_{FCV} = C_{FCVo} \text{ Exp } [a\ (t-2000)] \quad \textbf{(2)}$$

where, C_{FCVo} is the initial cost of a FCV in 2000.

The cost of the FCV's after 2005 could be influenced significantly as a result of intensive Research & Development and competitions between auto manufacturers to develop cost effective FCV's. Thus, this gives an annual growth rate a= 1/10 for FCV and -1/17 for ICV. The cost trends of these two types of vehicles in the ten years period were taken from the available literature.

COMMERCIALIZATION AND PUBLIC ACCEPTANCE OF FCV'S

Commercialization and public acceptance from the year 2000 to 2025, can best be described by Fisher and Pry model [11]:

$$F = \frac{1}{(1+\exp(-(t-b)))} (100\%) \quad \textbf{(3)}$$

where, F is the rate of commercialization and public acceptance and b is the year when F is reached 50% level. For various scenarios, b= 2010, 2015 and 2020. These are years where commercialization acceptance levels would reach 50%.

SAFETY AND DEPENDABILITY OF FCV's

It is safe to say that the commercialization and public acceptance are a strong function of safety and dependability of FCV's. Generally, hydrogen involves safety risks since it is highly flammable. Therefore, it should be handled with extreme caution.

Therefore, manufacturers of FCV's should take certain measures to optimize hydrogen compression, storage and system delivery for the daily usage. In addition, auto manufacturers should educate consumers about the risks involved with compressed hydrogen and its physical and chemical properties. This would result in higher public acceptance, safety and dependability, just as they did with the ICV's, which took almost a century to gain consumers confidence and achieve an optimum commercialization level.

The following power-law equation can best describe the safety and dependability of introducing a new technology namely FCV's from the year 2000 to 2025:

$$S_{FCV} = (t-1999)^{c}\,100\% - (100\%) \quad \textbf{(4)}$$

c= a conservative rate of 1/10 was chosen, to reflect worst case scenario, 1/7 for a moderate case scenario and 1/5 for an optimistic scenario.

The reason of using power-law equations (1) and (4) to describe the cost fraction and safety and dependability of a FCV is that these parameters could always be improved over a period of time but it cannot reach the 100% level. Conversely, equation (3) was used to describe the commercialization and public acceptance level, where it could possibly reach the desired level of 100% by the quarter of the century if proper and effective measures are taken to achieve this goal.

RESULTS AND DISCUSSIONS

Generally, commercialization and public acceptance of utilizing FCV's depend primarily on the capital cost, safety and durability of such vehicles. In the past decade, a significant improvement in the cost fraction of FCV's was detected in comparison to the ICV's, which demonstrated a decrease of 15% in 2000 from its initial cost, as depicted in Fig. **1**. For instance, the cost of a FCV declined from its initial average cost of $30, 400 in 1990 to $23, 900 in 2000 due to intensive R&D and competition among fuel cell developers. On the other hand, cost of an ICV increased from its initial average cost of $17, 300 in 1990 to 19, 900 in 2000 as a result of the features that were added to achieve a better safety (*i.e.* air-bags and antilock breaks), fuel economy and contribute to a lesser environmental pollution mainly NO_x, CO and CO_2.

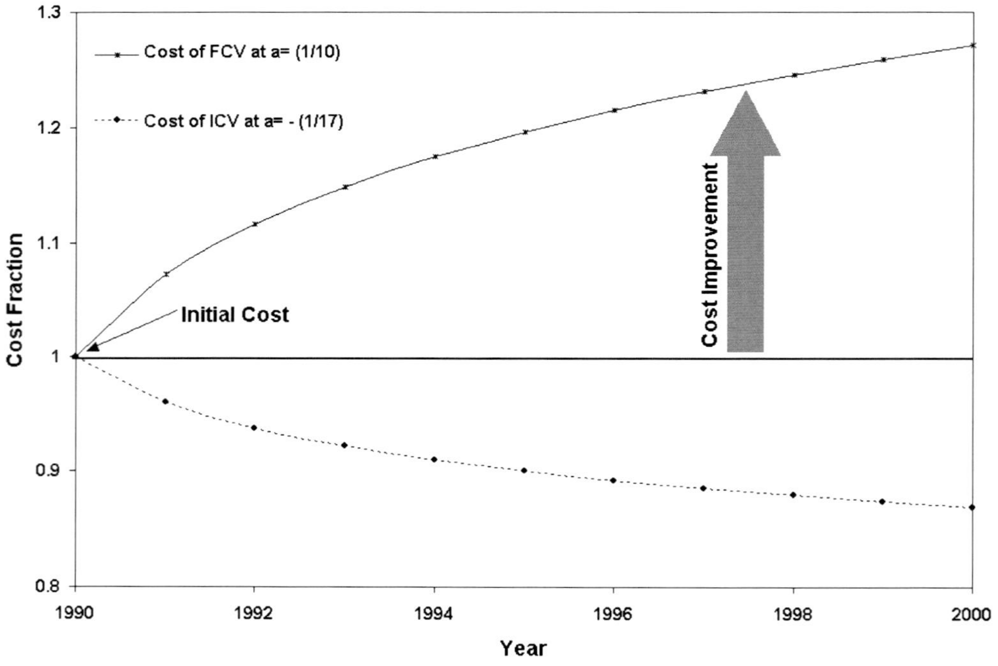

Figure 1: Cost fraction of ICV and FCV from 1990 to 2000.

It is anticipated that the average cost of FCV in the next twenty years will be reduced by approximately 25% if the normal trend is taken, as illustrated in Fig. **2**. However, the cost could take a more or less favorable trend depending on two major components namely cost of the fuel cell and cost of hydrogen storage system. Currently, the fuel cell cost could range from as high as 9, 000$/kW to a very optimistic rate of as low as 324 $/kW for mass production [12]. But eventually, the cost of the fuel cell will be lowered to a competitive rate in the future, as a result of intensive work in producing electrodes, membranes and specially the catalyst layers at attractive rates and at a mass production scale [13-15]. Moreover, cost of hydrogen storage must be reduced to a better rate than the current average rate of 1, 000-2, 000$/GJ in order to achieve the desired competitiveness in the automobile market [8, 9, 15, 16].

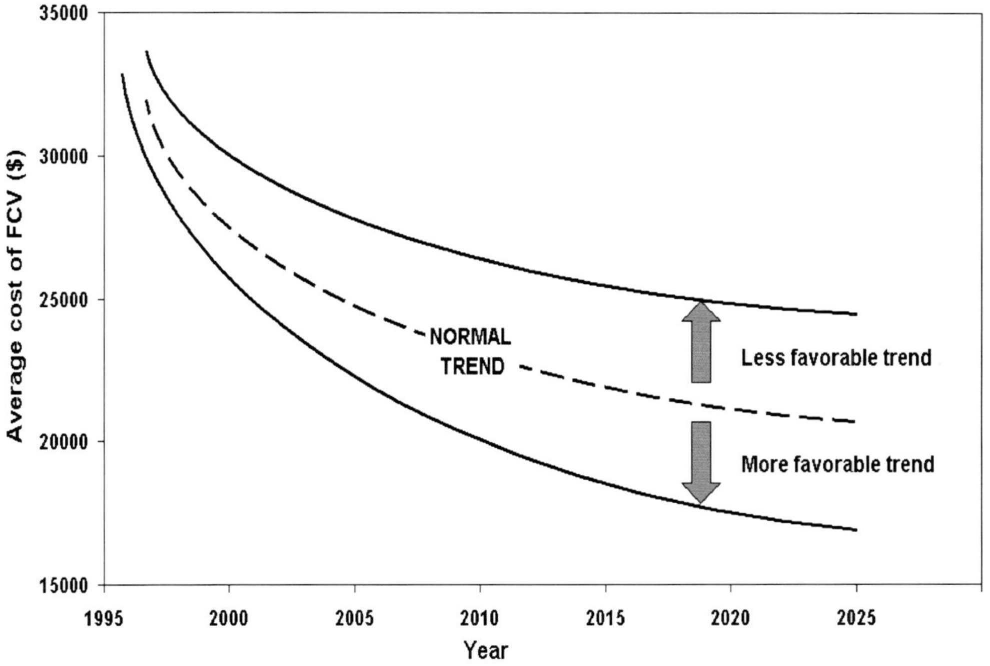

Figure 2: Estimated average cost of FCV in the next 25years.

With a current average end cost of hydrogen fuel in FCV's to be about 30$/GJ, a major obstacle that needs to be dealt with effectively in order to reduce the cost to below 10$/GJ rate. However it is anticipated that the end cost of hydrogen fuel will be drastically reduced to at least 20 $/GJ in 2008 and 10 $/GJ in 2015 [17]. Beside lowering the cost of hydrogen and FCV's, drastic measures in setting-up infrastructures for supplying hydrogen and enhance social awareness about alternative energy sources should be deployed in order for this technology to be realized on the road on a massive-commercial scale [18].

Commercialization and public acceptance in acquiring a new technology such as FCV's could take many years to achieve the desired level of 100%, as it was the case with ICV's. There are two key factors that could contribute significantly to the level of commercialization and public acceptance namely cost of FCV's and safety and dependability of such vehicles. Commercialization and public acceptance could take several possible scenarios where the acceptance level could reach 50% specifically in the year b=2010, 2015 or 2020 as illustrated in the Figs. **3** and **4**.

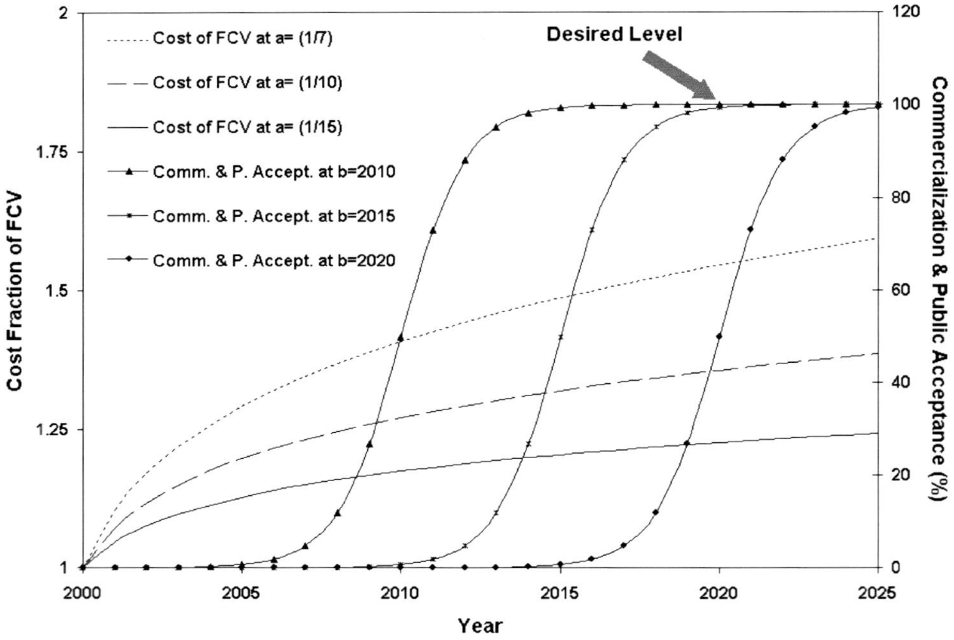

Figure 3: Commercialization and public acceptance *vs.* cost of a FCV at various scenarios.

Actually, better results can be demonstrated at b=2010 than the other two scenarios since the commercialization and acceptance level would reach the desired level of 100% well before 2015, which is at a much earlier stage than the other two scenarios. In addition, cost fraction of FCV's could take the same trend as it took in the 1990's with the exponent a= 1/10, or it could take a better trend (*i.e.* a=1/7) as shown in Fig. **3**. Definitely, the higher the cost fraction of a FCV the higher the commercialization and public acceptance will be accomplished. Moreover at a constant cost fraction, higher commercialization and public acceptance can be achieved if b is increased. For example, at a current trend of a=(1/10) the commercialization and public acceptance would increase by 12% if b is increased from 2010 to 2020.

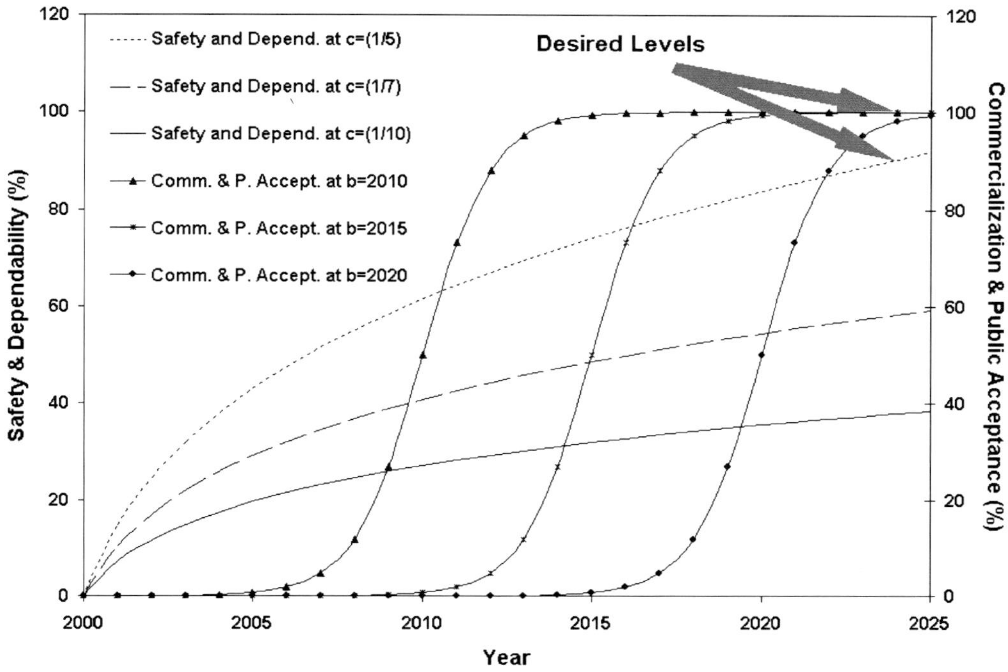

Figure 4: Commercialization and public acceptance *vs.* Safety and dependability of FCV's at various scenarios.

Similar conclusion can be drawn in the case of safety and dependability illustrated in Fig. **4**. The possible scenarios associated with safety and dependability of FCV's vary from the worst-case scenario where c= (1/10) to the best-case scenario where c= (1/5). However, an important point that should be kept in mind is that a higher safety is desired in order to get a higher acceptance level, which could be achieved with additional cost (*i.e.* lower cost fraction). Furthermore, safety policy should be surveyed and safety targets, legislation, standards should be set-up in order to build-up consumer confidence and prevent disasters and to mitigate any accidents [19].

It is essential that the industry must overcome major technical and institutional barriers in order to provide the customers with a clean, energy-efficient technology that performs as well as, if not better than, a commercially available product and at a comparable cost. However, the high cost and risk associated with overcoming these barriers such as creating a hydrogen infrastructure for fuel cell technology, makes the auto companies to be discouraged to invest in such high-risk technology. Therefore, in the absence of proper incentives, or a dramatic change in economics or availability of petroleum, the customer would have a very little reason to try an unproven new technology [20]. Finally, it is difficult to make adjustments to the energy system since it requires changes of technology choices, which would take decades to be realized. For instance, the ICV's took many decades from the beginning of last century to reach the level as it is now [21].

CONCLUSIONS

In this chapter, a general assessment of the commercialization and public acceptance of utilizing FCV's was conducted. The conclusions are summarized as follows:

- In the 1990's, a significant improvement in the cost fraction of FCV's was achieved in comparison with the cost fraction of ICV's. The future projected cost of the FCV's could take a more attractive trend if the costs of the fuel cell and the hydrogen storage systems are substantially reduced.

- Higher levels of commercialization and public acceptance of utilizing FCV's can be achieved at a higher cost fraction and higher safety and dependability.

- Although it is well desired to have a high level of commercialization and public acceptance of utilizing a new technology such as FCV's, it is difficult to make adjustments to the energy system since it requires changes of technology choices, which would sometimes take decades to be realized.

REFERENCES

[1] G. Hoogers, *Fuel Cell Technology Handbook* CRC Press, LLC, 2003.
[2] A.C. Gillet, "From water to water. Hydrogen as a renewable energy vector for the future" Proc. World Renewable Energy Cong., Malaysia, 1999, pp. 513-517.
[3] T. Lipman, J. Edwards and D. Kammen, "Fuel cell system economics: comparing the costs of generating power with stationary and motor vehicle PEM fuel cell systems" *Energy Policy* Vol. 32, pp. 101-125, 2004.
[4] J. Fernando Contadini, *Social Cost Comparison Among Fuel Cell Vehicle Alternatives*, Special report, Methanol Institute, 2003.

[5] J. Motavalli, "Power plays: fuel cells are reaching the market, in what could be a $100 billion industry", *Env. Magazine,* Jan-Feb, 2003.

[6] A. Kazim, "Introduction of PEM Fuel Cell Vehicles in the Transportation Sector of the United Arab Emirates", *Applied Energy,* Vol. 74, pp. 125-133, 2003.

[7] J.M. Ogden, T.G. Kreutz and M. Steinbugler, "Fuels for Fuel Cell Vehicles: Vehicle Design and Infrastructure Issues", *SAE Technical Paper Series* no. 982500, 1998.

[8] T. Johansson, H. Kelly, A. Reddy and R. Williams, "Renewable energy- sources for fuels and electricity", Island Press, USA, 1993.

[9] C. Thomas, B. James and F. Lomax, "Analysis of residential fuel cell systems & PNGV fuel cell vehicles", Proc. Hydrogen Program Rev. NREL/CP-570-28890, 2000.

[10] M. Dunn, "Environmental Vehicles: Competition for the car of the future", Proc. 17th World Energy Cong. in Houston, Texas, USA, 1998.

[11] J.C. Fisher and R.H. Pry, "A simple Substitution Model for Technological Change", *Tech. Forecasting Social Change,* Vol. 3, pp. 75-88, 1971.

[12] L. Schlecht, "Competition and alliance in fuel cell power train development", *Int. J. Hydrogen Energy,* Vol. 28, pp. 717-723, 2003.

[13] F. Pekhota, V. Rusanov and V. Fateev, "Progress in fuel cell development and possible market in Russia", Proc. 13th World Hydrogen Energy Conf., Beijing, China, pp. 156-160, 2000.

[14] T. Kosugi, A. Hayashi and K. Tokimatsu, "Forecasting development of elemental technologies and effect of R&D investments for polymer electrolyte fuel cells in Japan", *Int. J. Hydrogen Energy*, Vol. 29, pp. 337-346, 2004.

[15] F. Mitlitsky, A. Weisberg and B. Myers, "Vehicular Hydrogen Storage Using Lightweight Tanks", Proc. Hydrogen Program Review, NREL/CP-570-28890, 2000.

[16] S. Eaves and J. Eaves, "A Cost Comparison of Fuel-Cell and battery Electric Vehicles", *J. Power Sources,* 2004.

[17] S. Gronich and J. Garbak, "Energy Efficiency and Renewable Energy Hydrogen, Fuel Cells and Infrastructure Technology Program", Energy Eff. Renewable Energy Hydrogen, Fuel Cells and Infrastructure Tech. Program, FORS 5G-086, 2003.

[18] Agence France-Presse, "Cost, lack of infrastructure make fuel-cell cars beyond reach", ClariNet, 2003.

[19] T. Hijikata, "Research and development of international clean energy network using hydrogen energy (WE-NET)", *Int. J. Hydrogen Energy,* Vol. 27, pp. 115-129, 2002.

[20] P. Bak, "U.S. Department of Energy see widespread commercial hydrogen and fuel cell products in 2015 through private-public cooperation", Mat. DOE, 2003.

[21] K. Knapp, "Exploring Energy Technology Substitution for Reducing Atmospheric Carbon Emissions", *Energy Journal,* Vol. 20(1999), pp. 121-143.

APPENDIX: NOMENCLATURE

a	Annual growth rate of FCV and ICV
b	Year when F is reached 50% level
c	Annual growth rate of FCV

CF	Cost fraction of FCV and ICV
C_{1990}	Initial cost of the vehicle in 1990, $
C_{1990+i}	Cost of the vehicle after 1990, $
C_{FCV}	Projected cost of FCV, $
C_{FCVo}	Initial cost of the FCV in the year 2000
F	Rate of commercialization and public acceptance.
S_{FCV}	Safety and dependability of introducing FCV's
t	Time, year

CHAPTER 11

Philosophy for Controlling Auto-Ignition in an HCCI Engine

H. Machrafi*

University Paris 6 and ENSCP, Paris, France

Abstract: In the automobile industry, engines are mostly 4-stroke engines (intake stroke, compression stroke, combustion stroke and exhaust stroke) and generally there are two kinds of automobile engines: the Otto engine (Spark Ignition, SI) and the Diesel engine (Compression Ignition, CI). These two types of engines will be discussed as well as the most important pollutants they emit. In order to reduce the emission of these pollutants an alternative combustion process is discussed, called the Homogeneous Charge Combustion Ignition (HCCI). The advantages and problems are treated. One of the major problems to be solved is the ignition timing that is spontaneous in contrast with that of the spark ignition and diesel engines. Results from experiments, performed in a mono-cylinder engine, are analysed in order to discuss criteria that should be taken into account before the solution of controlling the auto-ignition can be discussed. The results are subsequently used in order to propose an outline for controlling the auto-ignition in an HCCI engine.

Keywords: Philosophic outline, compression ignition, auto-ignition, combustion control, parametric study, engine parameters, emission control, HCCI engine, EGR, influence chemical species, indicated efficiency, equivalence ratio, engine speed, auto-ignition characteristics, new technology, green energy.

INTRODUCTION

Two types of engines are mainly used today in vehicles. The petrol engine, also known as the spark-ignition engine (SI), often runs on gasoline fuel. In an SI engine a fuel/air mixture is obtained either by premixing, by injecting the fuel in the intake port or by injecting the fuel directly in the cylinder. The fuel mixes with the air that is sucked into the engine and, as a result, the cylinder is filled with a nearly homogeneous charge. When the charge has been compressed close to Top Dead Centre (TDC), a spark ignites the air/fuel mixture. The combustion process usually starts in the centre of the cylinder, after which the flame travels towards

***Address correspondence to H. Machrafi:** University Paris 6 and ENSCP, Paris, France; Tel.: +33 1 44276816; Fax: +33 1 44276813; E-mail: hatim-machrafi@enscp.fr the cylinder walls. This means that SI combustion is characterized by a flame propagation process. The produced heat from the oxidation reactions and the active intermediary reactions diffuse towards the adjacent gas layer in front of the flame front, causing further line ignition. Nowadays, the SI engines run on a stoichiometric mixture to best utilize the catalyst for exhaust after-treatment. Theoretically the fuel/air mixture is considered to be stochiometrical (equivalence ratio, $\phi = 1$) and the reaction complete, that is, producing only carbon dioxide and water. However, the equivalence ratio, describing a certain ratio of the fuel and the air in the inlet mixture, in most cases, does not have to be equal to one. Hence, the major species produced by the reaction for combustion of a hydrocarbon C_aH_b are CO_2, CO, H_2O, hydrocarbons and particulate matters. Using a fixed fuel/air ratio means that load control is only possible by controlling the mass flow of air into the engine. The throttle that is used for this purpose gives rise to pumping losses and a reduction in efficiency; the major disadvantage of the SI engine is its low efficiency at part load. The compression ratio in Otto engines is limited by knock and can normally be found in the range from 8-12 contributing to the low efficiency. At too high engine speeds even knocking takes place, which can destroy the engine. This is due to the fact that at high engine speeds, more fuel is injected in the cylinder at the same ignition delay time base [1, 2].

Diesel engines operate at higher compression ratios (12-24) than SI engines. In this type of engines, varying the amount of diesel fuel injected into the cylinder controls the load. In a Diesel engine the air and the fuel are separately introduced into the engine. Instead of ignition by a spark plug, the air-fuel mixture auto-ignites due to compression. The processes that occur from the moment the liquid fuel leaves the injector nozzles until the fuel starts to burn, are complicated; droplet formation, collisions, break-up, evaporation and vapour diffusion are some of the processes that take place. The rate of the combustion process is generally limited by these processes; a part of the air and fuel will be premixed and burn fast, but for the largest fraction of the fuel, the time scale of evaporation, diffusion, *etc.* is larger than the chemical time scale. Liquid fuel that does only partially burn results in soot formation. Together with NO_x, the emissions of soot characterize the Diesel combustion process. For present engines, a trade-off between these two emissions is observed, which poses a major challenge to reach future legislation for both emissions. The major advantages of the Diesel compared with the SI engine are the low pumping losses, due to the lack of a throttle, and a higher compression ratio, leading together to higher efficiency [1, 2].

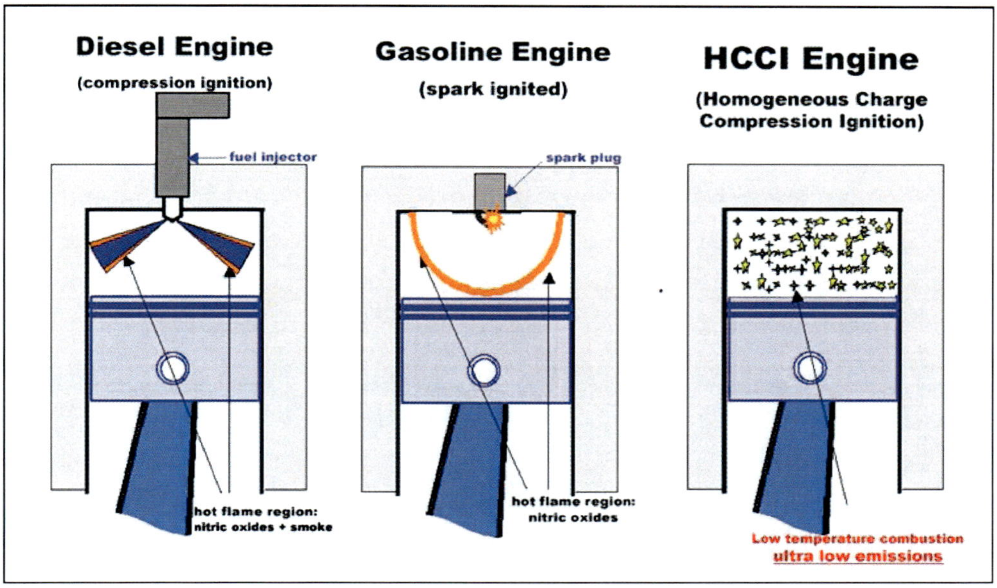

Figure 1: Comparison between the different combustion modes in an engine [4].

As said before, for the reasons put out above, a better alternative should be found. A major possibility that is proposed seems to be much promising: The Homogeneous Charge Combustion Ignition (HCCI). HCCI can be defined as a premixed, lean burn combustion process, preceded by a homogeneous air/fuel port-injection [3]. The HCCI engine generally runs on a lean, diluted mixture of fuel, air and combustion products, which is not ignited by a spark but by compression auto-ignition instead. In order to speed up the kinetics, the temperature of the charge at the beginning of the compression stroke can be increased. This can be done by heating the intake air or by keeping part of the warm combustion products in the cylinder (Integral Gas Recirculation). Both strategies result in a higher gas temperature after compression, which in turn speeds up the chemical reactions that occur in the (nearly) homogeneous mixture in the following cycle. In contrast with the SI engine, where a spark-plug is used to generate a propagating flame, and the CI engine, where the injection causes a diffusion flame, the ignition in the HCCI engine will occur when the temperature and pressure are high enough for fuel oxidation chemistry to become very fast. Auto-ignition is said to occur simultaneously at many locations, called *hot spots*, throughout the cylinder. Furthermore, it is claimed that turbulence is of minor importance for the combustion process, though more research is needed in this domain, and that flame propagation does not occur. It is generally agreed upon that the HCCI process is mainly driven by chemistry. Fig. **1** shows an example of the difference of the aforementioned types of engines.

While Otto engines are characterized by a low efficiency at part load and Diesel engines exhaust high concentrations of particulate matter and NO_x, the HCCI engine has neither of these problems: as the peak temperature in the cylinder is low for lean, diluted mixtures, NO_x concentrations are reported [5, 6] to be two orders of magnitude lower than for the Otto and Diesel processes and particulate emissions are negligible. Fig. **2** shows the zone of NO and soot formation as a function of the temperature and equivalence ratio as well as the operating zones for engines that would run in respectively the SI, CI and HCCI modes.

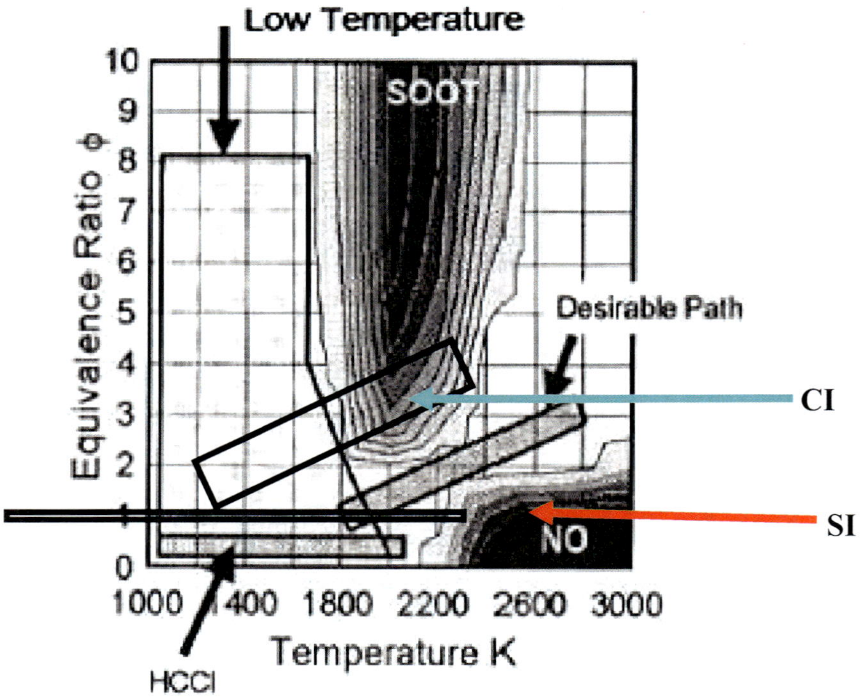

Figure 2: NO_x and particulate reduction at low temperatures with the HCCI method [7].

Due to its low emissions and a high efficiency, the HCCI process might in the future replace the Otto cycle during part-load operation. For high load operation another combustion process is necessary as a result of the low power density of HCCI engines. A major advantage of HCCI combustion, contrary to SI and CI engines, is that it allows for variable inlet conditions. Using techniques as a variable compression ratio or variable valve control, several fuels can be used. This opens the way of applying HCCI combustion to biofuels or even hydrogen.

Nowadays the HCCI concept cannot be implemented in automotive applications due to several problems: complex and expensive solutions (*e.g.,* variable valve train systems) may be necessary to make HCCI combustion possible and the operating window of smooth HCCI operation is still of limited size. But the major problem is the difficulty to control the moment of auto-ignition and the energy release rate. As the mixture in the cylinder is premixed and no spark plug is used, the chemical processes determine the onset of auto-ignition, as well as the fuel burn rate. In contrast to the SI and Diesel processes, HCCI combustion cannot be controlled directly; the spark-timing can be manipulated in spark-ignition engines, while the moment and amount of fuel injected can be controlled in Diesel engines. If we assume that the addition of fuel in HCCI engines is either by port-injection or early direct-injection, to obtain a near-homogeneous mixture, HCCI engines have no direct way of controlling the combustion process. At present, HCCI research has come to the stage that many seem to be working in an attempt to solve this problem. Increasing amounts of research is being published on extending the HCCI operating window [8, 9]. Some approaches used to control HCCI combustion are, split direct-injection in the negative valve overlap, variable valve timing and controlling the blend of two fuels. In this work, it is the idea that the solution of this problem may lie in the use of controlling some important physical, mechanical and chemical parameters as the inlet temperature, the equivalence ratio, the fuel composition and the compression ratio as well as EGR and certain additives. The variability of an engine running in HCCI mode, allows for many fuels to be used. However, the control of the timing and rate of combustion is still troublesome, the operating window for HCCI operation (load and engine speed) is small, the additional components making successful HCCI combustion possible lead to an increase in cost and the HCCI engine has relatively high friction losses due to the low power density. This poses a control problem, but is also a problem for the HCCI combustion. If misfire occurs, the gas mixture during the next cycle will be too cold for auto-ignition to occur (unless intake air heating is used) and the engine will stop. Therefore a profound analysis is needed to investigate the possibilities with which HCCI combustion can be controlled. For this purpose, results issued from [10] (where also the experimental set-up can be consulted) from an HCCI engine are used in this chapter. It is not the purpose of this chapter to discuss the results, but rather to use these results in order to propose a philosophy of controlling the auto-ignition in an HCCI engine.

EXPERIMENTAL COMPARISON SI ENGINE AND HCCI ENGINE

The purpose of understanding better the auto-ignition process is to contribute to solving the problem of controlling the auto-ignition in HCCI engines. This auto-ignition is intrinsically different from that of gasoline spark ignition or diesel compression ignition. To show an example of this difference the case of spark ignition combustion is taken. A pressure profile of a spark ignition combustion is compared to a pressure profile of an HCCI combustion. Since the spark ignition pressure profile is obtained from the HCCI engine, an optimal value is chosen for the spark timing, using classical conditions, such as an equivalence ratio of 1, 0, a compression ratio of 9, 5, an inlet temperature of 30 °C and gasoline as the fuel. The optimal value is based on the highest indicated efficiency and the lowest emissions for hydrocarbons and CO. The indicated efficiency could be defined as how the fuel-combustion-engine system is going to convert the introduced chemical energy into work on the piston. The indicated efficiency is calculated from the work on the piston, which can be defined by thermodynamics as follows:

$$W_{indicated} = -\oint PdV$$

Here, $W_{indicated}$ is the indicated work on the piston. The chemical energy (introduced power) that is introduced is defined as the specific energy of the fuel (J/kg) times the amount of fuel injected (kg/s). Combustion takes place every two turns of the crank shaft. The engine speed is 600 rotations of the crank shaft per minute or 10 rotations per second. So combustion takes place 5 times per second. So, to obtain the indicated power ($P_{indicated}$), the indicated work per second has to be multiplied by

N/(n*t), where N is the engine speed in (min^{-1}), n is the number of turns per combustion (in this case: 2) and t is 60 s/min. So the indicated efficiency is then:

$$\eta_{ind} = \frac{P_{indicated}}{P_{introduced}} = \frac{W_{indicated}N/120}{E_{fuel}F_{fuel}},$$

with E and F respectively the specific fuel energy and the injected fuel flow. Fig. **3** shows the results.

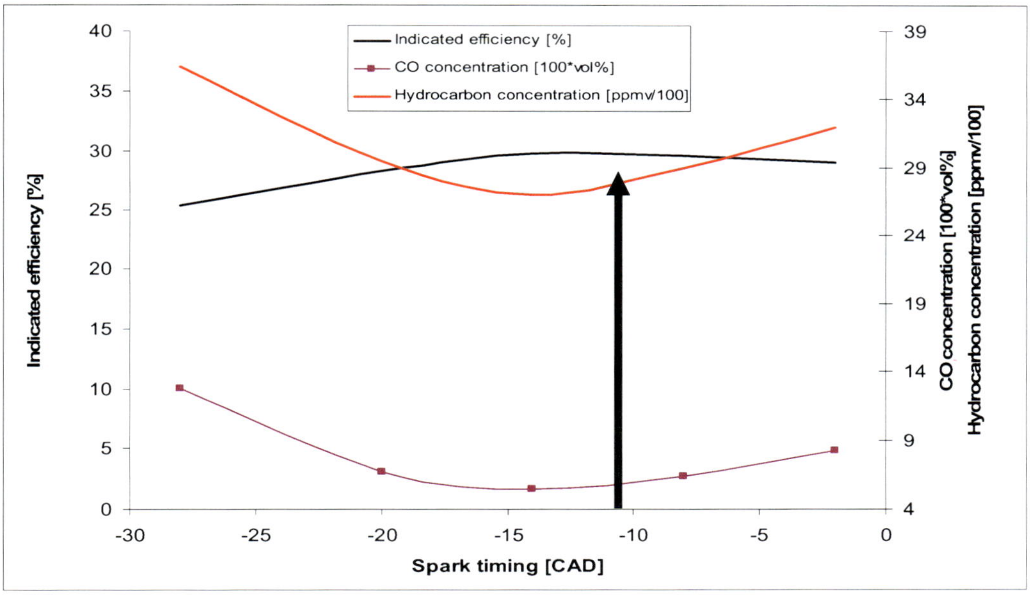

Figure 3: The indicated efficiency and the emissions of hydrocarbons, CO, CO_2 and O_2 as a function of the spark ignition advancement at an equivalence ratio of 1, 0, an inlet temperature of 70 °C and a compression ratio of 9, 5 with gasoline as the fuel.

Fig. **3** shows that the optimal value for the spark timing at the indicated conditions appears to be –13 CAD (with respect to the top dead centre (TDC)). At this value the pressure profile is presented in Fig. **4** for the case of spark ignition combustion with a spark timing of –13 CAD. As for the HCCI pressure profile, gasoline is also chosen as the fuel, at a compression ratio of 13, 5, an equivalence ratio of 0, 4 and an inlet temperature of 70 °C. As can be seen, the aim is not to compare the two modes quantitatively, but to assess the difference in the pressure profile, which is an overall resume of the auto-ignition process. The comparison is presented in Fig. **4**.

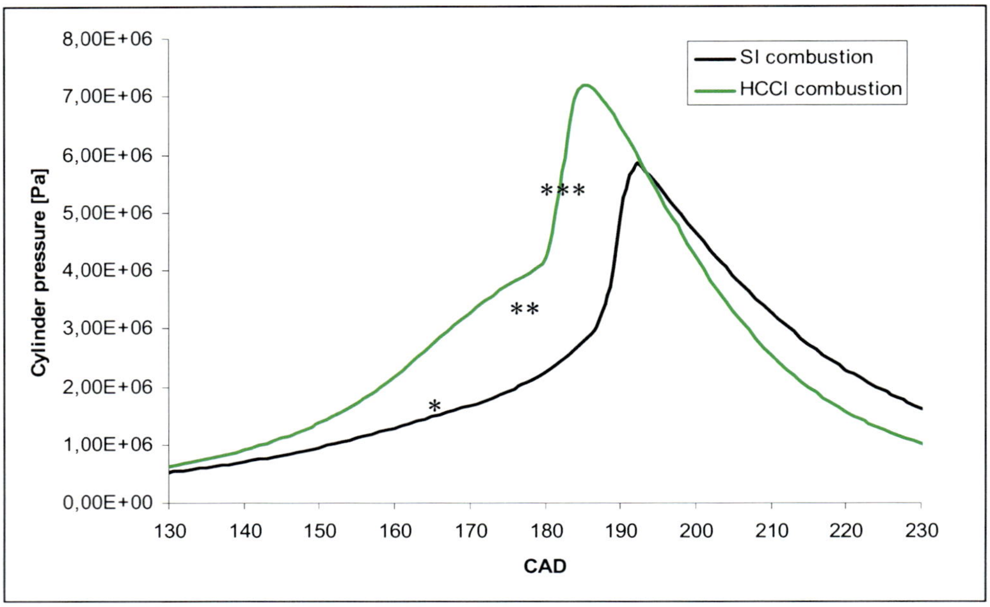

Figure 4: Qualitative comparison between SI combustion and HCCI combustion.

As Fig. **4** shows, the pressure profile is much different between the SI and the HCCI combustion modes. The SI combustion shows one ignition, which is issued after a propagation of the flame inside the combustion chamber, caused by the spark at –13 CAD in Fig. **3** or 167 CAD in Fig. **4**. Hereby the pressure gradient increases gradually until after the ignition. The HCCI combustion, however, shows a pressure gradient that increases (*) at first instance (cool flame), then decreases (**) (negative temperature coefficient (NTC)) and finally increases (***) until after the final ignition. Fig. **5** shows the pressure derivative (which is often used to represent the burn rate) as a function of the pressure to visualize better the combustion behaviour.

For the HCCI mode, the letter A, in Fig. **5**, represents the low pressure region from IVC to the cool flame, while the letter A' represents the NTC region. The letter B represents the high pressure region from the final ignition (around $dP/d\theta$ = maximum) to the maximum pressure*. These two take place at the compression stroke. The letter C represents the region after the maximum pressure back to the beginning that is the intake stroke. The letter D represents the exhaust and intake loops. For the SI mode, the letter A represents also the low pressure region before the ignition takes place. The letters B until D represent the same as for the HCCI mode. When looking especially to the region represented by letter A for both modes, one can observe that no maximum of the pressure derivative is observed for the SI mode. Furthermore, it appears that even with a lower equivalence ratio, the HCCI mode represents a higher burn rate at the final ignition and a higher maximum pressure. It can be seen from Figs. **4** and **5**, that the auto-ignition process for HCCI combustion is complex and intrinsically different from that of SI combustion. This suggests that controlling the HCCI combustion is different. Moreover, unlike the SI combustion, many parameters play an important role in the control of the HCCI combustion. This makes one think that more and more emphasis will be laid on the incorporation of physical-chemical coupled models in an on-board computer for the purposes to control the HCCI combustion. This, since the HCCI combustion is more sensitive to the parameters that control its auto-ignition than for the other combustion modes. This remaining of this chapter presents a brief discussion on the outline of how one may choose the parameters that seem to be the most appropriate to control the HCCI combustion.

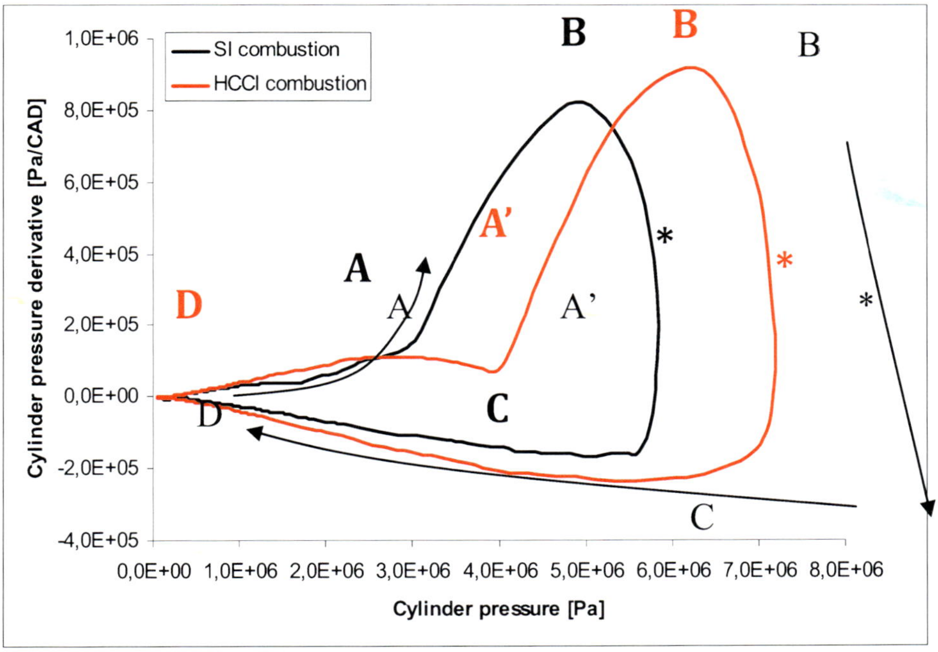

Figure 5: The pressure derivative as function of the pressure at an inlet temperature of 70°C and a compression ratio of 13, 5, comparing the auto-ignition process of gasoline95 in SI mode with equivalence ratio 1, 0 and in HCCI mode with equivalence ratio 0, 4.

OUTLINE TO CHOOSE THE MOST APPROPRIATE PARAMETERS FOR THE AUTO-IGNITION CONTROLLING PROCESS

When one wants to use a parameter for the control of the auto-ignition, the change in the parameter should not affect the work done on the piston and, importantly, the efficiency of the engine. Therefore, the effect of the same parameters on the indicated efficiency has been investigated and will be included with the discussion concerning the usefulness of the parameters. Two criteria will be used to determine whether a parameter is useful or not. The first criterion is, whether a parameter can be used in a wide range keeping low CO and hydrocarbon emissions (and a low as possible emission of CO_2), while the highest indicated efficiencies are influenced but slightly. The second criterion is that at the same time the parameter should have a strong effect on the ignition delays. This makes a slight change of a parameter that affects significantly the auto-ignition, result into a quick response to the requested demand in an engine and thus needs less time and stabilisation. Furthermore, a slight influence on the indicated efficiency means that changing this parameter can be done freely without the concern of losing efficiency in an engine. Also a change in a parameter that gives good results in terms of indicated efficiency but produces too much pollution cannot be acceptable to be used in an engine for the control of auto-ignition. The parameters that are used in this analysis are those of which the influence on the auto-ignition process is investigated in this work. Furthermore, the conclusions are only drawn on the investigated range of the parameters. In this light, three types of parameters could be mentioned. The parameters (type 1) that do not seem to be interesting to investigate for the control of auto-ignition, the parameters (type 2) that are interesting for this purpose and the parameters (type 3) that are not intended to be used for the control of the auto-ignition, but are fixed values that are used to have an optimal of indicated efficiency and emission control. Hereby, the findings are not definitive, since the HCCI engine does not necessarily represent a real engine, but is merely used as a tool. Moreover, this analysis does not provide definitive conclusions of the results, but rather presents an outline or philosophy of how to choose appropriate parameters for the control of the auto-ignition process, using the results of this work.

In [10] the parameters have been divided into three types. The concerned parameters are: dilution by CO, Exhaust Gas Recirculation temperature, dilution by N_2 and CO_2, addition of acetaldehyde, formaldehyde and NO. These parameters are chosen, since they represent important characteristics of Exhaust Gas Recirculation, which seems to be a promising controlling parameter for the auto-ignition in an HCCI engine. Other typical engine parameters that are also evaluated are the inlet temperature, equivalence ration and compression ratio. Finally, the choice of a fuel (compositions of n-heptane/iso-octane/toluene mixtures, gasoline and diesel) is also taken as important. For more information about the results of these experiments and these parameters, one may consult [10]. In the next, the mentioned parameters will be treated as acquired.The first one are the parameters of type 1, which do not seem to be interesting to be analysed further. The effect of CO and the Exhaust Gas Recirculation (EGR) temperature in the range discussed in [10], appeared to be of less interest than the other EGR parameters. Therefore, the addition of CO and the EGR temperature will not be considered in the analysis in this chapter. The effect of acetaldehyde appeared to pose some difficulties to be assessed. This renders its reactivity indeterminable in this work. Therefore acetaldehyde cannot be used as a parameter to be analysed further. The parameters of type 2 are the following. The effect of dilution of N_2 and CO_2 seems to be interesting [10]. and is worth investigating. The effect of NO and CH_2O (formaldehyde) also seem to be interesting, following the experimental results [10]. The equivalence ratio in an engine varies constantly and is also an interesting tool for the control of the auto-ignition process. For the sake of reducing the fuel consumption and CO_2 production it is desired to have an equivalence ratio as low as possible. However, to ensure an efficient functioning of the engine, the equivalence ratio should not be too low. So changing the equivalence ratio should be in a small range and should affect the auto-ignition largely. So a small optimal permissible range should be chosen for the equivalence ratio. The parameters of type 3 are the following. The inlet temperature has a clear effect on the auto-ignition delays and could be interesting to investigate, though it is most likely to be constant in an engine, since it has a large inertia. It is highly probable that also the compression ratio is to be chosen constant and an optimal value should be found, but in the near future, this parameter could be used satisfactorily to influence the auto-ignition process, with advancing technologies. In an engine the fuel composition is likely to be fixed, so the fuel compositions of n-heptane/iso-octane and n-heptane/toluene cannot be used for the control of HCCI auto-ignition. However, it can give an indication what fuel composition gives the best indicated efficiency. To assess the optimal values of the compression ratio, the inlet temperature, the fuel composition and the equivalence ratio, an extra discussion should be performed to determine what values are the most advantageous, with respect to low emissions and high efficiency. The parameters of type 1 are left out of the discussion. To determine the optimal conditions of the HCCI combustion in the HCCI engine, first the parameters of type 3 are discussed. Each parameter in the rest of the chapter's results will be accompanied by the equivalence ratio and the results will be presented in "iso-contours" in the form of mappings. The equivalence ratio is chosen for this purpose, since it is a parameter that influences the most the combustion efficiency. The slopes in the iso-contours allow determining the sensitivity of the indicated efficiency, the emissions or the auto-ignition delays as a function of the several parameters.

DETERMINATION OF THE OPTIMAL VALUES OF THE PARAMETERS OF TYPE 3

In this section, the parameters that are not likely to change in an engine are investigated to determine their optimal value. So the influence of the compression ratio, inlet temperature and the fuel composition on the emissions of CO, CO_2 and the hydrocarbons will be discussed. The fuel compositions that are studied in this work are n-heptane/iso-octane and n-heptane/toluene. A parameter value is considered optimal, when at that value, not only the indicated efficiency is high and the emissions low, but also having weak sensitivities at that value. The indicated efficiency is considered acceptable when it is higher than 30 %, based on present diesel engine performances. However, for the emissions, no concrete values can be given, since the HCCI engine in question is not a realistically optimised engine, but merely a tool to investigate the experimental auto-ignition process. Therefore, no optimal functioning of the engine can be presented and the values should be taken relatively. In this light, the important features that should be observed are the slopes of the iso-values in the figures, presented in this chapter. Figs. **6** and **7** show the influence of the equivalence ratio and the compression ratio, on the indicated efficiency and the emissions of CO, CO_2 and the hydrocarbons for respectively the fuels n-heptane and PRF40. These two fuels are chosen to show that the influence of the compression ratio can be different as a function of the fuel used.

A: Indicated efficiency

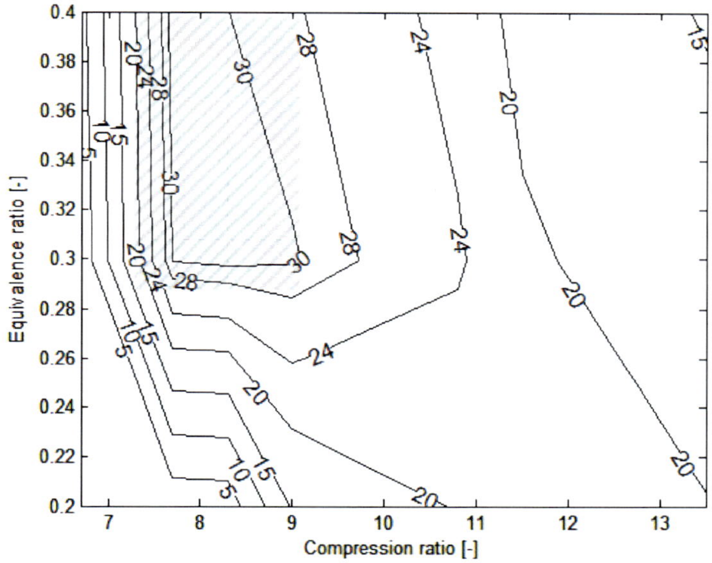

B: Hydrocarbon emission

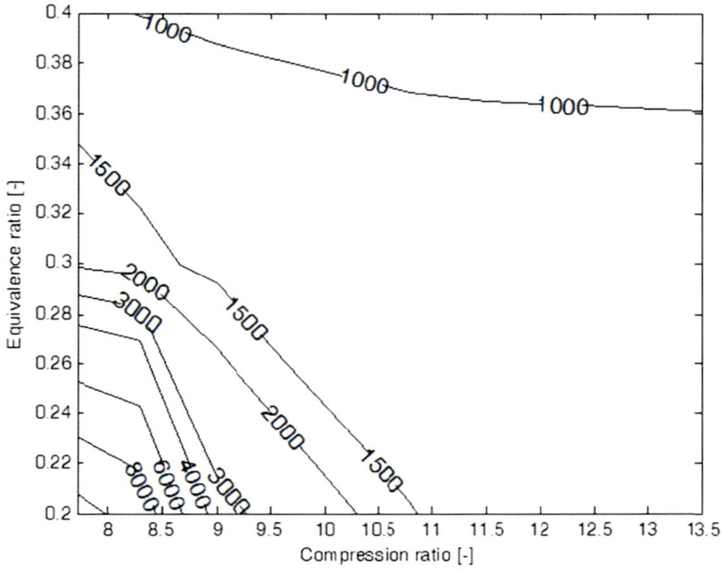

C: CO emission

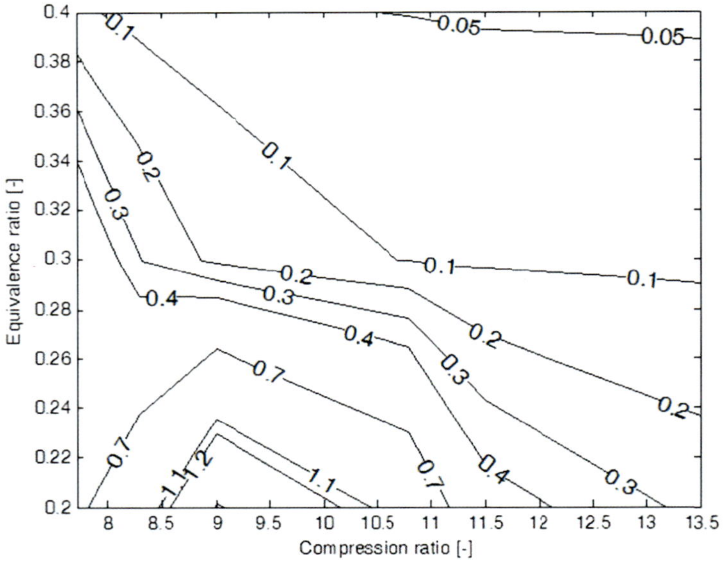

D: CO_2 emission

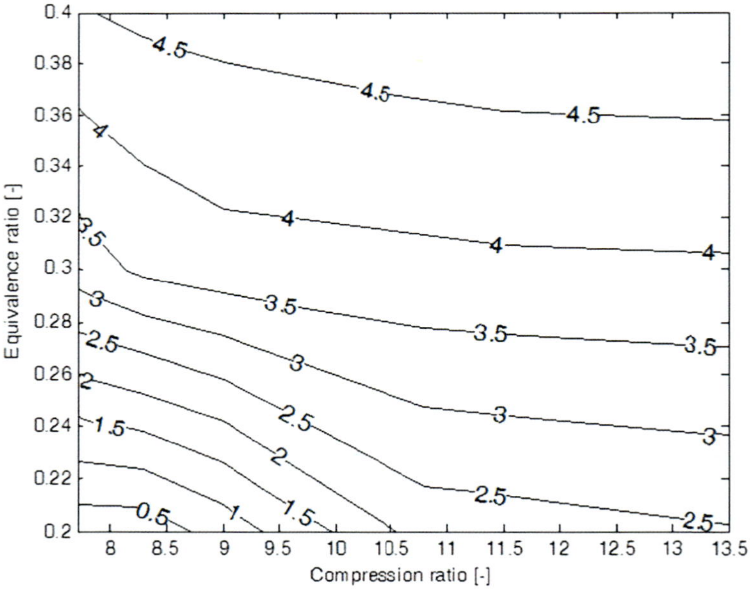

Figure 6: Iso-indicated efficiencies (A in %), hydrocarbon emissions (B in ppmv), CO emissions (C in vol%) and CO_2 emissions (D in vol%) as a function of the equivalence ratio and the compression ratio, for an inlet temperature of 70°C with n-heptane as the fuel.

A, Indicated efficiency

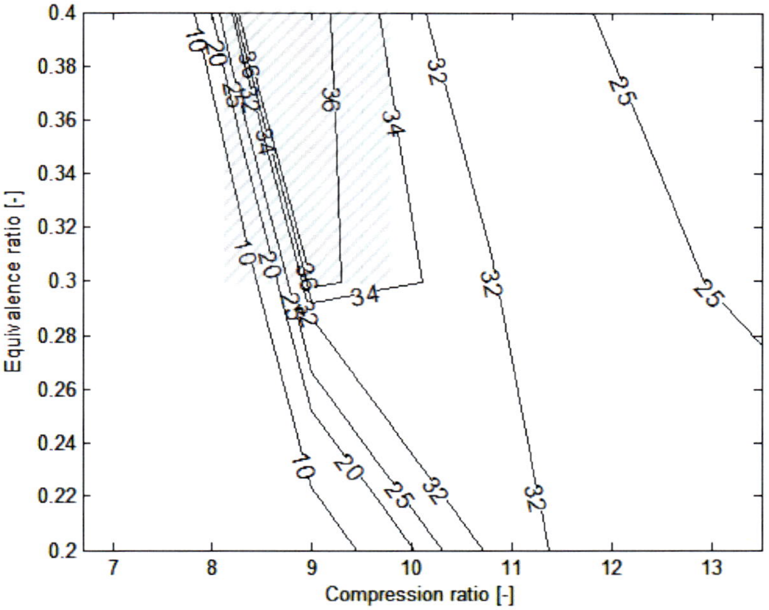

B, Hydrocarbon emission

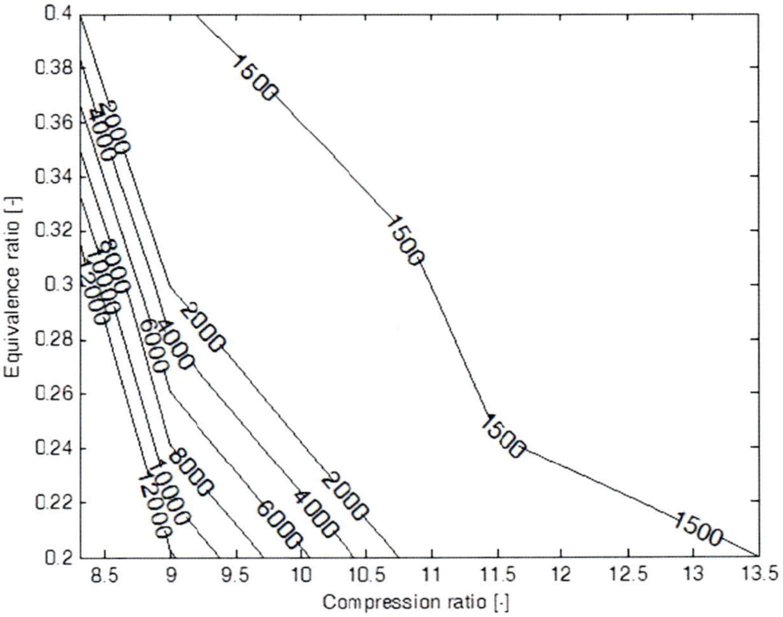

C, CO emission

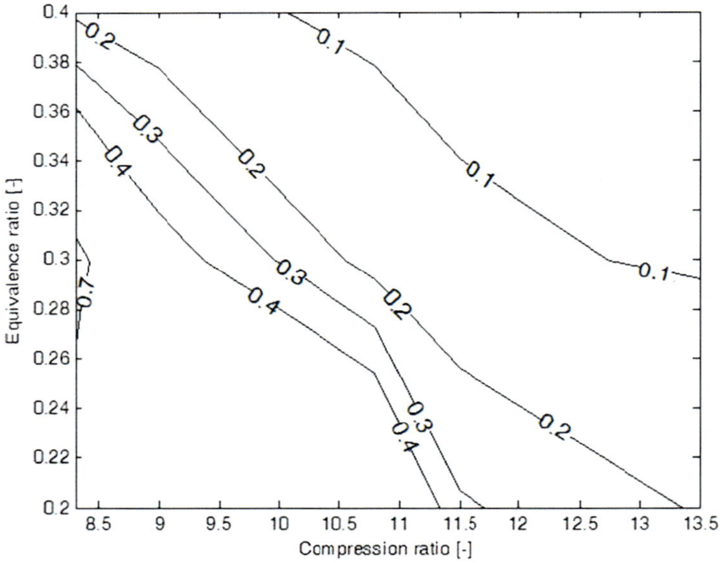

D, CO_2 emission

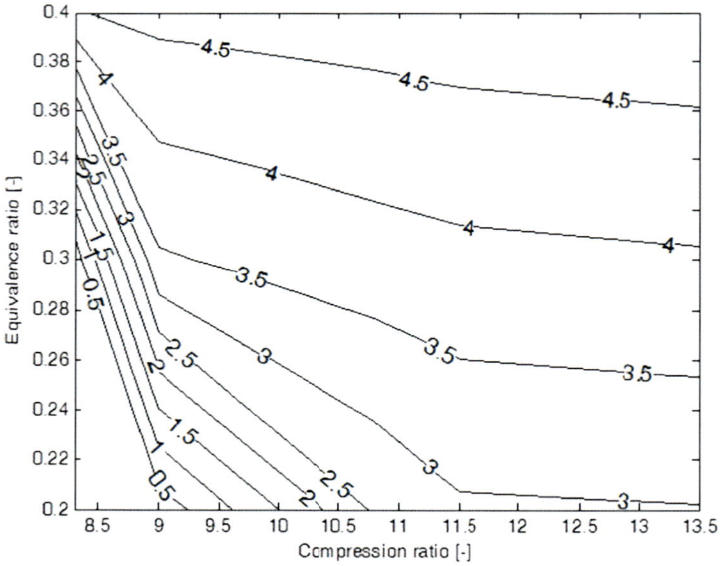

Figure 7: Iso-indicated efficiencies (A in %), hydrocarbon emissions (B in ppmv), CO emissions (C in vol%) and CO_2 emissions (D in vol%) as a function of the equivalence ratio and the compression ratio, for an inlet temperature of 70°C with "60 vol% n-heptane/40 vol% iso-octane" as the fuel.

It is clear, from Figs. **6** and **7** that an acceptable indicated efficiency (30 %) is attained when the equivalence ratio is higher than about 0, 3. It is not the objective to choose a particular equivalence ratio, but rather a range of equivalence ratios, where the indicated efficiency is as high as possible and the emissions of CO and the hydrocarbons as low as possible. What compression ratio is then suitable? A certain critical compression ratio (CCR) is needed for a fuel to burn in an engine cylinder. Furthermore, at increasing equivalence ratios higher the sensitivity of the emissions as a function of the compression ratio is less. This can be seen by the slope of the iso-values. The change in the slope is, for pure n-heptane, at a compression ratio higher than 8, whilst for "60 vol% n-heptane/40 vol% iso-octane" this is 10 and higher. The optimum of the indicated efficiency, at an equivalence ratio of at least 0, 3, is for n-heptane between 7, 5 and 9 and for "60 vol% n-heptane/40 vol% iso-octane" between 9 and 11, 5. Depending on the fuel composition, having as few emissions as possible and an as high indicated efficiency as possible, the compression ratio should be about 8 for n-heptane and about 10 (10, 2 to be precisely) for "60 vol% n-heptane/40 vol% iso-octane". The fuel plays apparently a role on the optimal compression ratio. The fuel composition is thus an important criterion for this analysis. Figs. **8** and **9** show the influence of the fuel compositions of mixtures of respectively n-heptane/iso-octane and n-heptane/toluene.

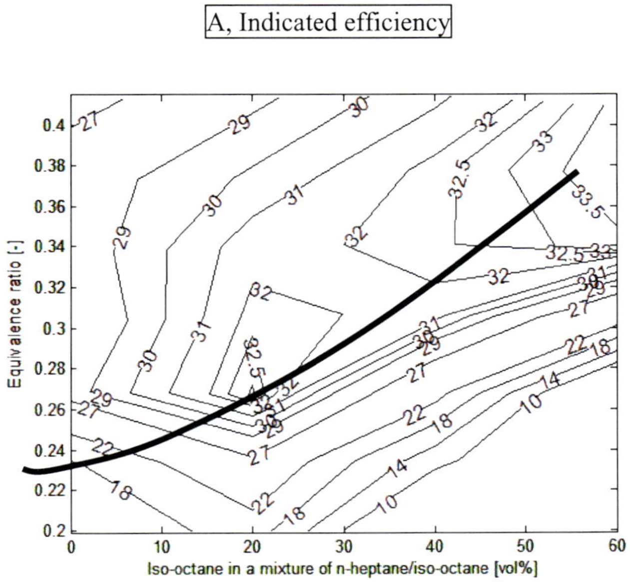

B, Hydrocarbon emission

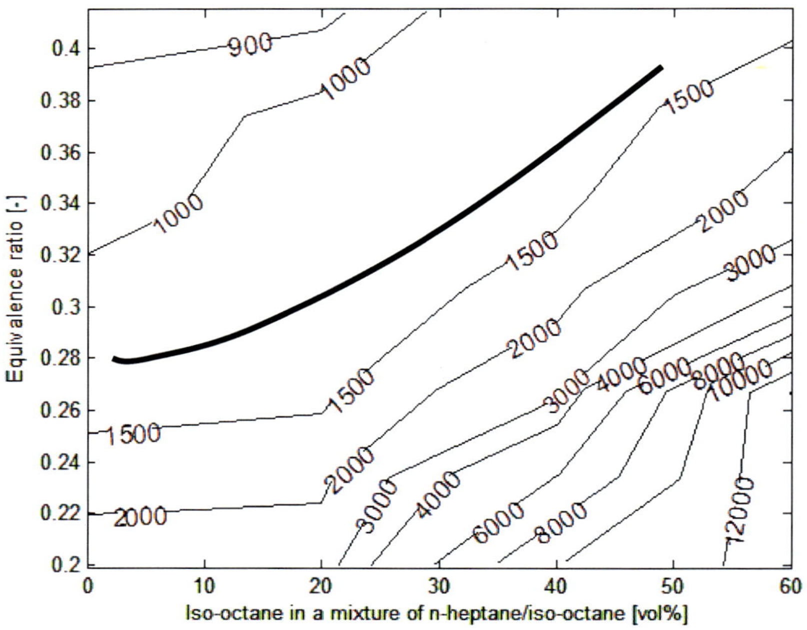

C, CO emission

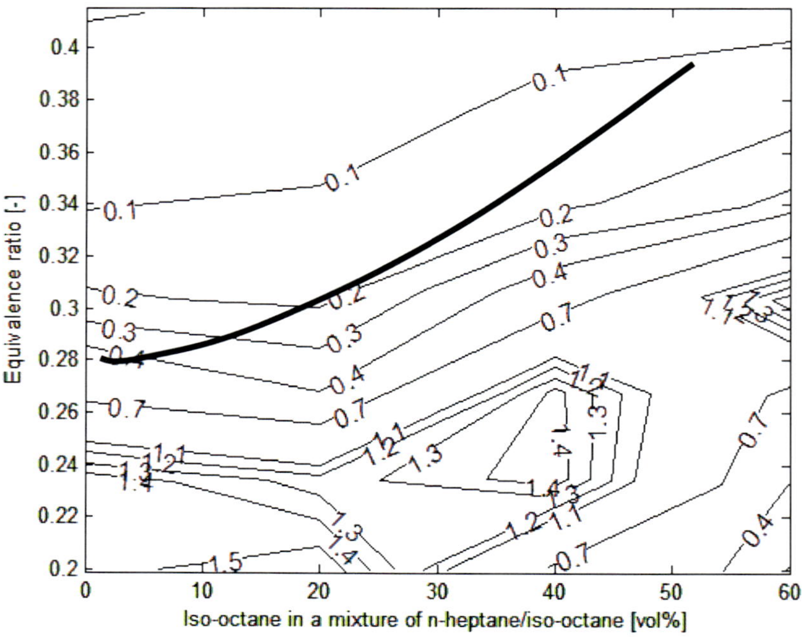

D, CO_2 emission

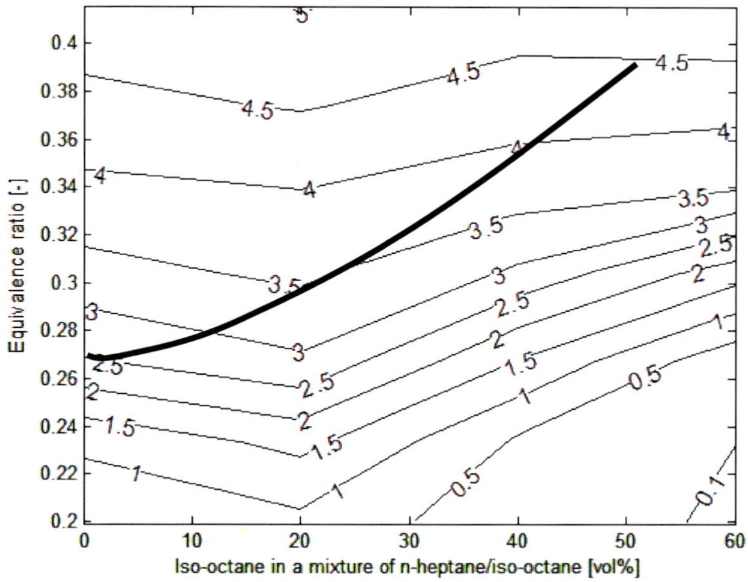

Figure 8: Iso-indicated efficiencies (A in %), hydrocarbon emissions (B in ppmv), CO emissions (C in vol%) and CO_2 emissions (D in vol%) as a function of the equivalence ratio and the fuel composition of n-heptane/iso-octane, for a compression ratio of 10, 2, an inlet temperature of 70°C.

A, Indicated efficiency

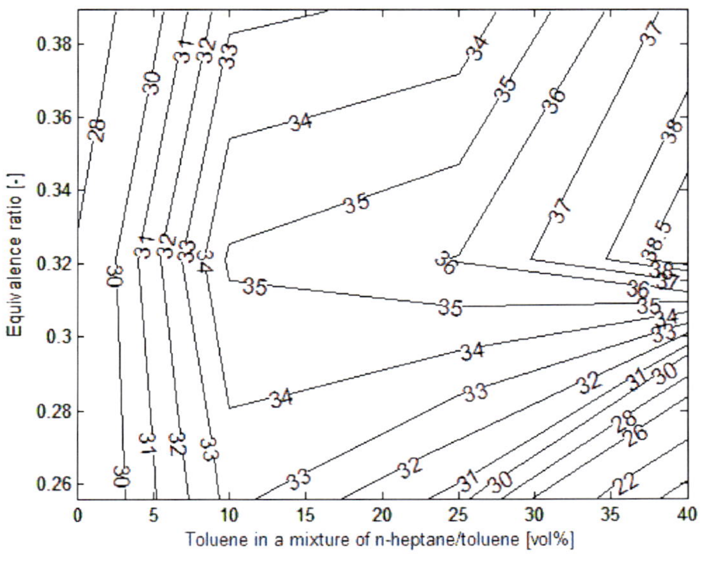

B, Hydrocarbon emission

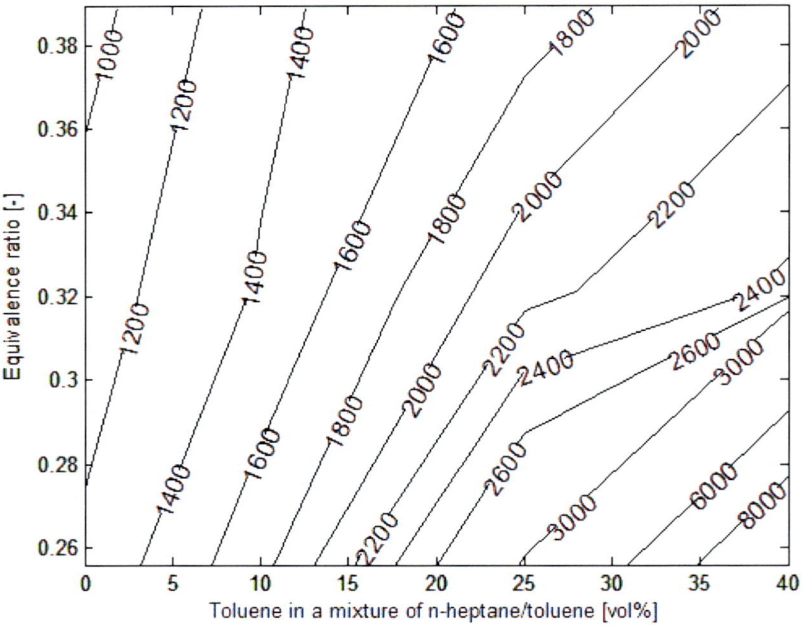

C, CO emission

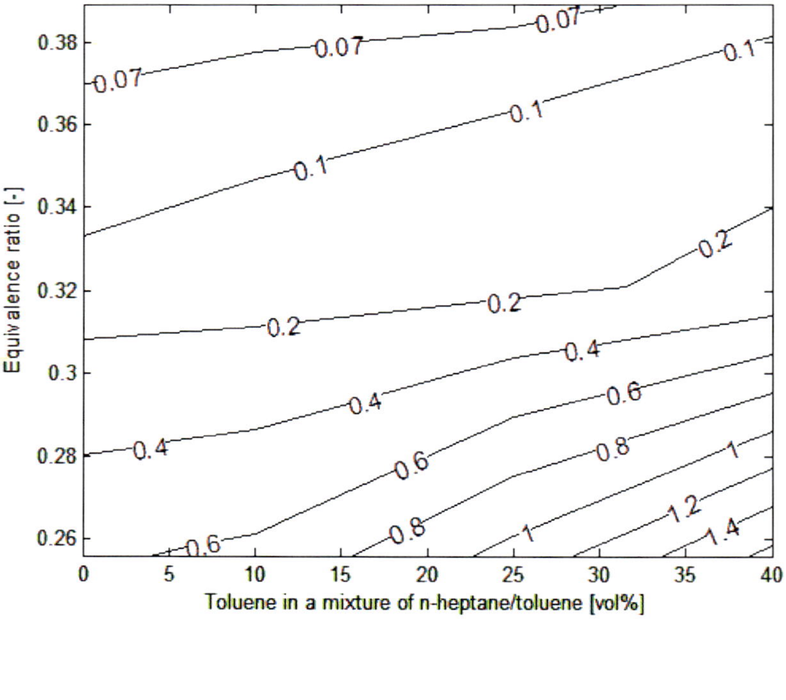

D, CO_2 emission

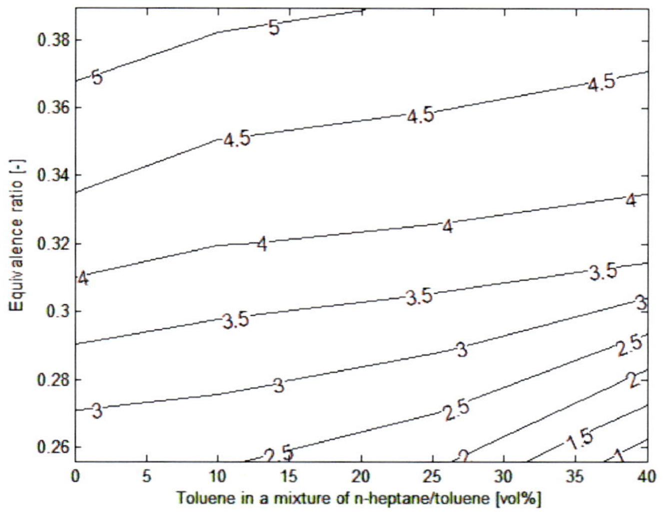

Figure 9: Iso-indicated efficiencies (A in %), hydrocarbon emissions (B in ppmv), CO emissions (C in vol%) and CO_2 emissions (D in vol%) as a function of the equivalence ratio and the fuel composition of n-heptane/toluene, for a compression ratio of 10, 2, an inlet temperature of 70°C.

First of all it seems, from Figs. **8** and **9**, that the indicated efficiency is very much influenced by the equivalence ratio. So this parameter has a narrower range for the control of the auto-ignition than the other parameters. The purpose is to have an equivalence ratio as low as possible for a low fuel consumption (being obviously also a function of the load), emissions that are as low as possible and an indicated efficiency as high as possible. It seems that a higher equivalence ratio is needed to have the same CO emissions for a fuel containing more iso-octane (Fig. **8**). The same goes for the hydrocarbon emissions. This implies that the fuel plays an important role on the choice of the equivalence ratio range. The highest indicated efficiency is indicated in Fig. **8** by a solid line. The same solid line is superposed over the emissions of the hydrocarbons, CO and CO_2. In order to have lowest equivalence ratio, for the same indicated efficiency, the fuel with 20 vol% iso-octane is the best. However, there, the CO emission is about 0, 4. When increasing the iso-octane content, following the solid line at iso-indicated efficiency, the CO_2 and hydrocarbon emissions do not increase until after an addition of 40 vol% of iso-octane, having, though, less CO emitted. Increasing furthermore, will increase the hydrocarbon emissions and especially that of CO_2.

Therefore, the most appropriate fuel seems to be "60 vol% n-heptane and 40 vol% iso-octane". There, the equivalence ratio is somewhat higher, but the sensitivity of the indicated efficiency due to a change in one of the parameters is much less at these conditions, which is a compromise. A same discussion can be lead for the fuel consumption of n-heptane/toluene. For this fuel, the maximum indicated efficiency is found to be 0, 32, at any fuel composition, the highest being for 40 vol% toluene. However, for this fuel composition, the emissions of CO and the hydrocarbons are at their highest. The indicated efficiency is, thus constantly in battle with the hydrocarbon and CO emissions. The slopes do not seem to change at any range of the fuel composition of n-heptane/toluene. Since the HCCI engine is not optimised, it is not possible to attribute a certain maximum allowable emission to designate the fuel composition. Furthermore, the fuel containing 40 vol% toluene, presents more knocking characterisitics that the fuel containing 40 vol% iso-octane. This makes one prefer to use the fuel "60 vol% n-heptane and 40 vol% iso-octane" (PRF40), for the conditions in this study. This choice is not definitive, but is issued from the presented analysis and could be subject to discussion. But for the sake of the discussion about the control of the auto-ignition, the PRF40 will be used for the rest of this section. The parameter that is left to discuss in this section is the inlet temperature. This is presented in Fig. **10**.

A, Indicated efficiency

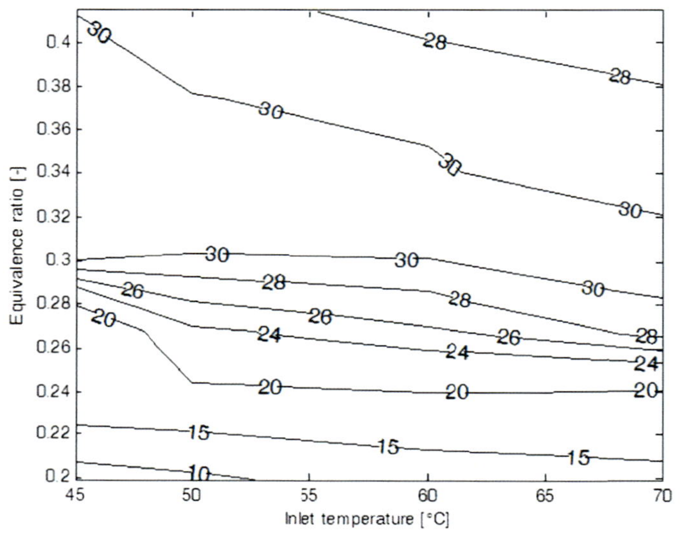

B, Hydrocarbon emission

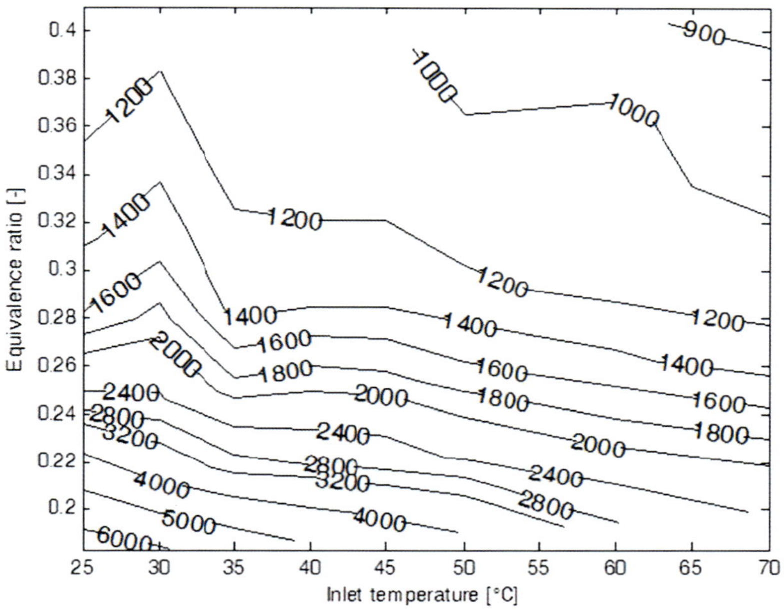

C, CO emission

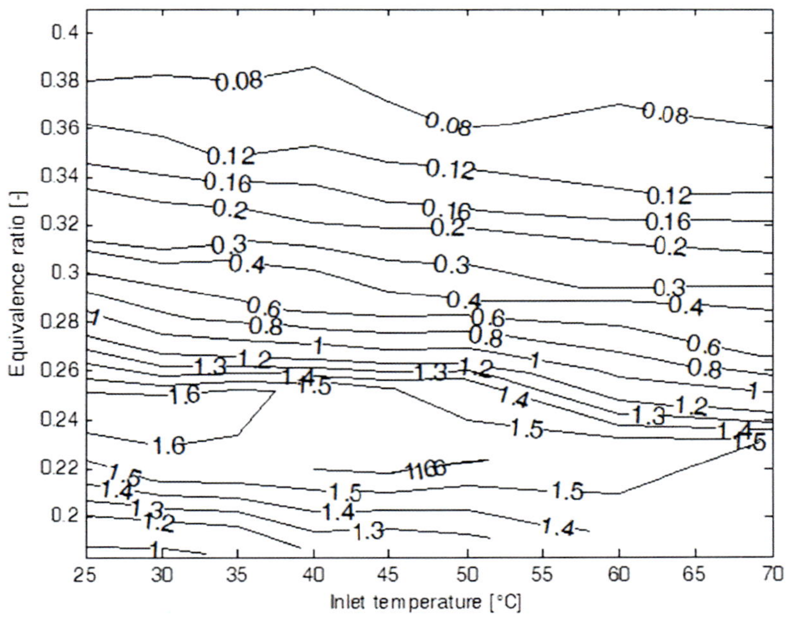

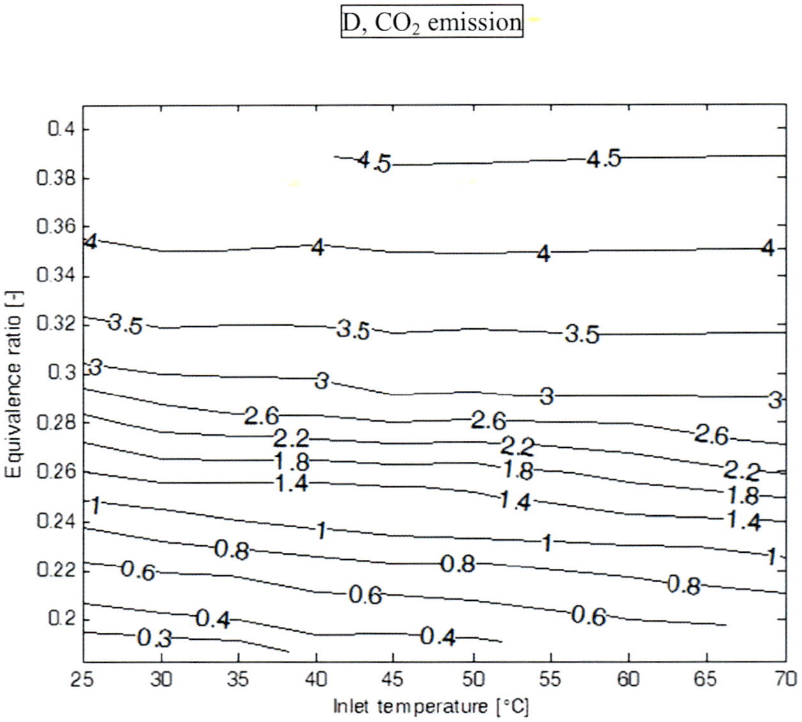

Figure 10: Iso-indicated efficiencies (A in %), hydrocarbon emissions (B in ppmv), CO emissions (C in vol%) and CO_2 emissions (D in vol%) as a function of the equivalence ratio and the inlet temperature, for a compression ratio of 10, 2 and n-heptane as the fuel.

Fig. **10** shows that the inlet temperature does not influence the emissions of CO and CO_2 significantly nor does it affect much the indicated efficiency and the emissions of the hydrocarbons, though a slight decrease is observed for the emission when increasing the inlet temperature. For n-heptane, the optimal efficiency seems to be attained at almost all the inlet temperatures that were investigated. So the inlet temperature seems to be a useful parameter, since, it influences the ignition delays, without seeming to influence much the emissions and neither the indicated efficiency. However, in practice in an engine the inlet temperature will technically speaking be difficult to control, due to its inertia. For this purpose the inlet temperature is not a good option to control the auto-ignition process and no further analysis will be performed concerning the sensitivity of the ignition delays due to a change in the inlet temperature. To minimize the emissions, the maximum inlet temperature will be used. A maximum inlet temperature, that is realistic for an engine is chosen to be 70 °C.

It is not needless to say that the optimal conditions, concerning the indicated efficiency and the emissions could also be attained by choosing the optimal values for other parameters such as the quantity of EGR, its temperature or by the chemically active species that are present in EGR. Nevertheless, the inlet temperature, the compression ratio and the fuel are kept constant at their optimal values, that result into an optimal indicated efficiency and thus a relatively optimal functioning of the engine being the following: an inlet temperature of 70 °C, a fuel composition of "60 vol% n-heptane/40 vol% iso-octane" and a compression ratio of around 10, 2. This, in order to isolate the effects of the EGR parameters and the equivalence ratio and their contribution to control the auto-ignition process. Note that the equivalence can be changed, be it at least 0, 32, but this would increase the fuel consumption. Nonetheless, if the equivalence ratio appears to be a good option for controlling the auto-ignition, by only changing it slightly it still can be used for this purpose.

To control the auto-ignition, other parameters that nearly do not effect the indicated efficiency but change easily the auto-ignition delays should be investigated. From the parameters that were investigated in this work, the most interesting that are left for discussion are the effects due to the equivalence ratio (being a major parameter) and the EGR, constituting of dilution effects (N_2 and CO_2) and chemical effects (NO, CH_2O and CH_3CHO).

DETERMINATION OF THE MOST USEFUL PARAMETERS OF TYPE 2 FOR THE AUTO-IGNITION PROCESS

The options for the control of the auto-ignition that are left in this work are the equivalence ratio, N_2, CO_2, NO and CH_2O. The influence of these parameters on the indicated efficiency and the emissions of CO, CO_2 and the hydrocarbons is studied in this section. Also, whenever the sensitivity of the indicated efficiency and the emissions is low with respect to the parameter, this influence of this parameter on the auto-ignition delays can be looked in. When the latter is strong, the parameter is useful to be used for the control of the auto-ignition process. Fig. **11** presents the results as a function of dilution by N_2 and the equivalence ratio.

A, Indicated efficiency

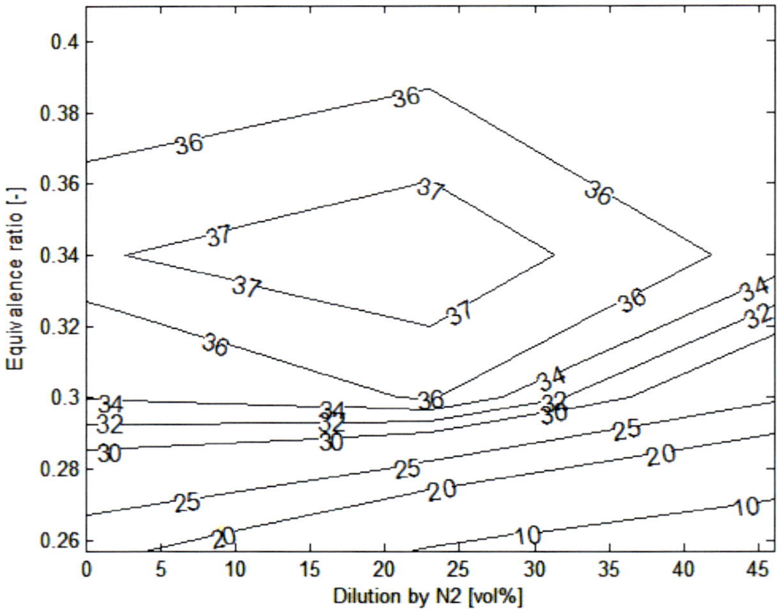

B, Hydrocarbon emission

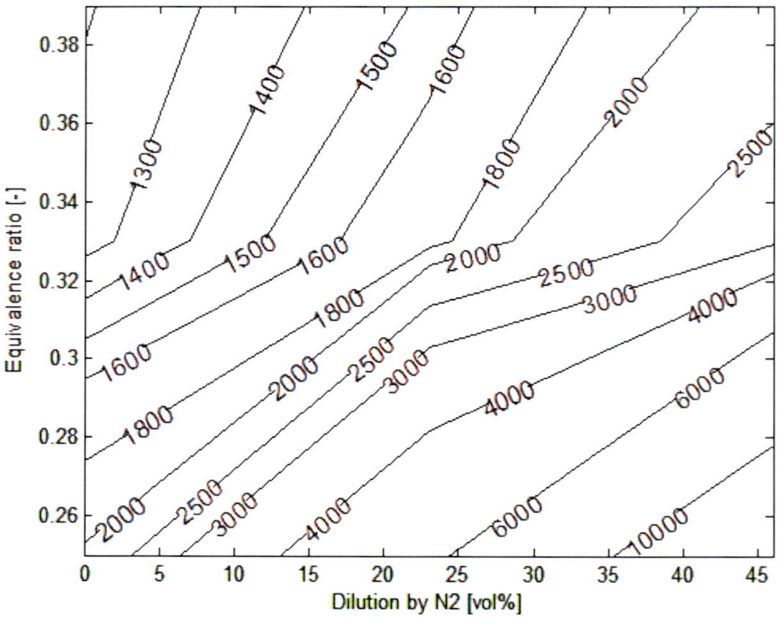

C, CO emission

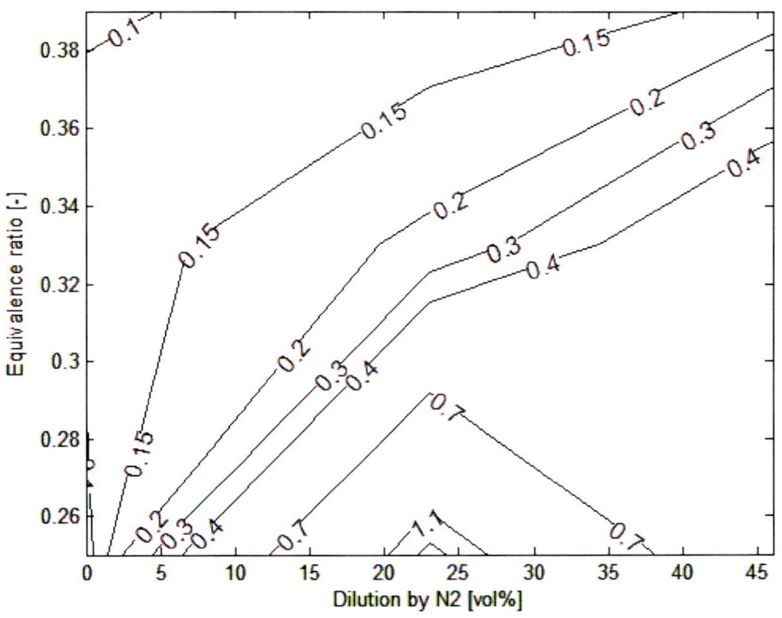

D, CO_2 emission

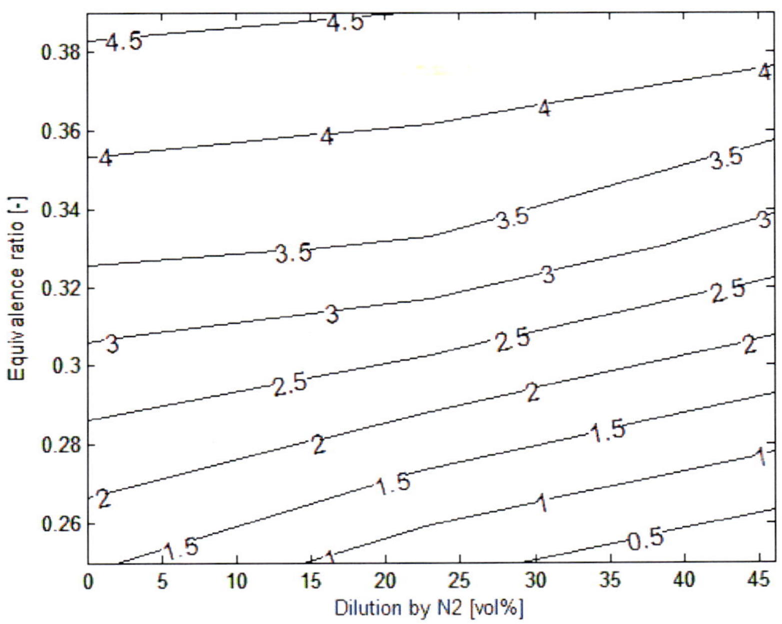

E, Cool flame delay

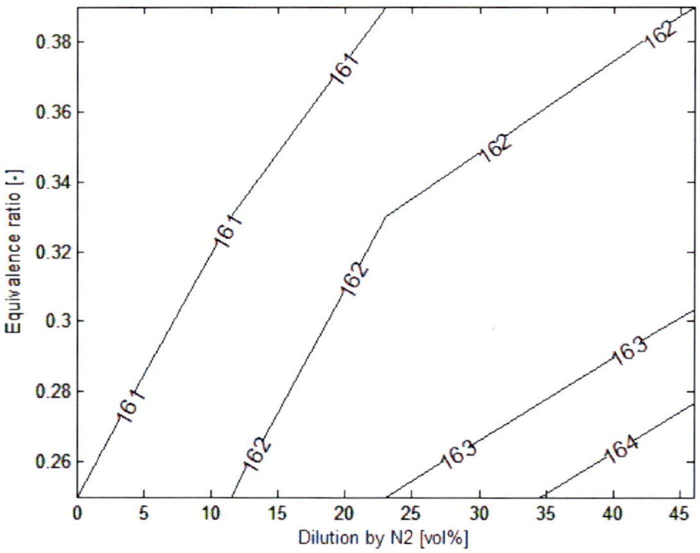

F, Final ignition delay

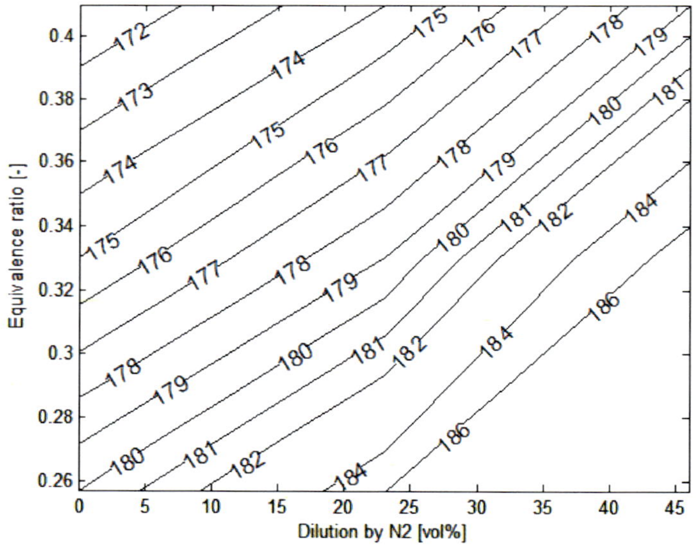

Figure 11: Iso-indicated efficiencies (A in %), hydrocarbon emissions (B in ppmv), CO emissions (C in vol%), CO_2 emissions (D in vol%), cool flame delay (E in CAD) and final ignition delay (F in CAD) as a function of the equivalence ratio and the addition of N_2 to the inlet mixture, for a compression ratio of 10, 2, an inlet temperature of 70 °C and "60vol% n-heptane/40vol% iso-octane" as the fuel.

Fig. **11** shows that the addition of nitrogen in the form of EGR changes but slightly the indicated efficiency at equivalence ratios higher than 0, 32. The change in the slopes for the hydrocarbons, for CO and CO_2 are not significant for

A, Indicated efficiency

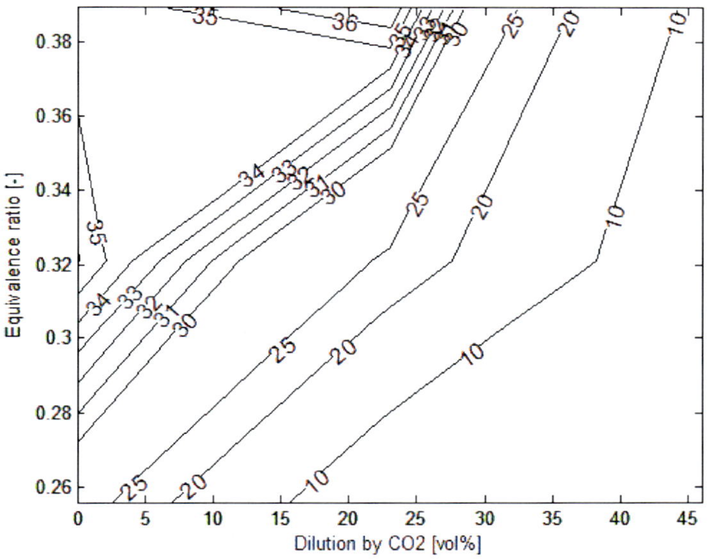

B, Hydrocarbon emission

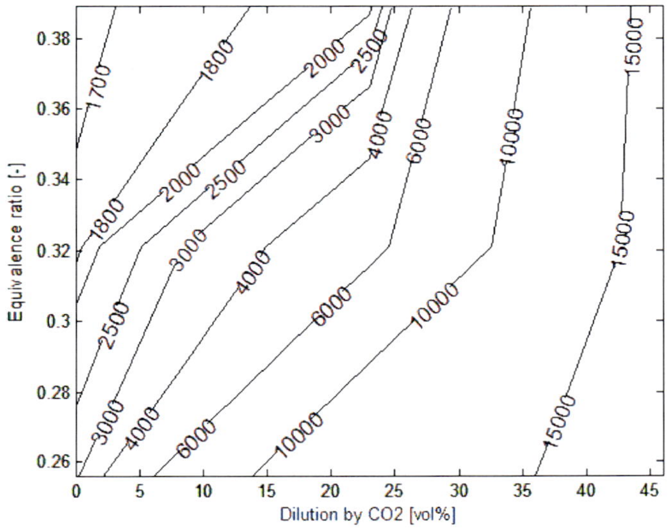

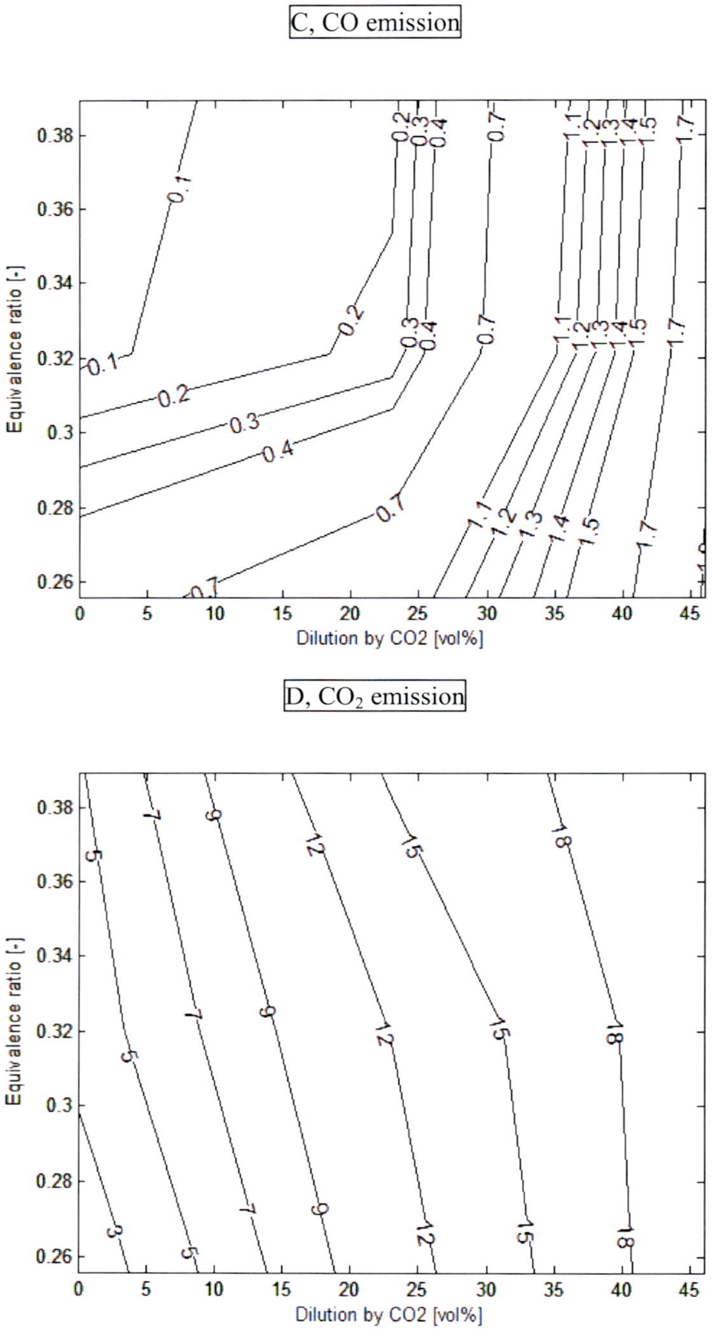

Figure 12: Iso-indicated efficiencies (A in %), hydrocarbon emissions (B in ppmv), CO emissions (C in vol%), CO_2 emissions (D in vol) as a function of the equivalence ratio and the addition of CO_2 to the inlet mixture, for a compression ratio of 10, 2, an inlet temperature of 70 °C and "60vol% n-heptane/40vol% iso-octane" as the fuel.

the sensitivity of their emission, at equivalence ratios higher than 0, 32. When looking over the whole range of N_2 dilution, at an equivalence ratio of 0, 32, it appears that the cool flame delay increases about 2 CAD and the final ignition delay about 4, 5 CAD, which is quite important. This makes the dilution by N_2 interesting to be used for the auto-ignition process. Another diluting species is investigated, CO_2, of which the results are presented in Fig. **12**.

Fig. **12** shows that the range for an acceptable indicated efficiency is rather limited. For an equivalence ratio higher than 0, 32, the CO_2 dilution factor should not exceed 13 vol% of the inlet mixture. The sensitivity of the indicated efficiency is rather important with respect to N_2. Since EGR ratios can go up to 50 vol% and N_2 shows a pronounced less influence on the indicated efficiency and the emissions of CO and the hydrocarbons, the dilution by CO_2 has less interest of further analysis. Not the least, the emission of CO_2 is rather high. Since the sensitivity on the emission is quite high for CO_2, it is not interesting to check the auto-ignition delays. Remain the results corresponding to the chemical species in EGR, which seem to be most interesting. The first one to discuss is the chemical species NO, of which the results are shown in Fig. **13**. Fig. **13** shows that the

A, Indicated efficiency

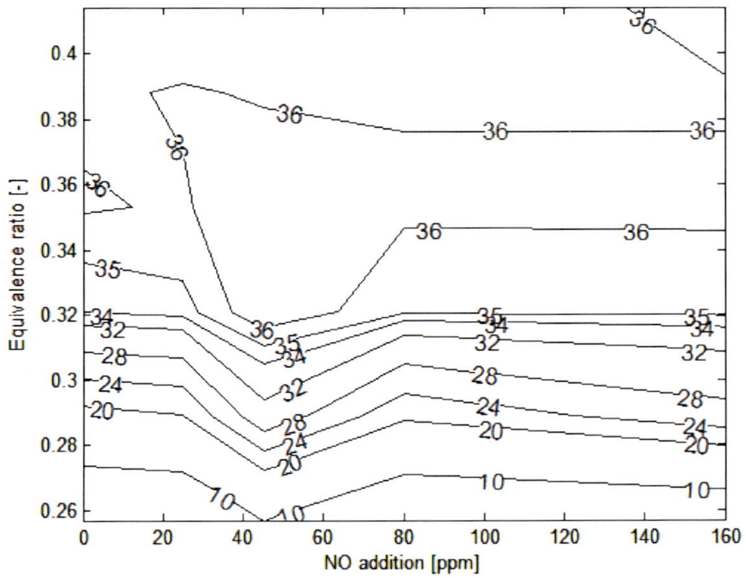

B, Hydrocarbon emission

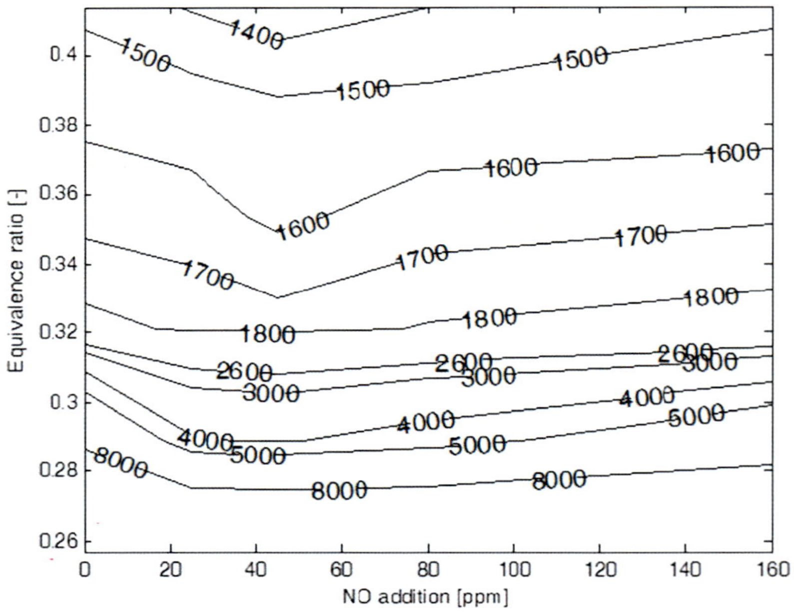

C, CO emission

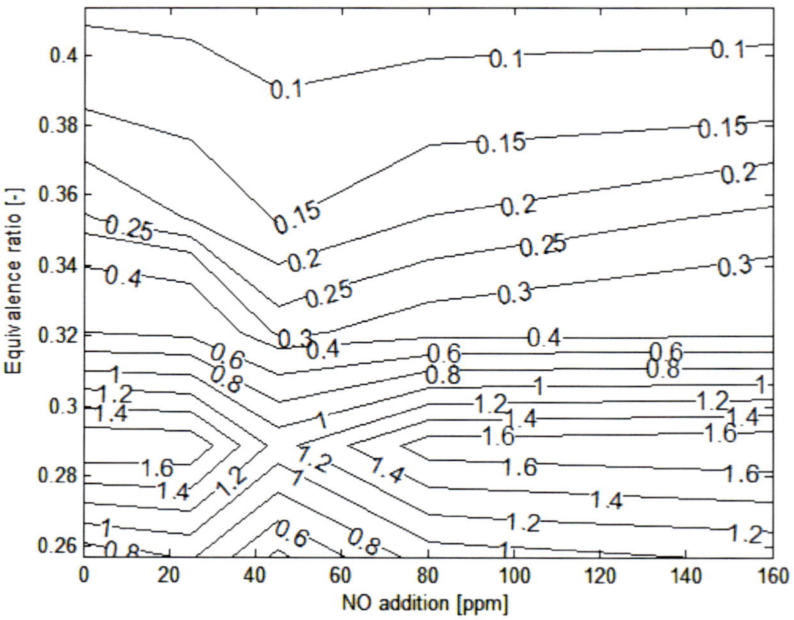

D, CO_2 emission

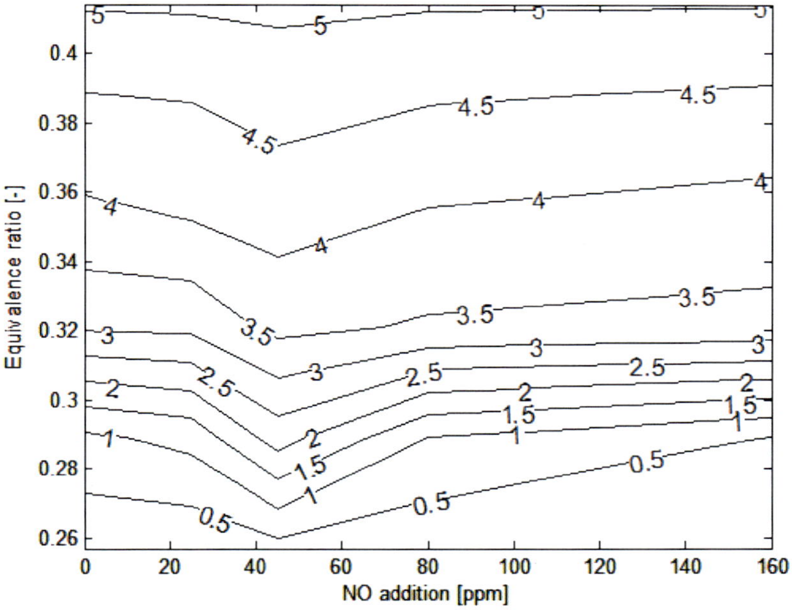

E, Cool flame delay

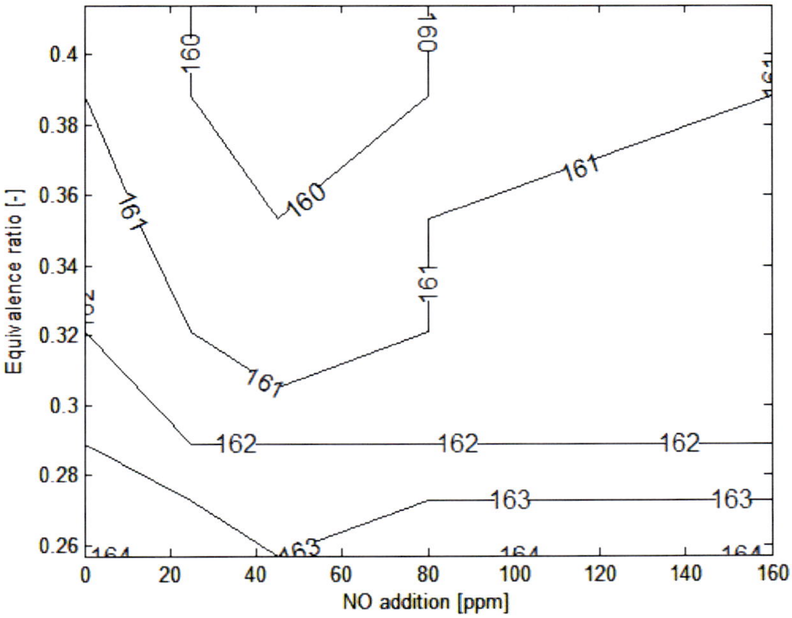

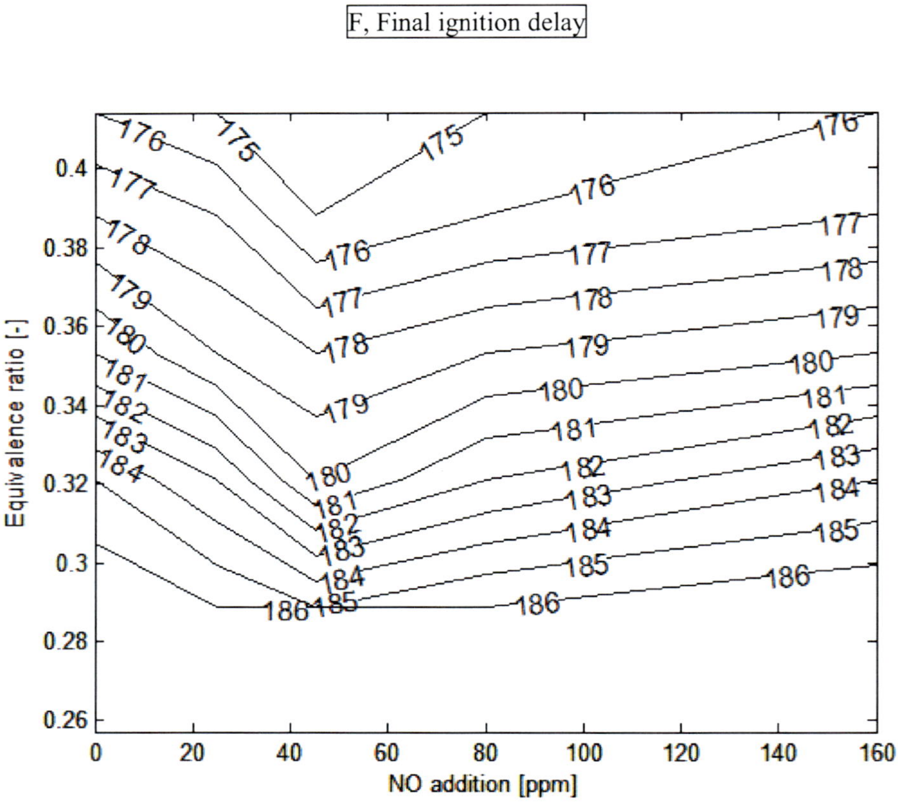

Figure 13: Iso-indicated efficiencies (A in %), hydrocarbon emissions (B in ppmv), CO emissions (C in vol%), CO_2 emissions (D in vol%), cool flame delay (E in CAD) and final ignition delay (F in CAD) as a function of the equivalence ratio and the addition of NO to the inlet mixture, for a compression ratio of 10, 2, an inlet temperature of 70 °C and "60vol% n-heptane/40vol% iso-octane" as the fuel and at a EGR ratio of 23 vol%.

addition of NO in the form of EGR changes but slightly the indicated efficiency nor the emissions of CO, CO_2 and the hydrocarbons. This means that NO can be fully used for auto-ignition control, without changing significantly the indicated efficiency or any of the treated emissions. During the experiments, the addition of NO resulted into a maximum NO content in the exhaust gas of 80 ppmv. So NO seems interesting for further investigation at the whole range, where it can be seen that the cool flame stays nearly constant, while the final ignition delay changes at an equivalence ratio of 0, 32 almost 5 CAD on adding only 45 ppm of NO. This makes the species NO a very interesting candidated for the auto-ignition control.

The next chemical species is formaldehyde (CH_2O), presented in Fig. **14**.

A, Indicated efficiency

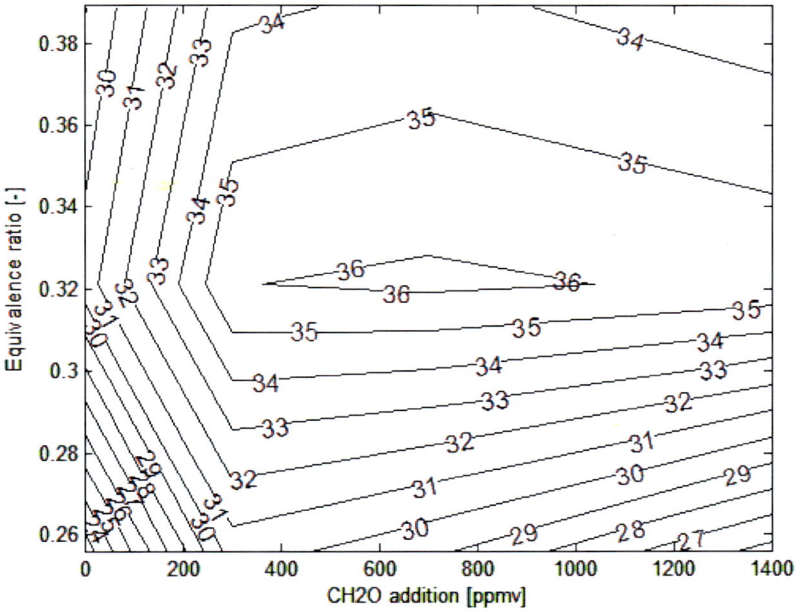

B, Hydrocarbon emission

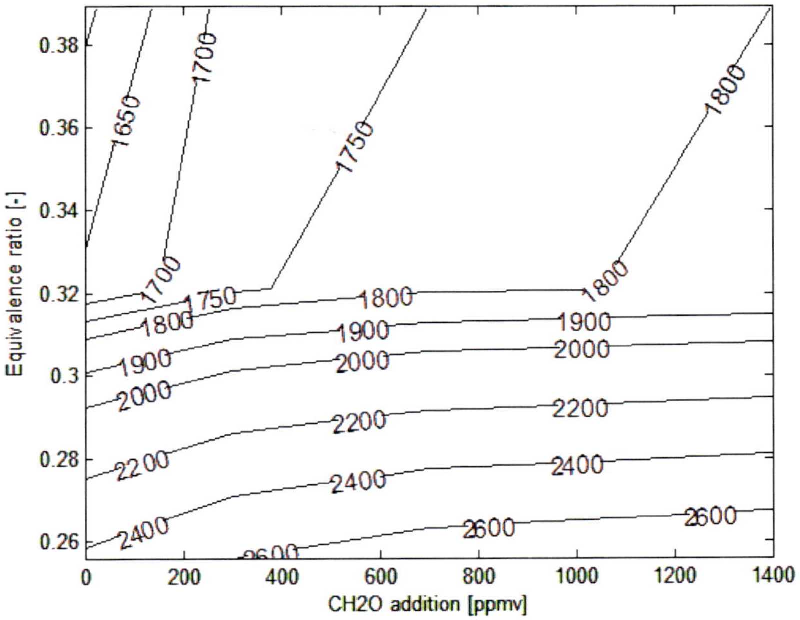

C, CO emission

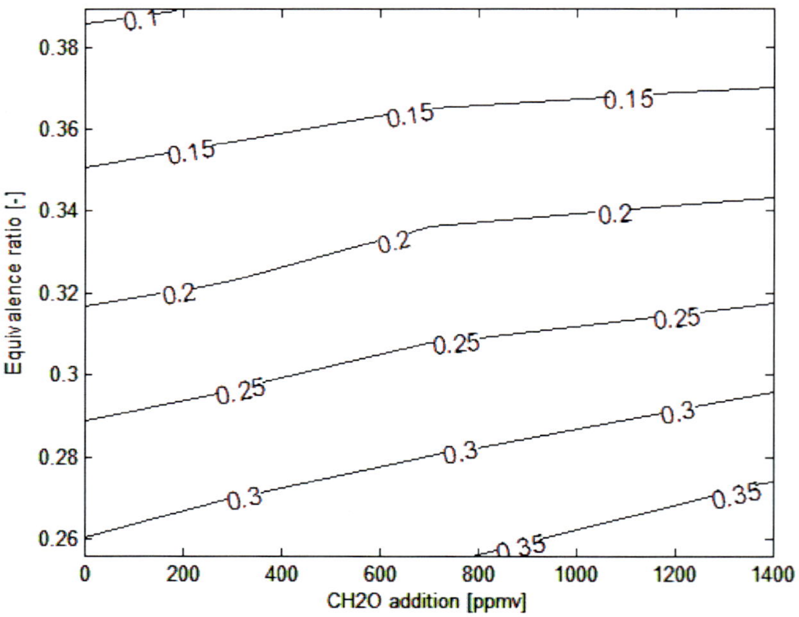

D, CO_2 emission

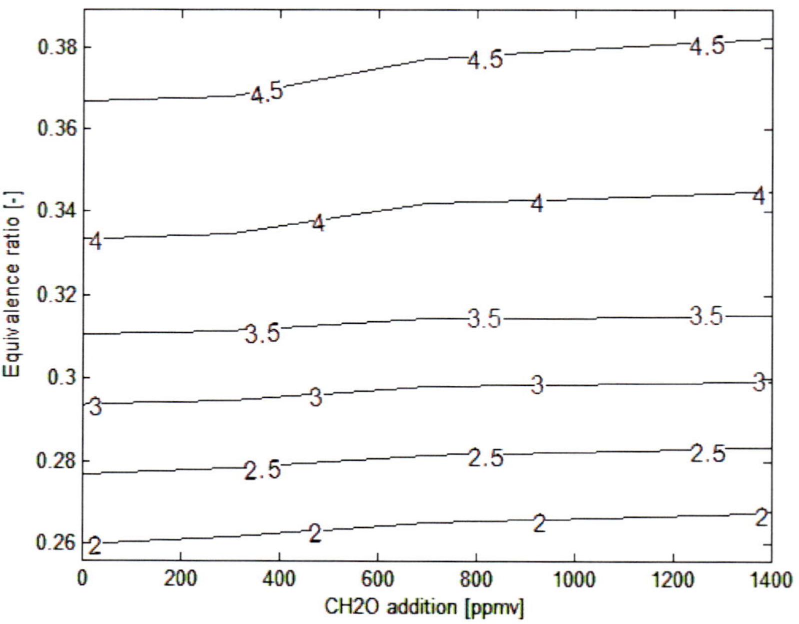

E, Cool flame delay

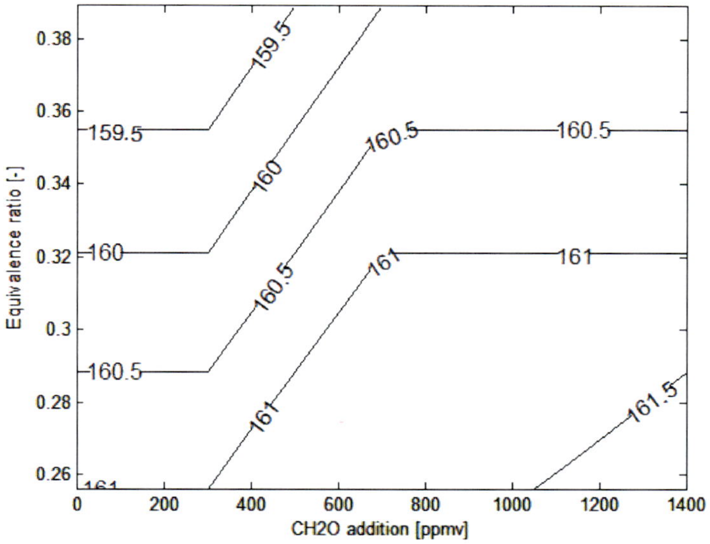

F, Final ignition delay

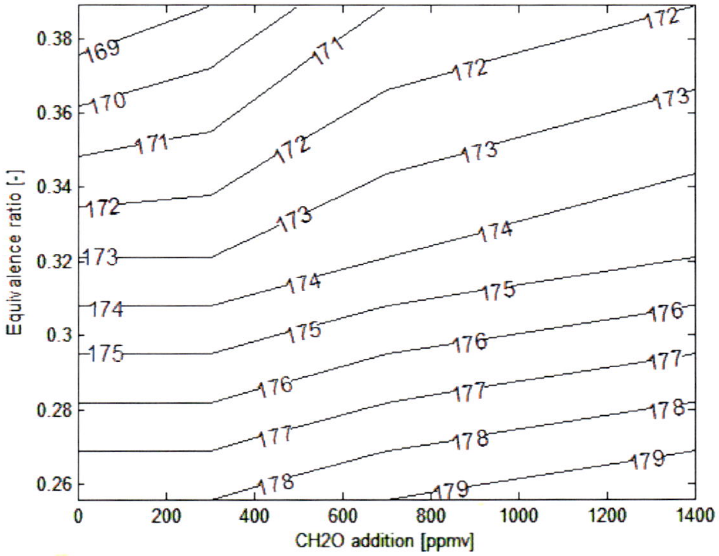

Figure 14: Iso-indicated efficiencies (A in %), hydrocarbon emissions (B in ppmv), CO emissions (C in vol%), CO_2 emissions (D in vol%), cool flame delay (E in CAD) and final ignition delay (F in CAD) as a function of the equivalence ratio and the addition of CH_2O to the inlet mixture, for a compression ratio of 10, 2, an inlet temperature of 70 °C and "60vol% n-heptane/40vol% iso-octane" as the fuel.

Fig. **14** shows that formaldehyde addition in this range does hardly effect the indicated efficiency. An indicated efficiency of 30 % and higher is obtained at all ranges for an equivalence ratio higher than 0, 32, the equivalence ratio at which the indicated efficiency is highest for all ranges of formaldehyde additions. As far as it concerns the hydrocarbon emissions, a strong difference in the slope is observed, though the difference between each contour is about 3 %. The sensitivity of CO and CO_2 to the change in the addition of formaldehyde is weak. Therefore the addition of formaldehyde seems not to evoke much difference in the indicated efficiency and the emissions. The cool flame delay changes only 1 CAD, while the final ignition delay, seem to change 2 CAD over the whole range. It is true that 2 CAD is not negligible, but when one sees that it needs 45 ppm of NO to invoke a change of 5 CAD in the final ignition and it needs 1400 ppm of CH_2O to invoke a change of 2 CAD, it is apparent that NO is more interesting than CH_2O. Furthermore, NO has the ability to both promote and delay the auto-ignition process, while formaldehyde only can perform the latter effect.

It is noticed that the most effective parameters, resulting into the lowest sensitivity for the indicated efficiency and emissions and the highest sensitivity for the auto-ignition delays, seem to be N_2 and NO or rather the combination of these two in EGR. Since the effectiveness of these parameters is intrinsically linked with the equivalence ratio, the equivalence ratio stays an important parameter to deal with and is combined with the parameters N_2 and NO.

CONCLUSIONS AND FINAL DISCUSSION

In this chapter the usefulness of the parameters that were discussed in order to be appropriate to control the auto-ignition process are determined. Some parameters were chosen to be constant representing an optimal value that resulted into an optimal indicated efficiency and relatively low emission values. It appeared that the optimal values for the compression ratio, fuel composition and inlet temperature, at the conditions of this work, were respectively 10, 2, the fuel composition of "60 vol% n-heptane/40 vol% iso-octane" and 70 °C. As far as it concerns the equivalence ratio, it appeared that it can be changed, as long as it is at least 0, 32. A minimum value for the equivalence ratio is requested to minimize the fuel consumption. However, the equivalence ratio should be high enough to ensure the requested engine power. The equivalence ratio appeared to have a double face. On one hand, it should be chosen thus that the engine functions optimally and on the other hand, it appears to be a good option for controlling the auto-ignition. The most interesting parameters that are interesting for the control of the auto-ignition seem to be N_2, NO and the equivalence ratio. The first represents a dilution factor and changes the gas temperature, having mainly a physical effect. The second changes the preferences for certain reaction pathways, by interacting strongly with the intermediate species, during the auto-ignition. The last parameter determines largely the energy supply and the fuel consumption. This means that one can control the power and the emission in the engine by primarily the equivalence ratio, while controlling the acceleration or deceleration by EGR, allowing only little change in the indicated efficiency and the emissions. This acceleration or deceleration is a result of the change in the final ignition delay, which is the most important factor to control. An example is shown in Fig. **15**, where the indicated efficiency and the final ignition delay are shown as a function of the equivalence ratio and the addition of NO at a constant dilution by N_2 of 23 vol%. The same results are presented for the influence of the dilution by N_2 in the same figure. In both cases the compression ratio is held at 10, 2, the inlet temperature at 70 °C and the fuel "60 vol% n-heptane/40 vol% iso-octane" is used.

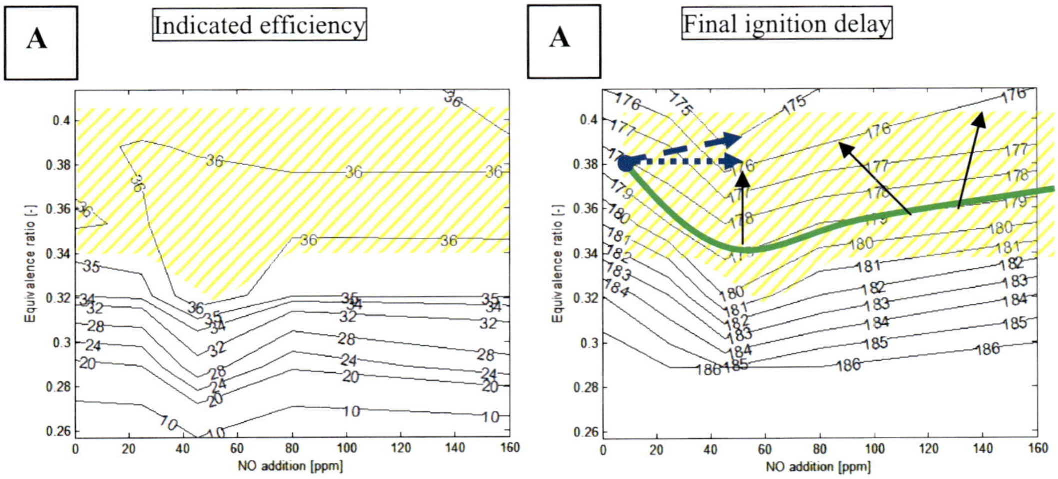

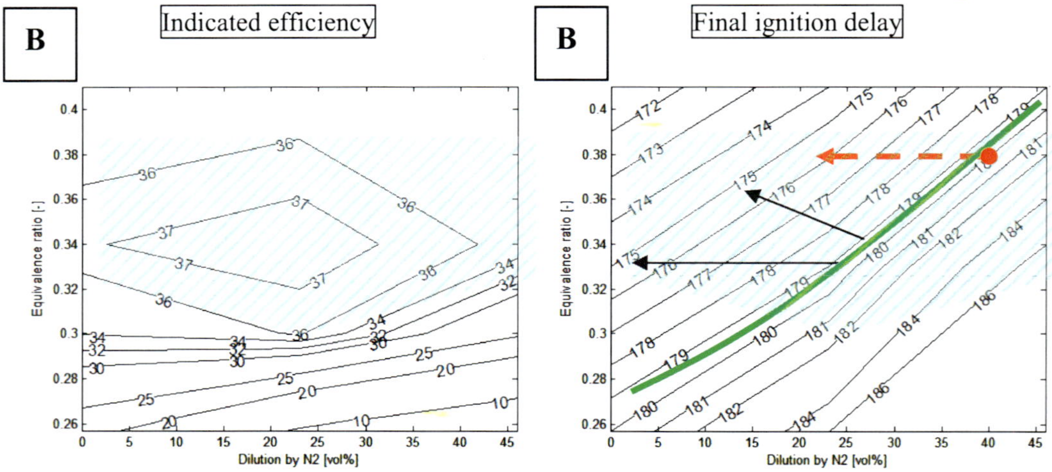

Figure 15: Iso-indicated efficiencies (left, in %) and final ignition delays (right, in CAD) as a function of the influence of respectively: **A)** the addition of NO (in ppm and at an EGR ratio of 23 vol%) and the equivalence ratio and **B)** the dilution by N_2 (in vol%) and the equivalence ratio. The fixed conditions are a compression ratio of 10, 2, an inlet temperature of 70 °C and "60vol% n-heptane/40vol% iso-octane" as the fuel, with an engine operating at 600 rpm.

Fig. **15** does not present a concrete way to control, but proposes a possible way of thinking to control. Therefore, this figure is merely demonstrative for the sake of the discussion. In Fig. **15A**, it can be seen that the highest indicated efficiency (36 %), is situated in the equivalence ratio range between 0, 34 and 0, 4 for all NO additions, with an exception of an NO addition of 45 ppm, where also an equivalence ratio of 0, 32 results into the same indicated efficiency (indicated in yellow in Fig. **15A**). The same indicated efficiency (at least 36 %) in Fig. **15B** is attained at equivalence ratios between 0, 34 and 0, 39, with the same exception at dilution around 23 vol% by N_2, where an equivalence ratio of 0, 3 is possible as well (indicated in green in Fig. **15B**). As is shown in the previous sections, at these indicated efficiencies, the emissions were at their lowest. A logical pathway to control could be that one wants a certain maximum indicated efficiency, say 36 % as shown previously. In this example, both the addition of NO and the dilution by N_2 will be examined along with the equivalence ratio as is presented in Fig. **15.** Say an engine operates at 600 rpm, with a final ignition delay of 179 CAD. This is indicated by the green solid lines in Fig. **15.** The same operating ranges for the indicated efficiencies are superposed as well on the final ignition maps. To change the engine's speed, maintaining the same indicated efficiency, the final ignition delay can be changed by changing the parameters that influence the interaction of the chemical species during the auto-ignition process. In this case, this can be the addition of NO (Fig. **15A**) or the dilution by N_2 (Fig. **15B**) accompanied by the equivalence ratio. Initial operating conditions could then be an equivalence ratio of 0, 38 and no addition of NO in case A (indicated by a blue dot) or an equivalence ratio of 0, 38 and a dilution of 37 vol% N_2 in case B (indicated by a red dot). Say that the final ignition delay should be changed by 4 CAD towards 175 CAD. The equivalence ratio should not change much in order not to influence much the power for the engine or the emissions. These values are not given in this case, since it is the way thinking that is of interest in this section. The respectively blue and red dashed arrows indicate a possible change, resulting into an equivalence ratio of 0, 39 with an NO addition of 45 ppm for case A and an equivalence ratio of 0, 38 with a dilution by N_2 of 17 vol% for case B. In case A, the equivalence ratio had to be changed in order to realise the 4 CAD change, since without the intervention of the equivalence ratio, the maximum effect NO could have on the final ignition delay, at a constant equivalence ratio of 0, 38, would be 3 CAD, indicated by the blue pointed arrow in Fig. **15A**. Of course more possibilities are present as indicated by the various arrows in both case A and B, all within the high indicated efficiencies. This means that the "ideal" pathway to choose depends on other factors as the emission, fuel consumption, power output, to mention a few. This restricts also the change in the equivalence ratio, since a minimum is needed for the power output and a maximum is obliged to reduce fuel consumption and CO_2 formation. These factors are part of a engine optimisation study, which is not the subject of this study, but is a factor that is necessary to attain a proper control of the auto-ignition process.

Another crucial issue plays an important role when one uses this way of thinking about controlling the auto-ignition. The objective of changing the final ignition is obviously to change the engine speed. However, the engine speed can have a great and complex influence on the auto-ignition characteristics. In a control point of view one would like to have the same pressure curve, power, indicated efficiency, for instance, when changing the engine speed. Taking the case of the gasoline surrogate, one may take a point of view of choosing the initial conditions at, for instance, an engine speed of 600 rpm, such that at higher engine speeds, the same ignition delay, pressure and heat release would be obtained. An example of a possible idea to apply this thought, would be the following. Take at 600 rpm an auto-ignition of gasoline surrogate at 40 vol% EGR (N_2), a compression ratio of 18, an inlet temperature of 70 °C and an equivalence ratio of 0, 6. Pressure curve number 1 in Fig. **16** would be obtained. When, as put out previously, the conditions change such (higher equivalence ratio or change in addition of NO) that the engine speed is 1500 rpm. If the same pressure curve is to be obtained for the same conditions at an engine speed of 1500 rpm, the EGR ratio should be reduced to 15 vol% and pressure curve number 2 is observed. Doing the same to increase the engine speed up to 2200 rpm, would result into an EGR ratio of 5 vol%, obtaining pressure curve number 3.

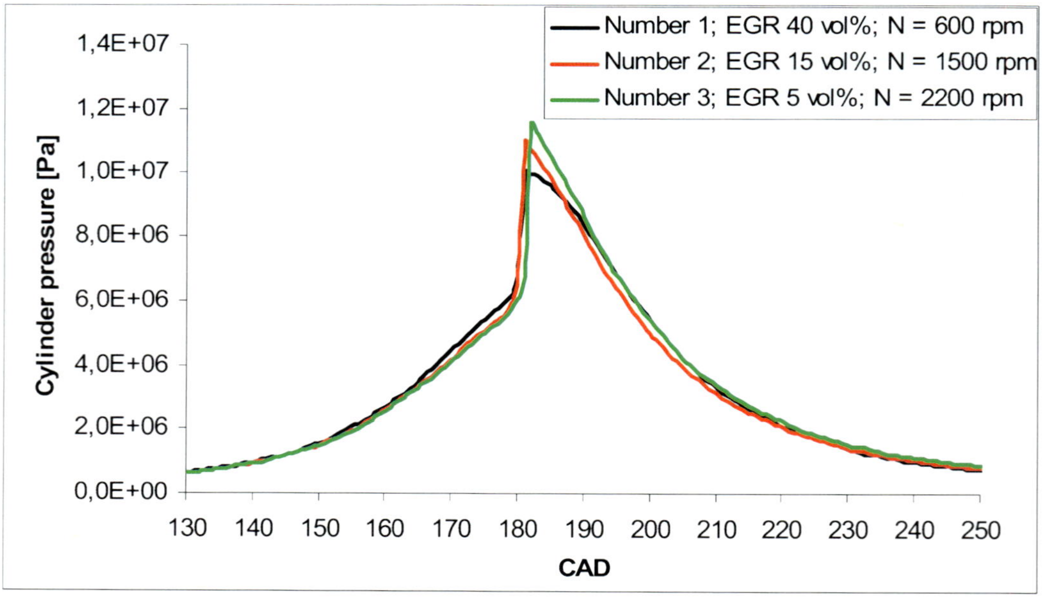

Figure 16: Example of changing the initial conditions to maintain the same auto-ignition characteristics at different engine speeds.

Fig. **16** presents a rather qualitative result. To adjust for the maximum pressure, heat release and other entities, in order to have the same pressure profile, some other parameters could be changed a bit, as for instance, the equivalence ratio. However, to arrive to such a control, one must understand the influence of the engine speed on the auto-ignition process. The effect of the engine speed can play a very important role in the auto-ignition process. In this work, no experimental investigation is done on the effect of the engine speed on the auto-ignition process. Nonetheless, since the surrogate mechanism, presented in [10] and [11] is used for the analysis of the auto-ignition process of several fuels, it is interesting to use it to provide for an estimation of the effect of the engine speed, lacking experimental data. For this purpose, two fuels with different burn rates and octane numbers are chosen for comparison: PRF40 (discussed previously in this chapter) and the gasoline surrogate used in this work. Fig. **17** shows an example where the cylinder pressure is presented for the fuels PRF40 and the gasoline surrogate, varying the engine speed.

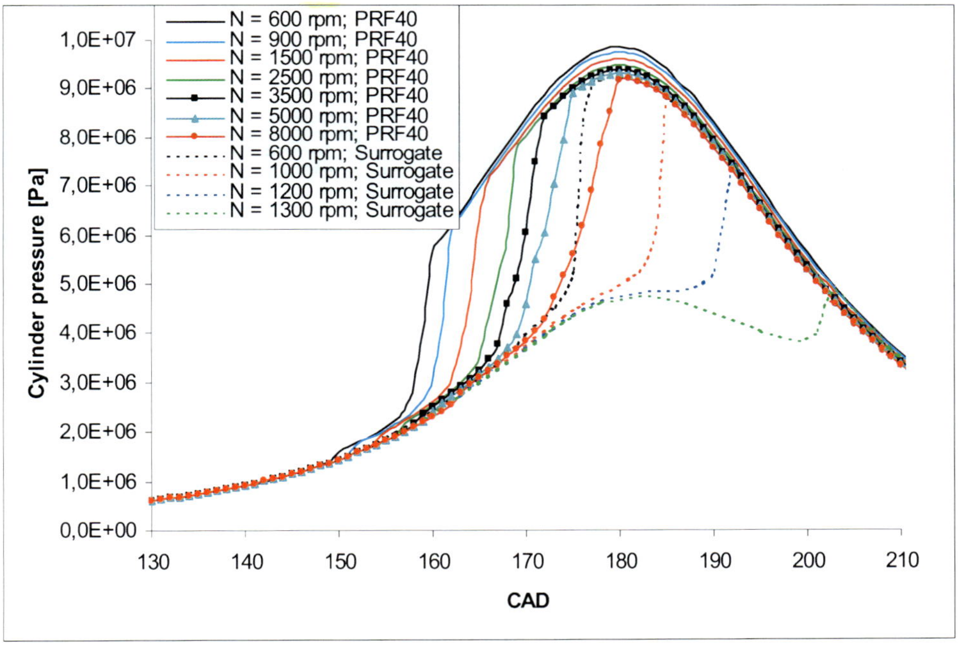

Figure 17: Influence engine speed (N) on the pressure profile for PRF40 and the (gasoline) surrogate at a compression ratio of 16, an equivalence ratio of 0, 4 and an inlet temperature of 385 K.

Fig. **17** clearly shows that the pressure profile changes considerably, delaying thereby the cool flame and the final ignition delay. This change is stronlgy dependent on the fuel. So, for the investigation of any control strategie, a fuel strategie research is important. It appears that the effects of engine speed on the auto-ignition process are a very complex phenomenon, which is directly involved with the control of the auto-ignition. As such this is subject to further investigation in a control-oriented research with experimental studies.

REFERENCES

[1] J.B. Heywood, "Fundamentals of Internal combustion", McGrawHill, 1988.

[2] C.H. Van Aken, "Zuigermotoren", *Faculteit der Toegepaste Wetenschappen, Universiteit Gent*, Belgium, 1996.

[3] R. Ogink, "Computer Modeling of HCCI Combustion", *PhD thesis: Chalmers University of Technology*, Sweden 2004.

[4] "Der Unterschied: Otto, Diesel, HCCI: Der HCCI Motor - Aus Zwei mach Eins: Unsere Motoren können alleine atmen": *#203478502 (http://www.motor-talk.de/bilder/der-hcci-motor-aus-zwei-mach-eins-g22780598/der-unterschied-otto-diesel-hcci-i203478502.html)*, accessed 2011.

[5] Y. Huang, C.J. Sung, J.A. Eng, "Dilution limits of n-butane/air mixtures under conditions relevant to HCCI combustion", *Combust. Flame*, vol. 136, pp. 457-466, 2004.

[6] B.M. Gauthier, D.F. Davidson, R.K. Hanson, "Shock tube determination of ignition delay times in full-blend and surrogate fuel mixtures", *Combust. Flame*, vol. 139, pp. 300-311, 2004.

[7] R. Aneja, B. Bolton, N. Hakim, Z. Pavlova-Mackinnon, "Attaining Tier 2 Emissions Through Diesel Engine and Aftertreatment Integration – strategy and experimental results", *Eighth Diesel Engine Emissions Reduction Conference* – Detroit, 2002.

[8] M. Christensen, B. Johansson, "Supercharged Homogeneous Charge Compression Ignition (HCCI) with Exhaust Gas Recirculation and Pilot Fuel", *SAE* 2000-01-1835 , 2000.

[9] J. Hyvonen, G. Haraldsson, B. Johansson, "Operating Range in a Multi Cylinder HCCI Engine Using Variable Compression Ratio", *SAE* 2003-01-1829, 2003.

[10] H. Machrafi, "Development and experimental validation of kinetic schemes for hydrocarbon mixtures for HCCI applications investigation of the auto-ignition process and the application to internal combustion engines", *PhD thesis, University Paris 6*, France, 2007.

[11] H. Machrafi, S. Cavadias, "Three-stage autoignition of gasoline in an HCCI engine: An experimental and chemical kinetic modeling investigation", *Combust. Flame*, vol. 155, pp. 557-570, 2008.

Index

A

B

C

D

F

Hatim Machrafi (Ed)

W

Made in United States
North Haven, CT
04 November 2026

21727492R00215